Applications of Image Processing and Soft Computing Systems in Agriculture

Navid Razmjooy
Independent Researcher, Belgium

Vania Vieira Estrela
Fluminense Federal University, Brazil

A volume in the Advances in Environmental Engineering and Green Technologies (AEEGT) Book Series

IGI Global
DISSEMINATOR OF KNOWLEDGE

Published in the United States of America by
IGI Global
Engineering Science Reference (an imprint of IGI Global)
701 E. Chocolate Avenue
Hershey PA, USA 17033
Tel: 717-533-8845
Fax: 717-533-8661
E-mail: cust@igi-global.com
Web site: http://www.igi-global.com

Library of Congress Cataloging-in-Publication Data

Library of Congress Cataloging-in-Publication Data

Names: Razmjooy, Navid, 1987- editor. | Estrela, Vania Vieira, editor.
Title: Applications of image processing and soft computing systems in agriculture / Navid Razmjooy and Vania Vieira Estrela, editors.
Description: Hershey, PA : Engineering Science Reference, [2019] | Includes bibliographical references.
Identifiers: LCCN 2018042435| ISBN 9781522580270 (hardcover) | ISBN 9781522580287 (ebook)
Subjects: LCSH: Agriculture--Data processing. | Image processing--Digital techniques. | Soft computing--Industrial applications. | Machine learning--Industrial applications.
Classification: LCC S494.5.D3 A66 2019 | DDC 338.10285--dc23 LC record available at https://lccn.loc.gov/2018042435

This book is published in the IGI Global book series Advances in Environmental Engineering and Green Technologies (AEEGT) (ISSN: 2326-9162; eISSN: 2326-9170)

British Cataloguing in Publication Data
A Cataloguing in Publication record for this book is available from the British Library.

Advances in Environmental Engineering and Green Technologies (AEEGT) Book Series

ISSN:2326-9162
EISSN:2326-9170

MISSION

Growing awareness and an increased focus on environmental issues such as climate change, energy use, and loss of non-renewable resources have brought about a greater need for research that provides potential solutions to these problems. Research in environmental science and engineering continues to play a vital role in uncovering new opportunities for a "green" future.

The **Advances in Environmental Engineering and Green Technologies (AEEGT)** book series is a mouthpiece for research in all aspects of environmental science, earth science, and green initiatives. This series supports the ongoing research in this field through publishing books that discuss topics within environmental engineering or that deal with the interdisciplinary field of green technologies.

COVERAGE

- Waste Management
- Radioactive Waste Treatment
- Biofilters and Biofiltration
- Green Transportation
- Contaminated Site Remediation
- Water Supply and Treatment
- Air Quality
- Electric Vehicles
- Sustainable Communities
- Pollution Management

IGI Global is currently accepting manuscripts for publication within this series. To submit a proposal for a volume in this series, please contact our Acquisition Editors at Acquisitions@igi-global.com or visit: http://www.igi-global.com/publish/.

The Advances in Environmental Engineering and Green Technologies (AEEGT) Book Series (ISSN 2326-9162) is published by IGI Global, 701 E. Chocolate Avenue, Hershey, PA 17033-1240, USA, www.igi-global.com. This series is composed of titles available for purchase individually; each title is edited to be contextually exclusive from any other title within the series. For pricing and ordering information please visit http://www.igi-global.com/book-series/advances-environmental-engineering-green-technologies/73679. Postmaster: Send all address changes to above address.

Titles in this Series

For a list of additional titles in this series, please visit:
https://www.igi-global.com/book-series/advances-environmental-engineering-green-technologies/73679

Green Finance for Sustainable Global Growth
Sang-Bing Tsai (University of Electronic Science and Technology of China Zhongshan Institute, China & Civil Aviation University of China, China & WUYI University, China) Chung-Hua Shen (Shih-Chien University, Taiwan) Hua Song (Renmin University of China, China) and Baozhuang Niu (South China University of Technology, China)
Business Science Reference • ©2019 • 397pp • H/C (ISBN: 9781522578086) • US $215.00

Building Sustainability Through Environmental Education
Lynn A. Wilson (SeaTrust Institute, USA) and Carolyn N. Stevenson (Purdue University Global, USA)
Engineering Science Reference • ©2019 • 335pp • H/C (ISBN: 9781522577270) • US $195.00

Unique Sequence Signatures in Plant Lipolytic Enzymes Emerging Research and Opportunities
Nihed Ben Halima (University of Sfax, Tunisia)
Engineering Science Reference • ©2019 • 197pp • H/C (ISBN: 9781522574828) • US $145.00

Handbook of Research on the Adverse Effects of Pesticide Pollution in Aquatic Ecosystems
Khursheed Ahmad Wani (Government Degree College Bijbehara, India) and Mamta (Jiwaji University, India)
Engineering Science Reference • ©2019 • 538pp • H/C (ISBN: 9781522561118) • US $295.00

Harnessing Marine Macroalgae for Industrial Purposes in an Australian Context ...
Göran Roos (Economic Development Board of South Australia, Australia) Anthony Cheshire (Balance Carbon Pty Ltd, Australia) Sasi Nayar (South Australian Aquatic Sciences Centre, Australia) Steven M. Clarke (South Australian Research and Development Institute, Australia) and Wei Zhang (Flinders University, Australia)
Engineering Science Reference • ©2019 • 291pp • H/C (ISBN: 9781522555773) • US $195.00

Advanced Methodologies and Technologies in Engineering and Environmental Science
Mehdi Khosrow-Pour, D.B.A. (Information Resources Management Association, USA)
Engineering Science Reference • ©2019 • 506pp • H/C (ISBN: 9781522573593) • US $295.00

For an entire list of titles in this series, please visit:
https://www.igi-global.com/book-series/advances-environmental-engineering-green-technologies/73679

IGI Global
DISSEMINATOR OF KNOWLEDGE

701 East Chocolate Avenue, Hershey, PA 17033, USA
Tel: 717-533-8845 x100 • Fax: 717-533-8661
E-Mail: cust@igi-global.com • www.igi-global.com

Editorial Advisory Board

Table of Contents

Detailed Table of Contents

The quality control of the agricultural products, which in many cases is through intuitive observation of the visible features of the product, plays a key role in the survival of the agricultural industry. For a long time, the qualitative categorization of these products has been performed by trained people who search products for the specific characteristics. On the other hand, hard and repetitive working can cause people to make some mistakes in computing the quality control errors. Hence, by entering the machine vision systems into this subject, they turned into a reliable, low-cost and real-time technology. Despite the existence of machine vision systems in this process, there are still major challenges in categorizing agricultural products in terms of quality, size, shape, and examination of defects. Potato is one of the most important agricultural products that is produced and has a high application. Unfortunately, it suffers from various types of diseases and defects. Hence, its quality control has a particular importance.

Visual examination of blood smears is an essential tool for analysis, prevention, and remediation of several types of maladies. The interest of computer-aided decision has been acknowledged in many medicinal instances (e.g., automatic ways and means are being explored to spot, classify, and measure visual items in hematological cytology [HC]). This chapter proposes an entirely automated blood smear diagnosis system for hemograms, which can lessen the time spent to scrutinize a slide. The present framework relies on morphological operations (MOs) and soft segmentation by means of the watershed transform (WT). Experiments demonstrate the method efficacy to count white blood cells (WBCs) and red blood cells (RBCs). Some considerations about implementations, design advice and possible variants, as well as improvements are discussed. The future of automated medical analysis is contemplated.

Identification of useful items that can be picked up from the damaged crops or batch of fruits/vegetables is a challenging task nowadays. Humans may fail to identify them correctly with their naked eyes due to strain. Image processing techniques can help to maximize the amount of the good agro-items easily by comparing the existing goods to templates. This chapter introduces an effective recognition method to spot good agro-items by extracting the local features using Gabor filter for orientation information. Another local information of that fruit/vegetable is extracted by speeded up robust features (SURF) algorithm. The extracted features are matched with their templates which results in the decision of individual feature extraction method. Finally, both local information is fused at decision level individually with AND operation (i.e., both algorithms will give correct decision to identify the good agro-item).

The disease occurrence phenomena in plants are season-based which is dependent on the presence of the pathogen, crops, environmental conditions, and varieties grown. Some plant varieties are particularly subject to outbreaks of diseases; on the other hand, some are opposite to them. Huge numbers of diseases are seen on the plant leaves and stems. Diseases management is a challenging task. Generally, diseases are seen on the leaves or stems of the plant. Image processing is the best way for the detection of plant leaf diseases. Different kinds of diseases occur because of the attack of bacteria, fungi, and viruses. The monitoring of leaf area is an important

development of computer science. This chapter focuses on collecting information and overcome the short comes of manual data handling and prediction of yield results of crop production. Data mining is a prominent agricultural research area for analysis of crop yield. These predictions are a very important in solving agricultural problems for crops.

Image evaluation procedures are widely employed in various domains to extract the useful information to make the necessary decision. This chapter implements a soft-computing tool to examine the benchmark plant/weed (BPW) pictures of computer vision problems in plant phenotyping (CVPPP2014) challenge database. The proposed work implements a hybrid image evaluation procedure based on social-group optimization algorithm (SGOA) and Shannon's entropy (SE)-based multi-thresholding and the Chan-Vese segmentation (CVS)-based extraction procedure. After extracting the crop/weed regions of BPW pictures, the superiority of the proposed tool is then assessed by implementing a comparative study between the extracted plant/weed region and its equivalent ground-truth. The results of this study substantiate that the proposed system is proficient in examining the BPW pictures and in future. This procedure can be considered to inspect the crop/weed pictures obtained with field supervising drones.

Rainfalls measured in a selected location in Ilorin, Nigeria and standard formula were used to fix the unknown parameters of the new numerical formula using Microsoft Excel Solver. The new numerical formula was used to estimate groundwater recharge from the rainfall. The accuracy of the new numerical formula was evaluated statistically and compared with the previous formulae in use using field groundwater recharge. Correlation between rainfall and estimated groundwater recharge was stablished. Annual cost benefit of groundwater recharge was computed. The study revealed that new formula provided the lowest relative error of 0.001%, the highest MSC of 17.747; the degree of accuracy of 99.999% correlation factor between rainfall and groundwater recharge using the new numerical formula was 0.1612 with correlation coefficient of 0.6079. The average annual cost benefit was1080.24 $ m-2 per year. It was concluded that modeling of groundwater recharge using the new numerical formula is a promising tool for estimating groundwater recharge with minimum error in water resources management.

With the advent of the cutting-edge technologies in information repository and communication, data storage is rapidly increased, and it is required to reduce the size of the data. Especially in the case of agricultural image data like types of plants, crops, seeds; kinds of diseases and their remedial pesticides; and the agricultural satellite; images require a huge volume of memory space to store. To avoid this problem, it is required to reduce the size of the data and redundancy of the data. To overcome this problem, this chapter proposes a compression method, based on an adaptive Gaussian Markov random field model for agricultural image data compression, where the images are assumed to be a Gaussian Markov random field. The parameters of the model are estimated, based on the Bayesian approach. The authors use arithmetic coding to store seed values and parameters of the model as it augments the compression ratio. They also have studied the use of the M-H algorithm, which updates the parameters and through which the image contents such as untexturedness are captured.

With the increasing computational power, areas such as machine learning, image processing, deep learning, etc. have been extensively applied in agriculture. This chapter investigates the applications of the said areas and various prediction models in plant pathology for accurate classification, identification, and quantification of plant diseases. The authors aim to automate the plant disease identification process. To accomplish this objective, CNN has been utilized for image classification. Research shows that deep learning architectures outperform other machine learning tools

Preface

Computer Vision (CV) in addition to Artificial Intelligence (AI) are potent practices for identifying and analyzing numerous physical, biological, chemical, mechanical and behavioral features of both agricultural products are farming practices. For example, before exportation the agricultural goods, they are habitually graded depending on shape, color, texture, size, and weight. Non-destructive tests can help to grade these products concerning the described characteristics, which greatly influences in the markets, the selection confidence intervals, and their applications.

Among the existing possibilities to automate agronomic production, there exist computer-based agricultural equipment with vision. Such mechanisms assure high quality benchmarks through high-speed categorization, optimization of electromechanical cutting, food recognition systems, packaging documentation and product tagging. Such automated agrisystems can also offer a dependable as well as standardized merchandise tracking framework, and other technology matters that an agro-company makes available to the consumers as well as to assist the production.

Insufficient technological knowledge about CV by the engineers, agronomists, geographers, agribusiness entrepreneurs and farmers along with the nonexistence of acquaintance with the cost-effective rationalization and methodologies relying on CV and that created incredulity, and, in some circumstances, there are negative responses. Despite this, the CV has turned out to be increasingly more used. Its evolution has also been impressive. The image processing applications boil down to comparison of numerical sets in contrast to a threshold to resolve if the produce is excluded or accepted.

Soft Computing (SC) has been introduced into agricultural image processing as a consequence of its effectiveness in treating uncertainties intrinsic to the acquired agronomic image data and relying on these procedures are less exhaustive concerning computational burdens.

In Chapter 2, a simple SC technique is explored to provide the backbone for low medical blood tests.

Quality control of agrarian products are examined in Chapter 3.

SC alludes to a relatively novel set of numerical computing practices in engineering, computer science, economy, AI, machine learning, and several other valuable applications that attempt to evade the restrictions of Hard Computing (HC) methods. All these areas need to study, model, validate and analyze very intricate phenomena with accurate scientific methods that were unsuccessful analytically, and mathematically to discover feasible resolutions in the past.

Compared with SC methods, the old HC optimization methods have only thrived in modeling and scrutinizing somewhat simple systems. More sophisticated concerns like the ones associated to biological models, environmental studies, medicine, economy, humanities, pharmacy, administration, and so forth remained outside the focal realm of HC approaches as well as from strict mathematical and analytical techniques. Some recent SC instances are fuzzy connectedness methodologies to image segmentation like fuzzy clustering approaches for yields like plant leaf disease diagnostics via images, optimization-based merchandises packing, neural network built agrarian products' examination systems, etc.

The work by Oke in Chapter 8 reflects about the issues mentioned earlier regarding the dichotomy hard versus soft computing.

One of the traditional control problems is that complex systems cannot be precisely depicted by mathematical paradigms, and are consequently arduous to control using existing methods. SC, on the other hand entails the partial truth, vagueness, and approximation to elucidate complex problems. SC tolerates imprecision/uncertainty, and incomplete facts to lower the computational cost, accomplish tractability, increase robustness, while acquiring knowledge about a problem. Tasks such as nonlinear programming, intelligent control, and decision-making have attracted increasing attention from different backgrounds and can also be accomplished with SC techniques.

It is becoming challenging to control the emergent complexity of recent machinery via ordinary control techniques. For example, several nonlinear and time-variant systems with enormous time delays cannot be straightforwardly controlled in addition to stabilized with customary techniques. One of the reasons for this is the shortage of an accurate model for the agrarian plant. SC is evincing as a resourceful way of regulating complex plants.

SC allows for combinations of several non-competitive or mutually exclusive methods that can result in variants that compliment to each other to unravel a given problem as shown in Chapter 7. It can be assumed that SC aims to unravel complex problems by handling the imprecision in the decision-making processes.

HC approaches involve exact system models for the problems under investigation and encompass traditional mathematical procedures to solve them while SC uses only approximate models, and the solutions hinge on inexact reasoning techniques.

Besides solving automation problems, SC has also been employed in design, manufacturing, commercialization, pattern recognition, and many additional applications.

This book overviews some SC applications in agribusiness and deals with some frequently employed schemes to elucidate multifaceted agronomic problems.

1. FUZZY LOGIC (FL)

Fuzzy Logic (FL) represents human knowledge impreciseness. The fuzzification rationale turns a crisp input into a fuzzy (or semantic) value because the inputs from sensors are always precise numerical values, which complicates processing. The inference engine generates fuzzy outputs employing IF-THEN rules. In a FL system, a crisp output stems from the defuzzification stage (Zadeh, 1994; Mousavi et al., 2014).

Some fuzzy rules appear in Chapter 1 where images from potatoes are analyzed using several methodologies.

2. GENETIC ALGORITHMS (GAs)

Genetic Algorithms (GAs) are SC strategies mainly employed to unravel optimization problems frequently encountered in everyday applications' models. The GA rationale mimics the natural selection to attain a good solution choice for a given application with some settings. A GA can find solutions when there are intricate search problems resulting from modeling real-life applications. For instance, they can explore several conceivable design scenarios and elements to unearth the finest combination of parameters, metrics and conditions that will yield more stable, more reliable and more affordable designs in overall. Their basic steps are:

1. Initialization to craft a preliminary population randomly;
2. Evaluation to examine each population member, with the corresponding assessment of the individuals' fitness grounded on their compliance with the desired requirements;
3. Selection to keep the ones that comply with the looked-for requirements only; and

4. Crossover to spawn new individuals by mingling the best characteristics of the present individuals. At the completion, the resultant individuals will be nearby to the wanted prerequisites.

The process restarts from the step (3) until a termination condition is reached as a final point (Mousavi et al., 2014; Hemanth & Estrela, 2018).

3. PARTICLE SWARM OPTIMIZATION (PSO)

To facilitate optimization, a Particle Swarm Optimization (PSO) procedure can deliver a large number of candidate solutions that must be evaluated to choose the best sets. PSO is a heuristic method relying on the insect swarms' motion (Moallem, & Razmjooy, 2012; de Jesus et al., 2018). PSO generates an initial random population swarm (aka a particle collection) that propagates towards an optimal point in the n-dimensional design space, and achieves the global optimum after some iterations. Each swarm particle explores the n-dimensional design space using the knowledge from the previous best particles to generate a velocity vector designating a search direction headed for an auspicious design point, and updates the particles' locations. In Chapter 11, Singh and Vinod explore the problem of controlling and performing path planning of a surveillance agribot intelligently.

4. MACHINE LEARNING (ML)/SUPPORT VECTOR MACHINES (SVMs)

The variety and profusion of qualitative features of agronomic products have been the foremost reason for the advance of Non-Destructive Techniques (NDTs). Chapter 5 by Kateris et al. delivers a very interesting discussion on such applications with an SC rationale. Quality control is a paramount task in both industrial and rural plants as control and automation come to be more widespread, and new methodologies provide prospects for production through novel understandings. The quality and quantity of products intensify the usage of advanced automated equipment. Machine tools grow into more automated while less reliant on human elements. These days, CV technology is widely used in the agrarian industry, for the most part in product quality control, robotic units supervision and self-guided mechanisms.

Several ML and SVM implementations are examined in Chapter 1 by Navid Razmjooy et al. to perform NDT of potatoes.

Chapter 6 written by Singh and Sharma performs data mining in agrarian settings to analyze crop production.

Chapter 4 applies SC for plant leaf disease detection.

An application involving image compression of agricultural imageries appears in Chapter 9.

5. DEEP NEURAL NETWORKS (DNNs)

Artificial Neural Networks (ANNs) mimic the manner the nervous systems and the brains work. They accomplish their optimum performance when the correlation between the inputs as well as outputs is extremely non-linear and they are highly appropriate for unravelling problems where there are no clear-cut algorithms or a specific collection of rules to determine a viable solution.

An ANN contains interconnected elements where each neuron accomplishes some operations and the overall task involves the weighted addition of these elementary operations.

An ANN is trained with an identified set of inputs responsible for producing the desired output signals. Training typically establishes the learning network patterns that impacts weighting functions and some previously fixed learning rules. They are twofold: supervised, or unsupervised. The ANN under study is trained with inputs and the equivalent output patterns, i.e., the outcomes are known for specific inputs in supervised learning. Insofar as unsupervised learning goes, the network output is trained to react to input patterns.

An ANN has chiefly the input, hidden, in addition to output layers. Each layer consists of an amount of nodes. The Backpropagation (BP) procedure is employed in the majority of ANNs to train it. The ANN output is appraised against the sought after output. If the outcomes do not agree with what was anticipated, then the weights associated are amended and the computational process is reiterated until the error remains below a threshold (precision).

Many specialists resolve complex agribusiness problems via Deep Neural Networks (DNNs), including the difficult servo problems. Chapter 10 uses a DL-based mobile application for plant disease diagnosis that has as a case study the Tomato cultivation.

6. FINAL COMMENTS ON THE SCOPE OF THE BOOK

Intelligent systems together with SC practices are more relevant as the computer processing capabilities upsurge and their cost decreases. Intelligence is required for complex decisions and deciding on the best result from many likely solutions, using complex algorithms. These solutions require fast processing and enormous storage space, which is accessible to many research centers, and all types of schools at low cost in recent times.

The need for SC techniques and constructing clever systems are more essential than ever and must incorporate the Internet of Things (IoT) concept. Nowadays, cheap but super-fast microcontrollers can handle most SC applications efficiently.

Every day, many SC systems appear in many industrial and agronomic applications, such as searching engines, drones, and biomedical devices their use is expected to raise alongside the usage of the Industry 4.0, and IoT devices.

The appearance of agricultural yields plays an imperative part when it comes to marketing price. Characteristically, factors such as size, texture, color, origin, freshness, and novelty, certifications, and the absence of defects. An understanding of the appearance of agronomic production concerning different image modalities is obligatory, e.g., to explore hereditary traits, to depict the assortment, vegetal categories, and consumer buying choices. In this book, the focal uses of SC techniques in agricultural goods, such as connections between form and genetics, product features, and merchandise sorting, are scrutinized, in addition to other insights brought in by CV. In this regard, CV and SC algorithms have advanced dramatically over time which prompts the demand for quality metrics encompassing the external features of agronomic products. Since the SC practices can assist producers to handle surveys, the apt investigations about computational intelligence in agronomy can stimulate the development of more sophisticated works in the future.

The use of SC methods in labor-intensive agribusiness decisions is extremely relevant and losses and mistakes can be circumvented by taking advantage of automatic operations. Different stages agrarian production, comprising cultivation and harvesting, can use more sophisticated storage and processing for quality measurement, macroeconomic decision-making, management/ maintenance of equipment, performance estimation, etc.

Navid Razmjooy
Independent Researcher, Belgium

Vania Vieira Estrela
Fluminense Federal University, Brazil

REFERENCES

de Jesus, M. A., Estrela, V. V., Saotome, O., & Stutz, D. (2018). Super-resolution via particle swarm optimization variants. In J. Hemanth & V. Balas (Eds.), *Biologically Rationalized Computing Techniques For Image Processing Applications* (Vol. 25). Cham: Springer. doi:10.1007/978-3-319-61316-1_14

Hemanth, D. J., & Estrela, V. V. (2018). *Deep Learning for Image Processing Applications*. Amsterdam, The Netherlands: IOS Press.

Moallem, P., & Razmjooy, N. (2012). Optimal threshold computing in automatic image thresholding using adaptive particle swarm optimization. *Journal of Applied Research and Technology*, *10*, 703–712.

Mousavi, B. S., Soleymani, F., & Razmjooy, N. (2014). Semantic image classification by genetic algorithm using optimised fuzzy system based on Zernike moments. *Signal, Image and Video Processing*, *8*(5), 831–842. doi:10.100711760-012-0311-7

Zadeh, L. A. (1994). Fuzzy logic, neural networks, and soft computing. *Communications of the ACM*, *37*(3), 77–84. doi:10.1145/175247.175255

Chapter 1
A Survey of Potatoes Image Segmentation Based on Machine Vision

Navid Razmjooy
https://orcid.org/0000-0002-0102-1482
Independent Researcher, Belgium

Vania Vieira Estrela
https://orcid.org/0000-0002-4465-7691
Fluminense Federal University, Brazil

Hermes Jose Loschi
State University of Campinas, Brazil

ABSTRACT

The quality control of the agricultural products, which in many cases is through intuitive observation of the visible features of the product, plays a key role in the survival of the agricultural industry. For a long time, the qualitative categorization of these products has been performed by trained people who search products for the specific characteristics. On the other hand, hard and repetitive working can cause people to make some mistakes in computing the quality control errors. Hence, by entering the machine vision systems into this subject, they turned into a reliable, low-cost and real-time technology. Despite the existence of machine vision systems in this process, there are still major challenges in categorizing agricultural products in terms of quality, size, shape, and examination of defects. Potato is one of the most important agricultural products that is produced and has a high application. Unfortunately, it suffers from various types of diseases and defects. Hence, its quality control has a particular importance.

DOI: 10.4018/978-1-5225-8027-0.ch001

INTRODUCTION

Potato with the scientific name of Solanum tuberosum is one year plant from the Solanaceae family. Over thousands of species of potatoes are exist in the worldwide. The original name of the potato is batata that the Spaniards changed its name into potato it in the 16th century.

This product is one of the big challenges in agriculture; because it can be used as a meal and competes with the plant for resources. As a result of this competition for resources, its skin yields suffer from the blemishes (defects). Defects have a significant effect on the potato products qualification (Ahmadvand, Mondani, & Golzardi, 2009).

Several types of research have been introduced to detect the potato defects for qualifying and grading them (Ahmadvand et al., 2009; Moallem & Razmjooy, 2012; Moallem, Razmjooy, & Ashourian, 2013; Qi et al., 2018; Salaman, 2014).

The first step in quality inspection of potatoes in computer vision is to correct segmentation of the potato from the background.

In other words, image segmentation is the process of separating a digital image into some classes. The number of these classes can be changed based on the subject purpose. Indeed, the main purpose of image segmentation is to simplify and change the representation of an image into something that is more meaningful and easier to analyze (Hau, 2015).

Recently, the application of the image segmentation in agricultural applications is increasing. The quality inspection of the agricultural products is getting to be smarter with the automatic computer vision based robots. Hence, designing a profitable algorithm for solving the segmentation problem for these robots is a challenging subject for the researchers and scientists.

In this study, a review of different methods for potato tuber segmentation to separate the potato tuber image from the background image has been presented.

There are different methods for segmentation of potato images; in the following, the methods and their characteristics are discussed.

BI-LEVEL IMAGE SEGMENTATION

In order to obtain a simple and fast system, monochrome (grayscale) images are the best choice (Russ, 2016). Using grayscale images reduces the complexity and speeds up the system. There are different works of using this kind of system. The general algorithm of this approach is as follows:

1. Image acquisition
2. Convert image into the gray color space
3. Image thresholding
4. Performing the morphological operations
5. Two level segmented image including potato tuber and background

In brief, after reading the image, it should be first converted to the gray color space. Figure 1 shows the gray mode for a simple image.

In Figure 2, the gray image should be threshold by a threshold technique to achieve a bilevel image including the potato and background. For instance, consider the previous image; by applying a threshold method (for example Otsu method is analyzed (Xu, Xu, Jin, & Song, 2011)) and the morphological operations, the output image including a highlighted potato tuber separated from its background is achieved.

Unfortunately, while this approach has high speed on segmenting the potato tubers, it has some drawbacks; using this system is a success while using images with uniform intensity. Consider the images in Figure 3 with their shadow and uniform intensity which are segmented.

Figure 1. (A) Input image (RGB color space) and (B) gray conversion of the (A)

(A)

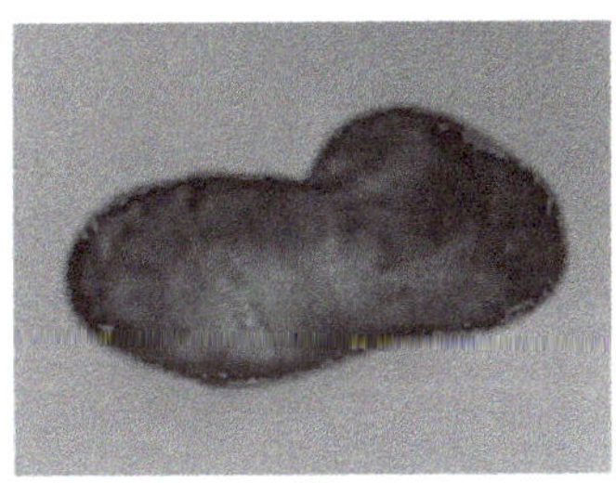

(B)

Figure 2. Grayscale image (A) and its threshold result after image threshold and morphological operations (B)

(A)

(B)

Figure 3. Inappropriate segmentation of potato tubers

From the above images, it is clear that presenting the shadow and non-uniform intensity makes the process unsuccessful.

COLOR IN IMAGE SEGMENTATION

Consider an input image with RGB color space. In order to the simplicity of RGB color space, it changes to a popular used through computer vision purposes. Red, green and blue are three primary additive colors and are represented by a three-dimensional, Cartesian coordinate system (García-Lamont, Cervantes, López-Chau, & Ruiz, 2018). Figure 4 shows this representation for some potato and background.

Figure 4. Analysis of Potato Colors in the RGB Color-Space: 3D-plot of Potato (blue), 3D-plot of non-Potato (red); the Potato colors approximately distributed in a linear fashion in the RGB color space

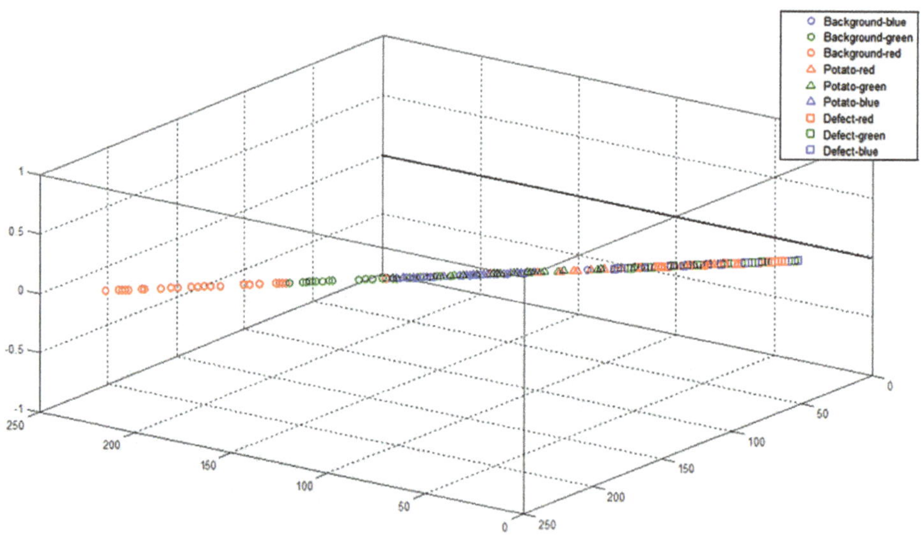

For increasing the accuracy and reducing the system's sensitivity, color-based approaches are the best choice. But it is important to note that selecting the best colorspace is itself a significant problem. A solution for selecting the best colorspace is to testify to different colorspaces.

AUTOMATIC SELECTION OF COLOR SPACES

Theory

The overall flowchart of the color space selection method is shown in Figure 5.

The first step in this approach is to eliminate the noise. A noisy image make the approach to has unfitted and even unpredictable results (Khmag, Ramli, Al-Haddad, & Kamarudin, 2018); this happens because of making some unbalanced and high frequencies in the clusters (which makes the cluster system to get very much clusters from its normal situation). Therefore using a proper noise removal system makes the fault effects to remove.

Figure 5. Diagram of the system overview

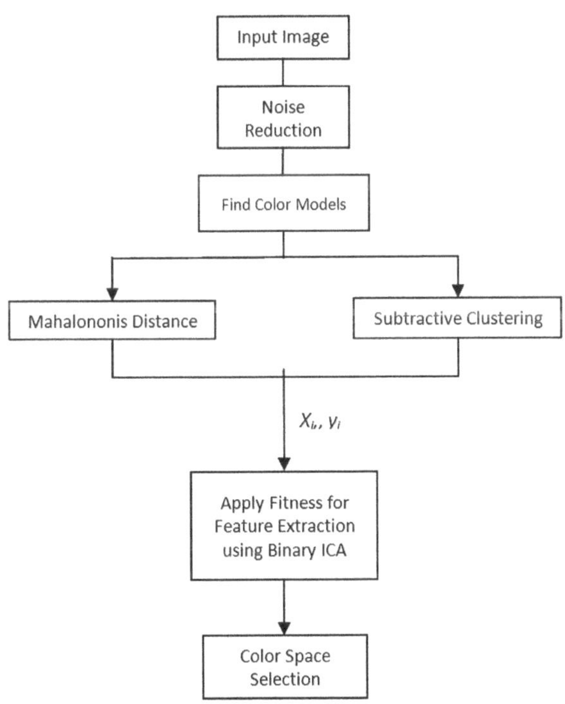

Gaussian low pass filter (GLPF) is utilized for this purpose (Dogra & Bhalla, 2014). Mathematically, a Gaussian filter develops the input signal by convolution with a Gaussian function; it is also known as the Weierstrass transform; the transfer function of a Gaussian low pass filter is given as below:

$$H(u,v) = e^{-a} \tag{1}$$

$$a = {(x^2 + y^2)}\big/{2r^2} \tag{2}$$

where, (x, y) is a local coordinate system, *r* represents in statistics the standard deviation sigma.

When this formula implemented into two dimensions, turn out a surface whose counters are concentric circles with a Gaussian distribution from the center point.

The consequences which are acceding from this distribution are used to build a convolution matrix, is implemented in the original image.

Each pixel's new result is set to a weighted mean of that pixel's neighborhood. The original pixel's value receives the heaviest weight (having the highest Gaussian value) and neighboring pixels receive little weights as their distance to the original pixel intensification (Boyat & Joshi, 2015; Khmag et al., 2018).

After noise removing, by assuming the RGB input, 10 model of another dimension of colors which have more application in machine vision are obtained. In Figure 6, the histogram (probability density function) of different color spaces for an input image is shown.

As it is clear from the above, two clusters are enough for the considered image; but because all of the colorspaces except *S* have two clusters; therefore we need another property to decrease the dimensions.

Mahalanobis Distance

Mahalanobis is a distance which depends on several variables with various correlation patterns prospering basically to illustrate sample sets. From a group of values with mean value:

$$\mu = (\mu_1 \mu_2 \cdots \mu_N)^T \tag{3}$$

Figure 6. Input potato image and the histogram of its color spaces

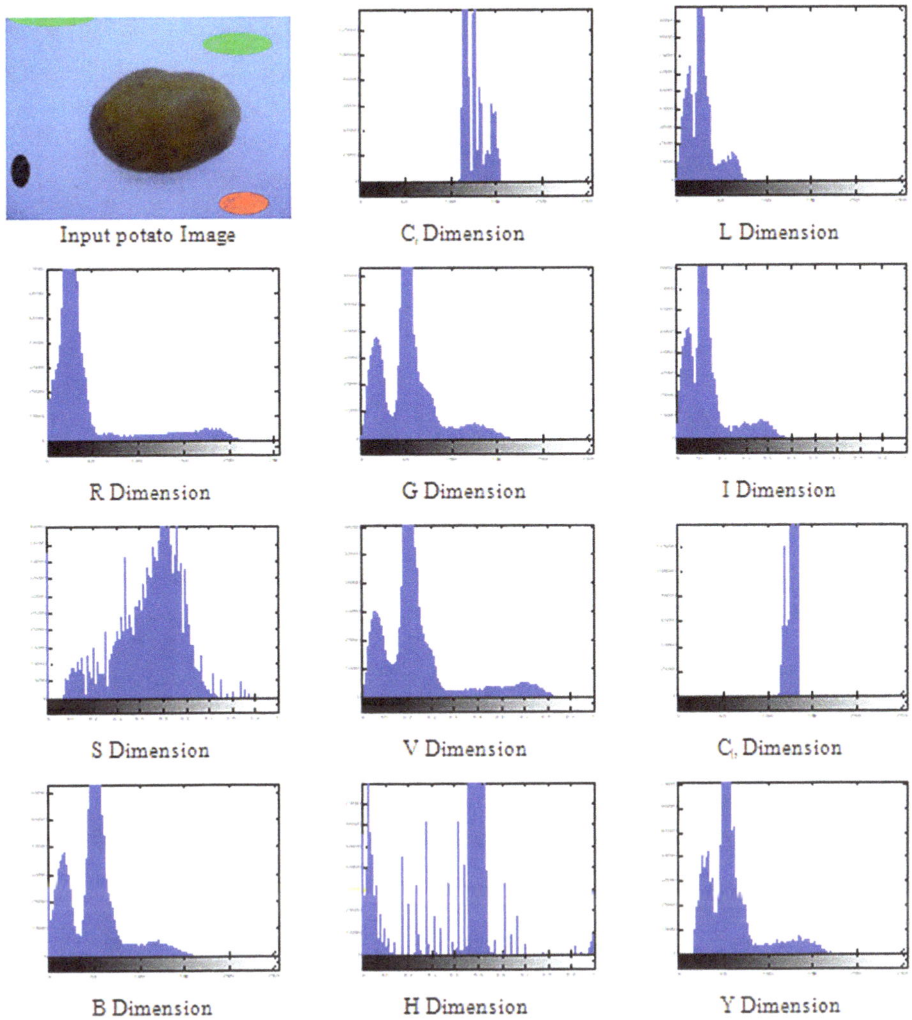

where N is the number of total pixels. Mahalanobis distance can be defined as a dissimilarity measure between two random vectors $\vec{x}$ and $\vec{y}$ of the same distribution with the covariance matrix (Yeager, Gregory, Key, & Todd, 2018):

$$d(\vec{x}, \vec{y}) = \sqrt{(\vec{x} - \vec{y})^T S^{-1} (\vec{x} - \vec{y})} \tag{4}$$

where S is the covariance matrix and is defined as,

$$S = Cov(x_i, y_i) = E[(x_i - \mu_i)(y_i - \mu_i)], \quad and, i = 1 \cdots N \tag{5}$$

To find the Mahalanobis distance, a mean value of the color spaces is selected as below:

$$\frac{1}{n}\sum_{i=1}^{n} x_i \tag{6}$$

where n is the number of color spaces and x_i is the value of the ith color space. Table 1 illustrates the Covariance matrix and mean vector for colorspaces of the input image; it can be seen that G dimension has the least distance from the input image and its covariance matrix is steady toward the others.

Figure 7 shows a query image and its results; it is clear from the image that the G dimension has the least difference from the input image (i.e. it is more close to

Figure 7. Different in Most distance images and least distance Images from the Mean Image: (A) Query Image, (B) Mean Value Image, (C) G dimension of the Query Image, (D) C_b dimension of the Query Image, (E) Different Result between (C) and (B), (F) Difference Result between (D) and (B).

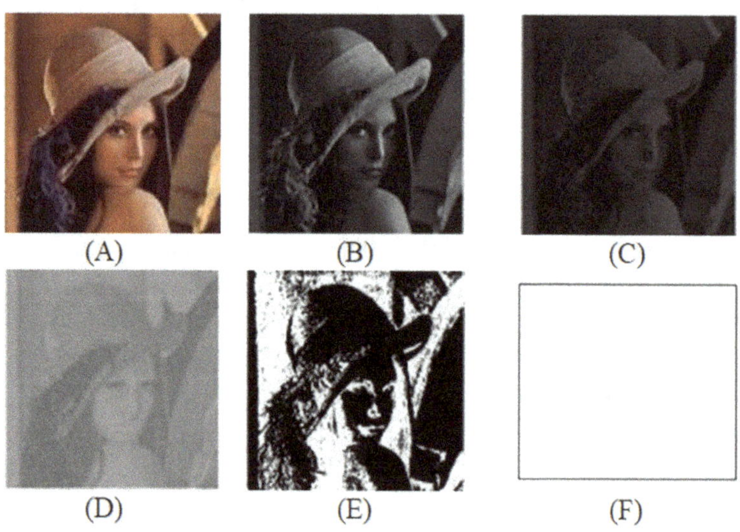

Table 1. Covariance matrix and mean vector for 13 dimensions of a query image

Dimension: R			Distance value: 27.8765	Dimension: G			Distance value: 7.2863
Covariance Matrix			Mean Vector	Covariance Matrix			Mean Vector
16.3333	15.1667	-3.5000	30.6667	14.3333	11.6667	-6.0000	50.3333
15.1667	16.3333	5.0000	29.3333	11.6667	10.3333	0	47.6667
-3.5000	5.0000	31.0000	28.0000	-6.0000	0	31.0000	47.0000
(a)				(b)			
Dimension: B			Distance value: 17.1569	Dimension: H			Distance value: 77.7395
Covariance Matrix			Mean Vector	Covariance Matrix			Mean Vector
13.0000	11.0000	-8.5000	57	0.0343	-0.0905	0	0.5421
11.0000	13.0000	2.5000	55	-0.0905	0.2387	0	0.5477
-8.5000	2.5000	31.0000	54	0	0	0	0.5449
(c)				(d)			
Dimension: S			Distance value: 77.8652	Dimension: V			Distance value: 78.3074
Covariance Matrix			Mean Vector	Covariance Matrix			Mean Vector
0.0017	0.0016	0.0001	0.4635	0.1999	0.1692	-0.1307	0.2235
0.0016	0.0016	0.0005	0.4684	0.1692	0.1999	0.0384	0.2157
0.0001	0.0005	0.0027	0.4850	-0.1307	0.0384	0.4767	0.2118
(e)				(f)			
Dimension: I			Distance value: 78.4825	Dimension: Y			Distance value: 13.5003
Covariance Matrix (10^{-3})			Mean Vector	Covariance Matrix			Mean Vector
0.0980	0.0854	-0.0410	0.1203	10.3333	8.0000	-5.0000	54.6667
0.0854	0.0889	0.0171	0.1150	8.0000	7.0000	0	53.0000
-0.0410	0.0171	0.2119	0.1124	-5.0000	0	21.0000	52.0000
(g)				(h)			
Dimension: C_b			Distance value: 153.2191	Dimension: C_r			Distance value: 127.6332
Covariance Matrix			Mean Vector	Covariance Matrix			Mean Vector
0.3333	0	0	133.6667	0	0	0	119.0000
0	0	0	134.0000	0	0.3333	0	119.3333
0	0	0	134.0000	0	0	0	119.0000
(i)				(j)			
Dimension: L			Distance value: 34.3280	Dimension: a			Distance value: 88.1940
Covariance Matrix			Mean Vector	Covariance Matrix			Mean Vector
2.7131	2.2921	-1.1176	26.5919	0.0024	0.0255	0.0007	-5.8017
2.2921	2.1627	0.1697	25.5525	0.0255	0.3302	0.0139	-5.1720
-1.1176	0.1697	5.9463	25.1925	0.0007	0.0139	0.0009	-5.4993
(k)				(l)			
Dimension: b			Distance value: 90.0839				
Covariance Matrix			Mean Vector				
0.1828	-0.0699	0.0289	-6.4256				
-0.0699	0.0315	-0.0162	-6.7137				
0.0289	-0.0162	0.0102	-6.6070				
(m)							

black) and difference between the most value (C_b) and the mean value is very high (i.e. it is more close to white).

After finding the Mahalanobis distance and subtractive clustering, two characteristics for each color spaces are achieved: distance and number of clusters. It is important to know that some of the extracted dimensions are useless in the segmentation process.

Therefore, we need to choose those color spaces which are actually important for image segmentation. In other words, eliminating the useless dimensions in the image is a key factor for reducing the image analysis complexity.

After achieving the presented information, a fitness function is introduced whereby minimizing it, the best colorspaces can be achieved:

$$Fitness = \sum_{i=1}^{n} a_i \left(\frac{x_i}{y_i} \right) \tag{7}$$

where i is the number of color spaces, x_i is Mahalanobis distance of ith dimension from the mean vector and y_i is the number of clusters center for i^{th} dimension. The fitness function can be minimized by an optimization algorithm. For example, consider the Figure 8 and its results (Table 2).

The original first image is shown in Figure 8(A) and its gray level and its histogram are shown in Figure 8(B) and (C). Figure 8(D) shows a proper selection of color space using the proposed method, and 8(E) shows its histogram, Figure 8(F-H) shows the thresholding operation by Kapur's method, Otsu's method and Rosin's method respectively using just gray level of the images, and Figure 8(I-K) illustrate the thresholding images using Kapur's method, using the proposed approach, respectively.

As contrasted with the described results, the lemon and blue plates are extracted completely by using the proposed approach and Rosin threshold.

Table 2. Some Result of Threshold Points after color selection

Lemon	Test Image
Hue (H)	Selected Color space
0.4531	Otsu
0.6510	Kapur
0.1333	Rosin

Figure 8. (A) Lemon Image, (B) gray color model (Blue), (C) Histogram of (B), (D) correct color model selection by proposed method (Hue), (E) Histogram of (D), (F) Threshold of (B) using Kapur, (G) Threshold of (B) using Otsu, (H) Threshold of (B) using Rosin, (I) Threshold of (D) using Kapur, (J) Threshold of (D) using Otsu, (K) Threshold of (D) using Rosi.

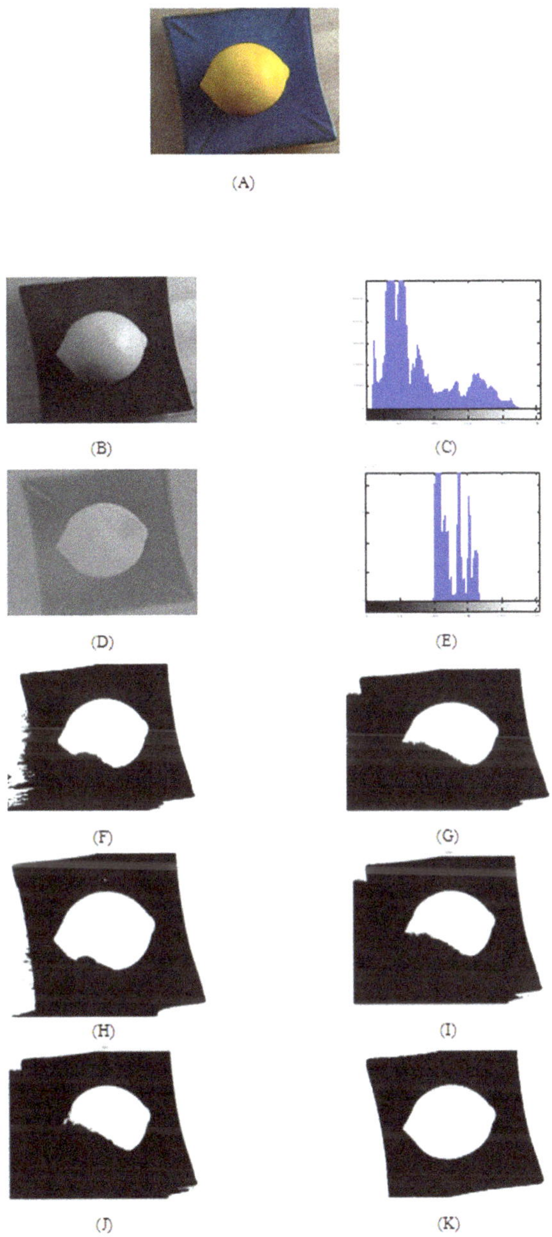

COLOR-BASED IMAGE SEGMENTATION

Simple Image Color Segmentation

Because of its ease of programming, this algorithm is a suitable method in terms of speed than the other color based methods. The general process steps are as follows:

1. Image acquisition
2. Convert image into a correct colorspace for evaluation
3. Image segmentation
4. Two level segmented image including potato tuber and background

As it is presented, the second part of the algorithm is very important (Garces & Reinhard, 2018). For instance, consider the following colorspace conversion based on the presented method before. By this technique, we select a 3d colorspace for segmentation as below:

- U = Red + Green - 2 × Blue
- L = Green + Intensity
- Y = RGB × L

The first step removes the darker parameters and softens the image; since in the previous step, the amount of darkness in the pot becomes darker, the next step brightens the image by adding some amount of brightness It does not go too far.

Note that if the amount of brightness was high, the entire image becomes white. Therefore, an empirical coefficient of 0.5 is considered for this colorspace.

In the last step, the RGB image is multiplied in order to obtain the final result; the results indicate that this algorithm is highly accurate and has a much faster speed. But, it should be noted that using the color space selection technique should be applied to all databased not for a single database

Figure 9 shows an example of the algorithm for the potato image.

Image Color Segmentation Based on Neural Networks

In this section, we consider using color image selection technique. Hence, the algorithm is as follows:

1. Image acquisition
2. Contrast enhancement
3. Converting the image into gray level

Figure 9. (A) Input RGB image (B) Image after the second stage with a coefficient of 0.5 (C) Image after the second stage with coefficient 1, (D) Final image

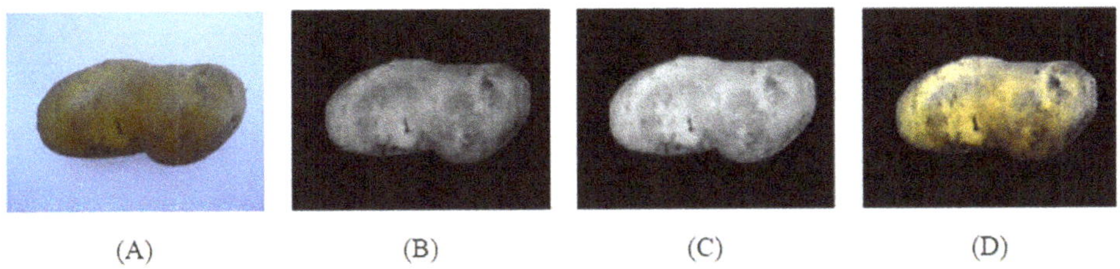

4. Applying the neural networks for image classification
5. Morphological operations
6. Two level segmented image including potato tuber and background

Consider an RGB input with non-uniform intensity. As it is explained in the previous section, the purpose of contrast enhancement is by highlighting the defect regions whilst leaving the unimportant background regions intact. This provides the defect detection point to get better locate and represent each defect in the image. By applying the contrast enhancement, we have the results shown in Figure 10.

After contrast enhancement, the next step is to convert the colored image into the gray level image.

Mean value of the RGB (Intensity) color model is the most useful descriptor and is selected as the graying method of potato image; therefore the original image in RGB color space is then converted to the Intensity dimension as below:

$$I = \frac{1}{3}(R + G + B) \tag{8}$$

It is important to know that for enhancing the accuracy of the image especially in non-uniform images, it is better to discard the third step.

The next step is to train the neural networks to segment the pixels whose related to the potato than the others. The input of the networks can be a gray level image or a colored image by a definite colorspace.

It is noted that the performance of the neural network is authentic upon to the assumption that we have enough number of training samples.

Let's consider a multi layer perceptron.

The backpropagation algorithm is a generalization of the least mean square algorithm that improves network weights to minimize the mean squared error between the desired and actual outputs of the network (Fan, Cong, & Wang, 2018).

Figure 10. (A) Input Image, (B) Histogram of (A), (C) Image after Contrast Enhancement, (D) Histogram of (C)

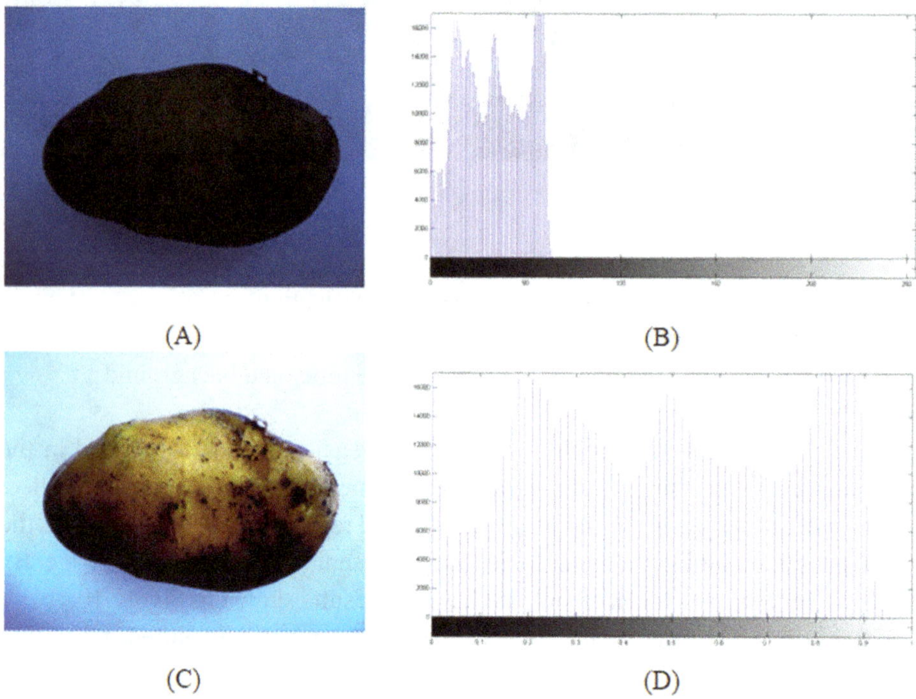

Backpropagation employs supervised learning in which the network is trained using data for which inputs, as well as desired outputs, are known.

Once trained, the network weights are vapid and can be used to calculate the output values for new input samples. Model Standard *multilayer perceptron* (MLP) architecture consists of more than 2 layers; A MLP can have any number of layers, units per layer, network inputs, and network outputs.

MLP in image classification can be run in some steps as described below.

Start with randomly chosen weights.

While *MSE* is unsatisfactory and computational bounds are not exceeded, do for each input pattern:

1. Compute hidden node inputs
2. Compute hidden node outputs
3. Compute inputs to the outputs nodes
4. Compute the network outputs

5. Compute the error between network outputs, and desired output,
6. Modify the weights between hidden and output nodes
7. Modify the weights between input and hidden nodes
8. End

where R is the input vector, IW is the weight vector, and b is the bias vector (Fan et al., 2018).

A two-zone image segmentation (as potato or background) is used in here. For instance, for segmentation *of the* potato tubers based on *pixel-based*, 3 characteristics including the red, green and blue values of each image pixels are considered as the input of the network; this process classifies each pixel independently from its neighbors. The output is 1 dimension and is zero (background) or one (potato).

After all, morphological processing consists of filling holes, and opening followed by closing is applied to the result images (Razmjooy, Mousavi, & Soleymani, 2012).

Figure 11 shows a part of the dataset which is used to compare the methods.

The Present work has implemented Artificial Neural Network consists of three methods as follow: Standard three layer MLP with backpropagation algorithm, with features of image RGB color pixel values taken as input values, and detected potatoes or background as the output. Here we use log-sigmoid as the activation function. A batch steepest mode gradient descent because of the requirement to less update (that inputs and biases are applied to network before they updated) is selected, and other parameters are chosen as Learning rate= 0.01, Epochs= 1000, Target error goal: 10^{-4}.

For better comparison, a classic method (KNN) and another neural network (SVM) are also considered.

3 different kernels and Sequential Minimal Optimization are used for checking the best classifier for potato tuber segmentation. C (optimal parameter) is equal to 100, and end-accuracy is 10^{-4}. The evolutional situation was quite the same as each model. For the nearest neighborhood, k is assumed 4. The process is illustrated completely in Figure 12.

Figure 11. A simple part of trained dataset: (A) background, (B) Potato sets

(A) (B)

Figure 12. (A) Input image, (B) image after segmentation (this one is by MLP), (C) Filling operation on the (B), (D) Opening operation on the (C), (E) closing operation on the (D), (F) final segmented image

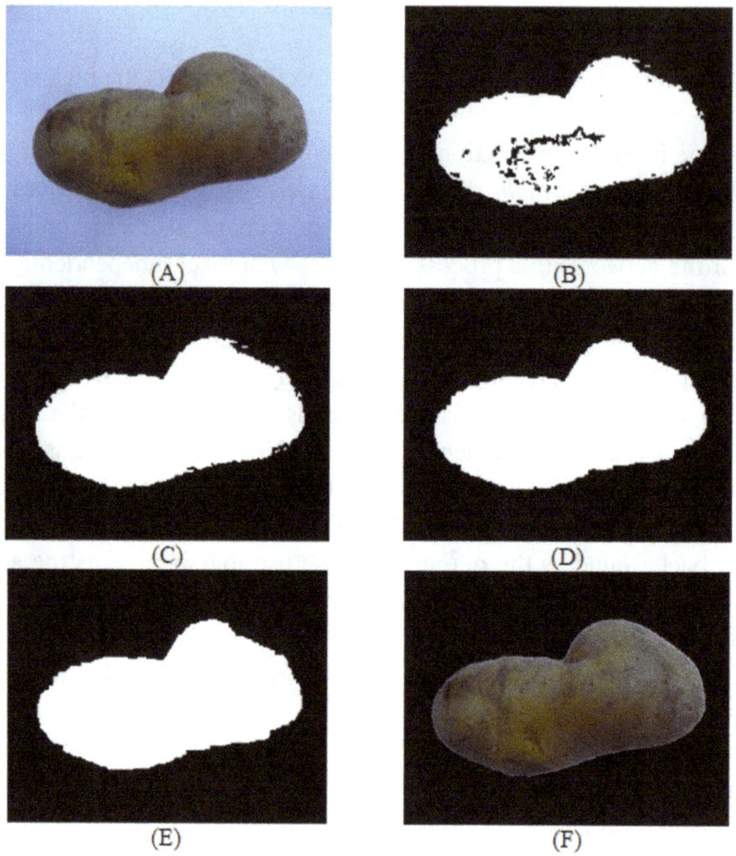

However, the morphological operations help to define the image contours and eliminate or reduce noise (undesired speckles outside the contour of the object) (see Figure 13).

To evaluate the performance of the system, three performance metrics are defined. The first metric is the correct detection rate **(CDR)** and is given in Eq. (9). The false acceptance rate **(FAR)** is the percentage of identification moments in which false acceptance occurs. The false rejection rate **(FRR)** is the percentage of identification moments in which false rejection occurs. The FAR and FRR are expressed in Equations (10) and (11), respectively:

Figure 13. Steps in segmentation based on neural networks

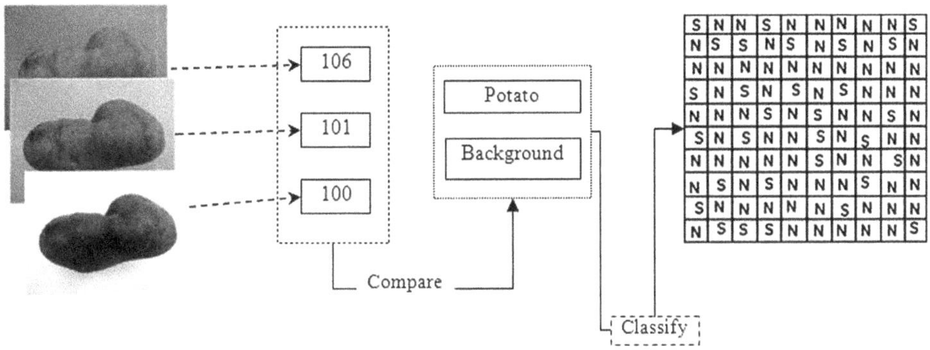

$$CDR = \frac{No.of.Pixels.Correctly.Classified}{Total.Pixels.in.the.Test.Dataset} \quad (9)$$

$$FAR = \frac{No.of.non - Potato.Pixels.Classified.as.Potato.PixelsClassified}{Total.Pixels.in.the.Test.Dataset} \quad (10)$$

$$FRR = \frac{No.of.Potato.Pixels.Classified.as.non - potato.PixelsClassified}{Total.Pixels.in.the.Test.Dataset} \quad (11)$$

Tables 3 and 4 present the performance of compared classifiers in accuracy and speed.

Table 3. Classification Comparison of Performance in the presented methods

Metric	Method					
	Knn	**MLP**	**SVM-Linear kernel**	**SVM-Poly Nominal kernel**	**SVM-Quadratic Kernel**	*SVM-SMO based*
CDR(%)	93	96	96	96	96	**96**
FAR(%)	5	2	2	2	2	***3***
FRR(%)	2	2	2	2	2	***1***

Table 4. Classification Comparison of speed in the presented methods

Method	Operation Time (seconds)
Knn	0.501
MLP	0.095
SVM-Linear kernel	0.057
SVM-Poly Nominal kernel	0.091
SVM-Quadratic Kernel	0.075
SVM-SMO based	***0.015***

A simple example of edge detection of the potato tuber on the image by the proposed method is shown in Figure 14.

It is obvious from the above that Support vector machine with sequential Minimal Optimization has better performance in both accuracy and time; Figure 15 shows the final results of the final image segmentation.

Image Color Segmentation Based on Optimized Neural Networks

Consider the previous method. In this method, after finding the proper color using the proposed operation, an optimized version of the neural network is applied for solving the potato segmentation problem.

The main idea in this technique is to focus on the weight optimization of ANN using an evolutionary algorithm to achieve a better system for segmentation; the evolutionary method can be GA, PSO or any other optimization algorithm. Here,

Figure 14. (A) the input image, (B) detected image after applying the proposed neural network, thresholding, morphological operations, and labeling

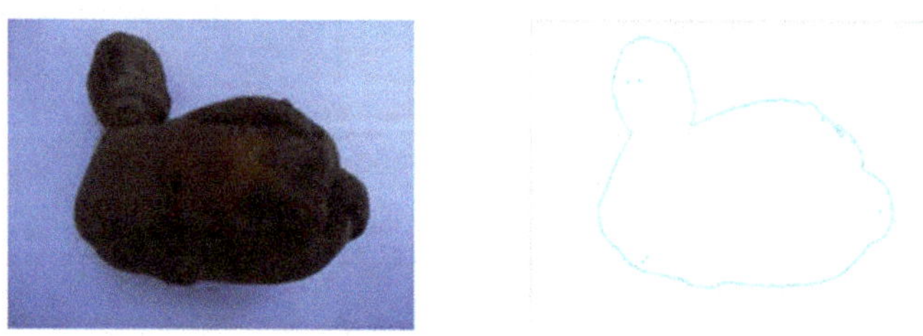

(A) (B)

Figure 15. Segmentation algorithm results: (a) original color Images, (b) Results with the proposed method

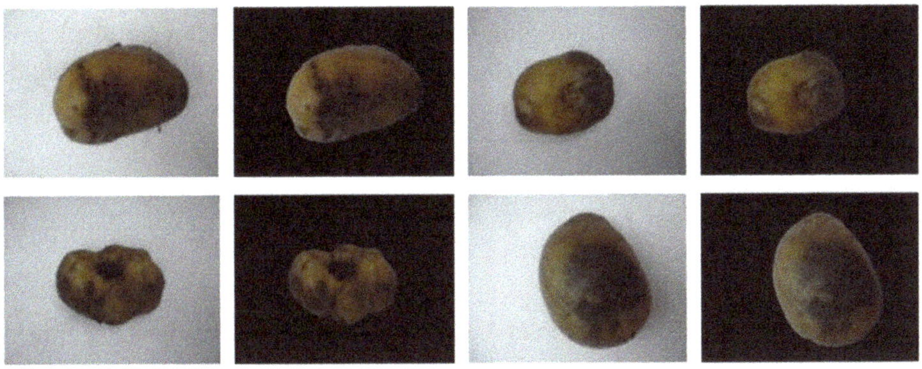

(a) (b) (a) (b)

let's consider the GA algorithm for optimizing the initial weights of the feed-forward neural network (Razmjooy, Mousavi, & Soleymani, 2013).

Good potato color pixel classification should provide coverage of all various potato types. This type of problem is well-suited to artificial neural networks, which have been proven as an efficient tool for pattern classification aims where decision rules are hidden in highly complex data and can be learned only from examples. The image is then classified by trying the performance for each pixel and making a decision about which of the signatures it resembles most.

Mathematically, a neuron's network function $f(x)$ is defined as a composition of other functions $g_i(x)$, which can additionally be defined as a composition of other functions. This can be conveniently represented as a network structure, with arrows depicting the dependencies between variables. A widely used type of composition is the *nonlinear weighted sum*, where

$$f(x) = K\left(\sum_i \omega_i g_i(x)\right) \tag{12}$$

The hyperbolic tangent. It will be convenient for the following to refer to a collection of functions g_i as simply a vector $g = (g_1 \ldots g_n)$. In such a state, Backpropagation (BP) is a common method which is used for feedforward networks. It computes the error on each of the training pairs and adjusts the weights to fit the desired output. This is done in several epochs until the total error on the training set become small enough or when the error stops to decline. Indeed, it employs supervised learning in which the network is trained using data for which inputs, as well as target outputs, are noticed. Once trained, the network weights are vapid and can be used to calculate output values for new given samples.

BP is a gradient descent algorithm on the error space, which most likely gets trapped into a local minimum, succeeded it entirely dependent on initial (weight) settings. This shortcoming can be removed by an exploration searching ability of the evolutionary algorithms such as GA.

A heuristic concerning the order of the interpolation is to put connection weights to the same node together. An evolutionary search of weights values can be formulated as follows:

1. Generate an initial population of *N* weight chromosomes. Evaluate the fitness of each EANN depending on the problem.
2. Depending on the fitness and using suitable selection methods reproduce a number of children for each individual in the current generation.
3. Apply genetic operators to each child individual generated above and obtain the next generation.
4. Check whether the network has achieved the required error rate or the specified number of generations has been reached then goes to step 2.
5. End.

In order to express the ANN, consider a two-layered network which is formulated as Equation (13):

$$\sum_{i=1}^{H} w_i \sigma\left(\sum_{j=1}^{d} w_j x_j + b\right) \tag{13}$$

where *H* denotes the number of neurons in the hidden layer, *w* denotes the weights of the network, *b* denotes the bias value and σ denotes the activation function of each neuron which in this case is considered as sigmoid.

The network is trained by performing optimization the value of the weights for each node interconnection and bias terms; until the values output at the output layer neurons is as close as possible to the actual outputs. The mean squared error of the network (MSE) can be defined as:

$$MSE = \frac{1}{2}\sum\nolimits_{k=1}^{g}\sum\nolimits_{j=1}^{m}\left(Y_j(k) - T_j(k)\right)^2 \tag{14}$$

where m is the number of output nodes, g is the number of training samples, $Y_j(k)$ is the desired output, and $T_j(k)$ is the real output.

Table 5 shows the Covariance matrix and mean vector for 10 dimensions of a query image; from the table, it is clear that G dimension has the least distance from the input image and its covariance matrix is steady toward the others. A sample of Mahalonobis distance is given in Table 5.

Using different datasets shows that C_r component of YC_bC_r color-space instead of gray level one presents the better result of potatoes segmentation. Therefore, The structure of the classifiers is a vector of 1×n pixel coefficient vectors from each image either potato or non-potato image which n is the number of neurons in the hidden layer: since the transfer function used is a sigmoid function (in this work MLP); it will be producing the output between 0 and 255(uint8 mode).

Thus, the output of the neural network needs to be modified so that it is either 0 or 1. In this work, a single threshold value is used in determining the potato and non-potato classes; if the MLP network output is higher than the threshold value, then the output is 1. Otherwise, the output is 0. After all, morphological processing

Table 5. Mahalanobis distance for a given sample

R	*G*	*B*	*H*	*S*
90.5947	128.6035	202.8550	118.4246	118.4669
V	*I*	*Y*	*Cb*	*Cr*
117.8777	118.5174	117.7406	137.3826	81.1532

Figure 16. Steps in supervised classification

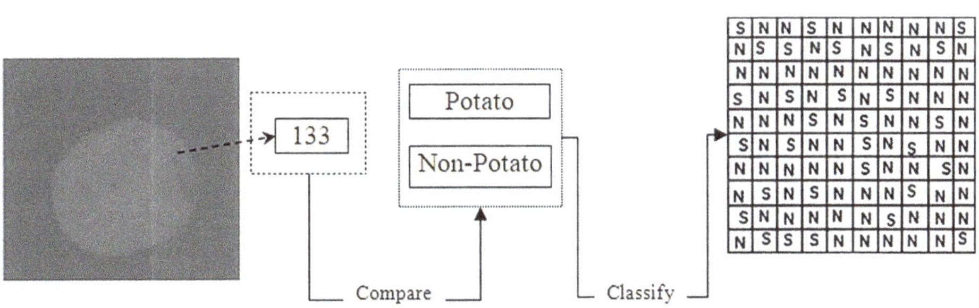

consists of filling holes, and opening followed by closing is applied to the result images. The effect of morphological processing on different datasets is shown in Figure 17.

Figure 18 shows a part of the dataset which is used as the input of the neural network

Image Color Segmentation Based on Fuzzy Systems

This technique uses the fuzzy rule-based system to segment the potato tubers images from the background.

In the proposed algorithm, after selecting appropriate color space, the distance between image pixel and real potato pixels is computed. Furthermore, this distance

Figure 17. output image after thresholding: (A) Input image, (B) optimized ANN segmentation, (C) after morphological operations, and (D) Output image

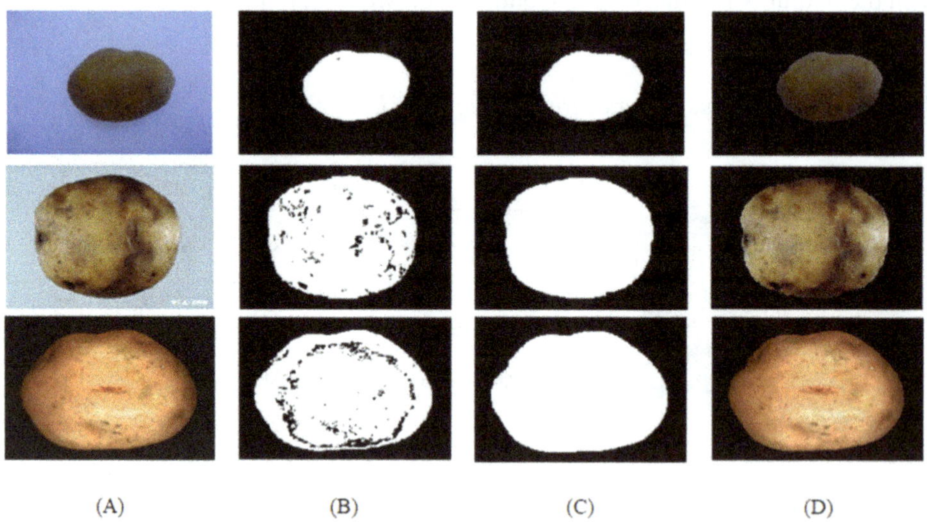

Figure 18. Test Set: (A) not potato-like sets, (B) potato-like sets

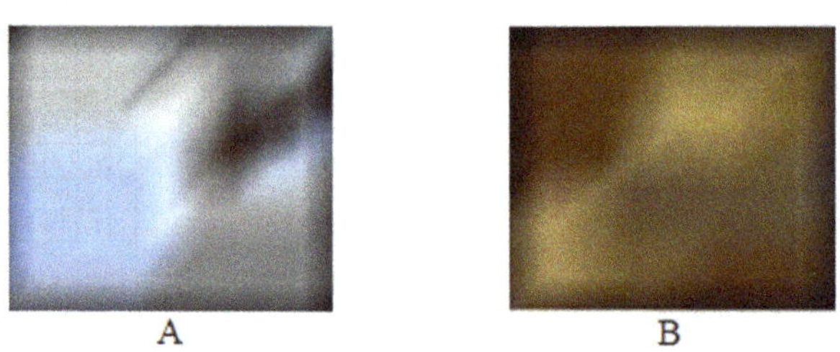

feeds to a fuzzy- rule-based classifier to extract potato candidate in the input image. A subtractive clustering can be also used to decide on the number of rules and membership functions of the fuzzy system. To minimize the design error in potato pixel selecting, membership functions shapes are can be also optimized by genetic algorithm (GA).

To segment potatoes in the input color image, an image thresholding is applied to the output of the fuzzy system, where the corresponding threshold is minimized based on GA. To improve the segmentation results, a sequence of some morphological operators are also applied to the output of the thresholding stage.

The overall schematic diagram of the proposed algorithm is shown in Figure 19.

From the previous sections, the concept of contrast enhancement, noise reduction, and color space selection are declared. Final colorspace which is obtained from the presented technique (for the Ardabil dataset1) is (H, S, (3/4) V).

The next step is to determine the potato pixels. For the proposed color-space, the fuzzy system is as a 1-input, 1-output system applying the normalized distance between the color of each pixel to the potato vector mean as input, and the likelihood of being potato pixel as an output.

Because of good results, Mahalanobis distance is selected; indeed Mahalanobis is a distance that depends on different variables with various correlation patterns prospering typically to illustrate sample sets. Formally, the

Mahalanobis distance of a multi-variable vector x which is represented below:

$$x = (x_1 x_2 \cdots x_N)^T \tag{15}$$

From a group of values with mean:

$$\mu = \left(\mu_1 \mu_2 \cdots \mu_N\right)^T \tag{16}$$

where N is the number of total pixels, and S is the covariance matrix and is defined as:

$$S = Cov\left(x_i, y_i\right) = E\left[\left(x_i - \mu_i\right)\left(y_i - \mu_i\right)\right]$$

$$and\, i = 1 \cdots N \tag{17}$$

Mahalanobis distance can be defined as a dissimilarity measure between two random vectors $\vec{x}$ and $\vec{y}$ of the same distribution with the covariance matrix:

Figure 19. Diagram of the system overview

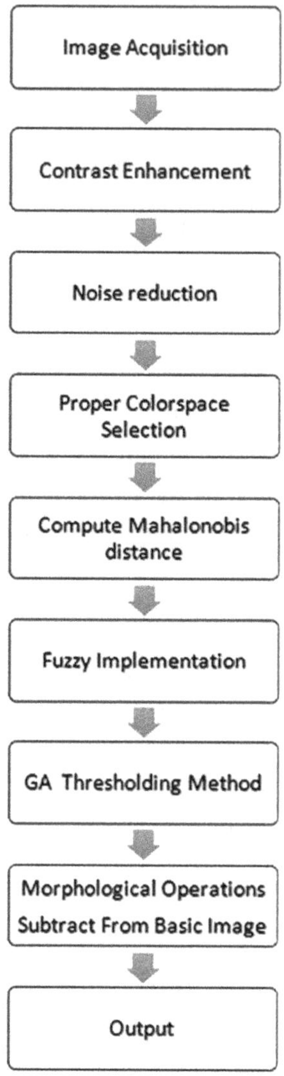

$$d\left(\vec{x},\vec{y}\right)=\sqrt{\left(\vec{x}-\vec{y}\right)^{T} S^{-1}\left(\vec{x}-\vec{y}\right)} \tag{18}$$

Before applying the fuzzy system, we try the Mahalanobis distance on the input image pixels.

Since *Mamdani* type is one of the most commonly used fuzzy interface methods, we used HSV type to fuzzification (Kyi & Zin, 2018). In order to optimize and decide on the number of membership functions (MF'S) and rules we use *subtractive clustering* that is applied to input space (containing more than 156000 potato and non-potato pixels); subtractive clustering is a fast, one-pass algorithm for estimating the number of clusters and the cluster centers in a set of data. Figure 20 shows initial clustering for the proposed Fuzzy system.

Implementing HSV method, gives us the best manner for system classification. That is four membership functions. The achieved rule in potato-color segmentation FIS is:

If Input is Z, Then Output is where Z ∈ *[Potato, Rather Potato, Low. p. Potato, Not Potato. Center-Of-Area* (COA) is then chosen which is the most widely used among all defuzzification approaches.

The *center-of-area* technique, COA, minimizes the following expression:

$$\left| \sum_{\inf y}^{coa(B')} B'(y) - \sum_{coa(B')}^{sup\ y} B'(y) \right| \tag{19}$$

where *inf* is the greatest lower bound, and *sup* is the least upper bound of the support of the fuzzy set *B*', respectively. The expression (10) gives numerical value $y_0 = y_{coa(B')}$, which divides an area under the membership function into two (approximately) equal parts. The value $y_{coa(B')}$ differs from the defuzzification value obtained by the COG technique which is given below:

Figure 20. Subtractive clustering for potatoes

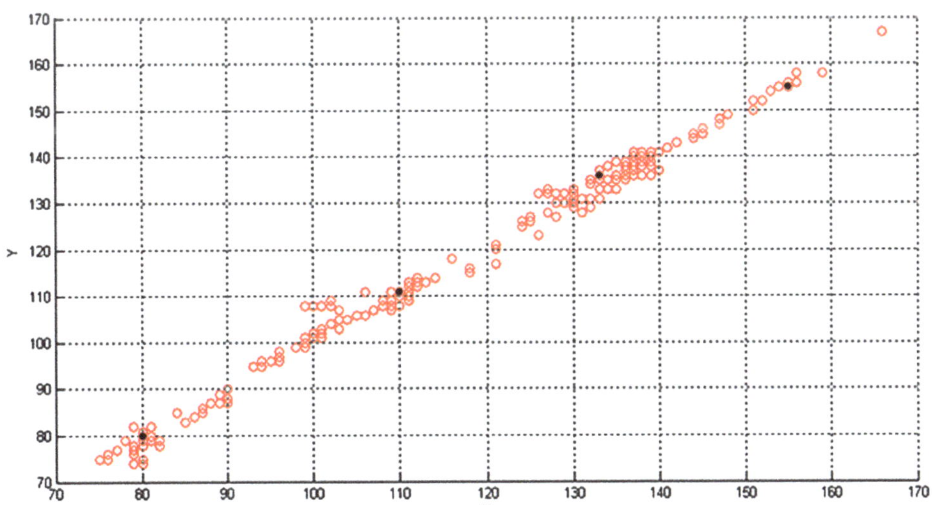

$$y_0 = defuzzifier\left(B'\right) = \frac{\sum_{i=1}^{N_q} B'\left(y_i\right) y_i}{\sum_{i=1}^{N_q} B'\left(y_i\right)} = COG \tag{20}$$

The optimum shapes of Membership functions (MFs) are achieved by the genetic algorithm.

Experiential knowledge is employed to diminish the searching space and to enhance the process. Since the MFs shapes are selected in order to a previously designed system and the parameters of MFs plus the threshold are utilized as the inputs of the GA, whose fitness function defines by comparing the whole detected potato pixels in the image (*D*) with the actual number of these pixels (*A*). The objective function, which should be minimized, is presented below:

$$Fitness = \left|1 - \frac{A}{D}\right| \tag{21}$$

Optimized Input and output membership function are shown in Figure 21.

Which *Input member function* is the Mahalanobis distance from potato means vector pixels and any color vector pixels and the *Output member function* is the likelihood to potato. The optimized output is between 0 and 1 and the output figure is the potato likelihood pixels probability. Figure 22 shows the result from the image fuzzification to a potato and as it can be seen, it is more probable for potato-like pixels to be selected; indeed they are closer to 1 toward the others.

Furthermore, some errors are engendered that were solved by thresholding [4]. To make a binary image from the top result and to separate the foreground and background of the image [20], [21] a proper threshold should be selected. It obtains about 80% manually.

Figure 23 shows the binarization on the prior potato manually; It is clear that the custom HSV method by manual threshold has some errors a little of the shadow is segmented as a potato.

For optimizing and relieving HSV method, a Genetic Algorithm (GA), by a fitness to make an optimized threshold was applied that depends on HSV histograms. The fitness function is empirically selected as follow:

$$Fitness = 1.223 - \mid (0.5 \times \left(mean\left(c\right) + mean\left(c'\right)\right) \mid \tag{22}$$

where 1.223 is found by trials and errors and *C* is the chromosome which shows the proper threshold point for our potato segmentation.

Figure 21. Input (A) and output (B) member function

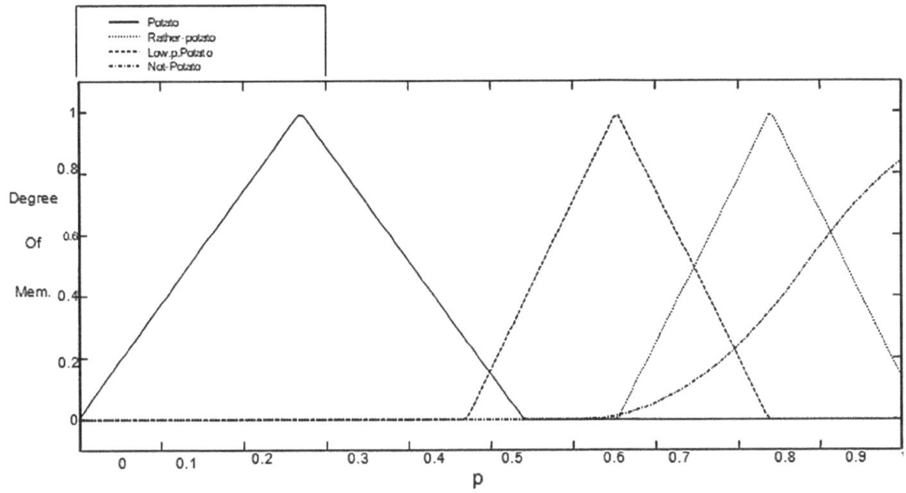

(A)

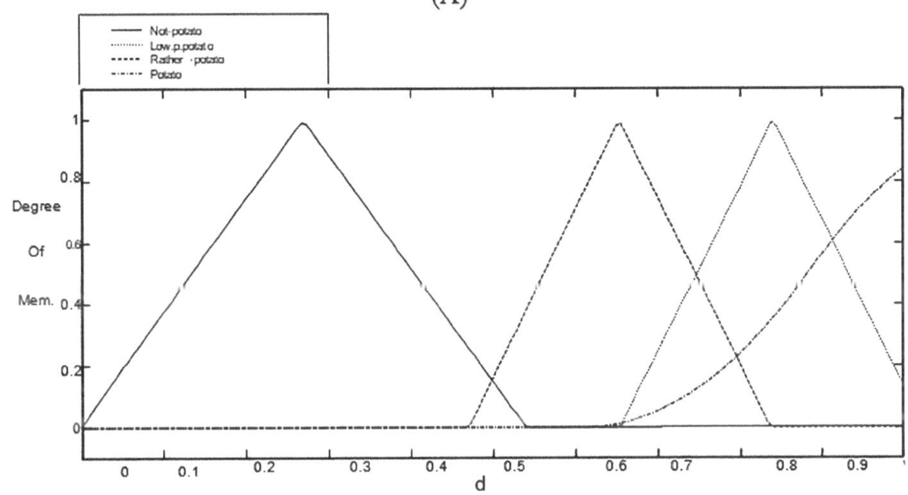

(B)

Initial population was selected 100, and crossover and mutation coefficients (mutation did non-uniform) are 0.8, and 0.04, respectively. Migration coefficient was selected 0.2. The threshold is the chromosome of GA.

After testing about 120 thousands of random potato color pixels from databases, the final value for threshold is got 0.863%. GA overcomes the manual ones and this

Figure 22. (A) The input image, (b) Probability image from the input image

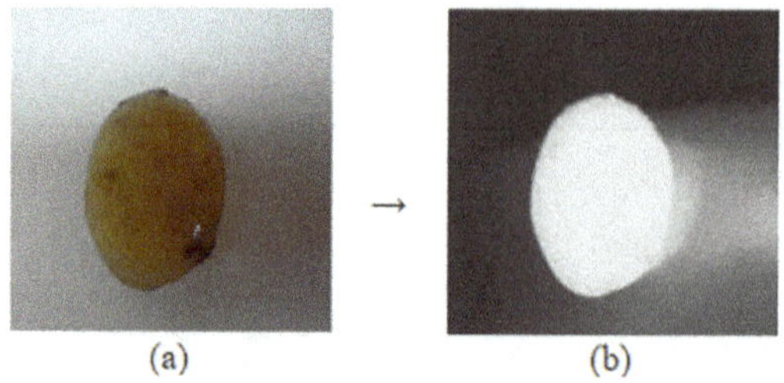

Figure 23. (A) Fuzzified image, (B) Manual binarization of the prior image

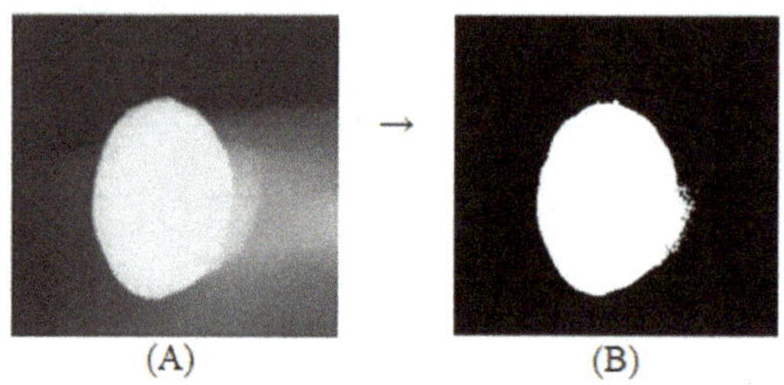

means that the pixels with potato-likelihood more than 0.863% are regarded as the potato-pixels (Figure 24).

After thresholding, morphological operations including filling holes, closing, and opening are utilized respectively. The results are shown in Figure 25.

For illustrating the system robustness for different datasets, some results are shown in Figure 26.

In Figure 27, the applications of the fuzzy method in industrial and outdoor environments are shown.

As it is cleared, some parts of potato tubers are segmented inaccurately; this is because we don't improve the system and its morphological operations based

Figure 24. (A) Probability image, (B) Binarization of prior image by GA

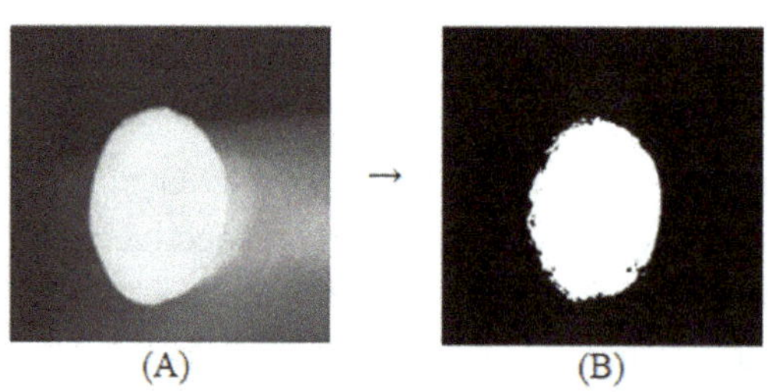

Figure 25. (A) Threshold image (B) Filling (A), (C) closing (B) and (D) opening (C)

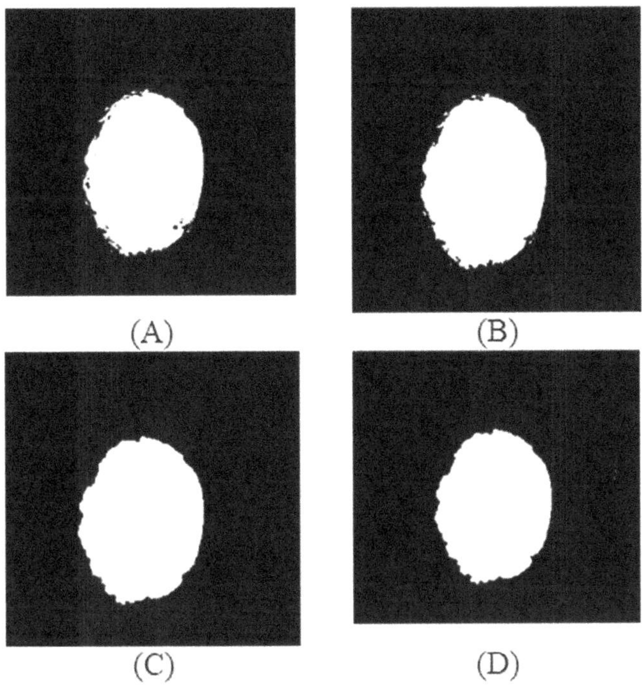

on different illumination which is occurred. Therefore, one important subject in segmentation quality inspection of the agricultural products (or any other objects) is to improve the image processing system based on environment illumination and disturbances.

Image Color Segmentation Based on Optimized Neuro-Fuzzy Systems

In this section, a hybrid method is proposed and implemented using fuzzy inference system in optimized color space. This system, which is designed by neuro-adaptive learning technique, apply a sample image as an input and can reveal the likelihood of being a special color for each pixel through the image. The intensity of each pixel shows this likelihood in the grey level output image.

After choosing threshold value, like before, a binary image is obtained, which can be applied as a mask to segment desired color in the input image.

Besides using fuzzy systems, optimizing color space for segmentation is another feature of the proposed method. This optimizing is implemented by genetic algorithms and influence on system accuracy (Jain, Kandel, & Teodorescu, 2017).

Figure 26. (A) input image, (B) image after fuzzification and GA thresholding, (C) image after morphological operations and (D) output image

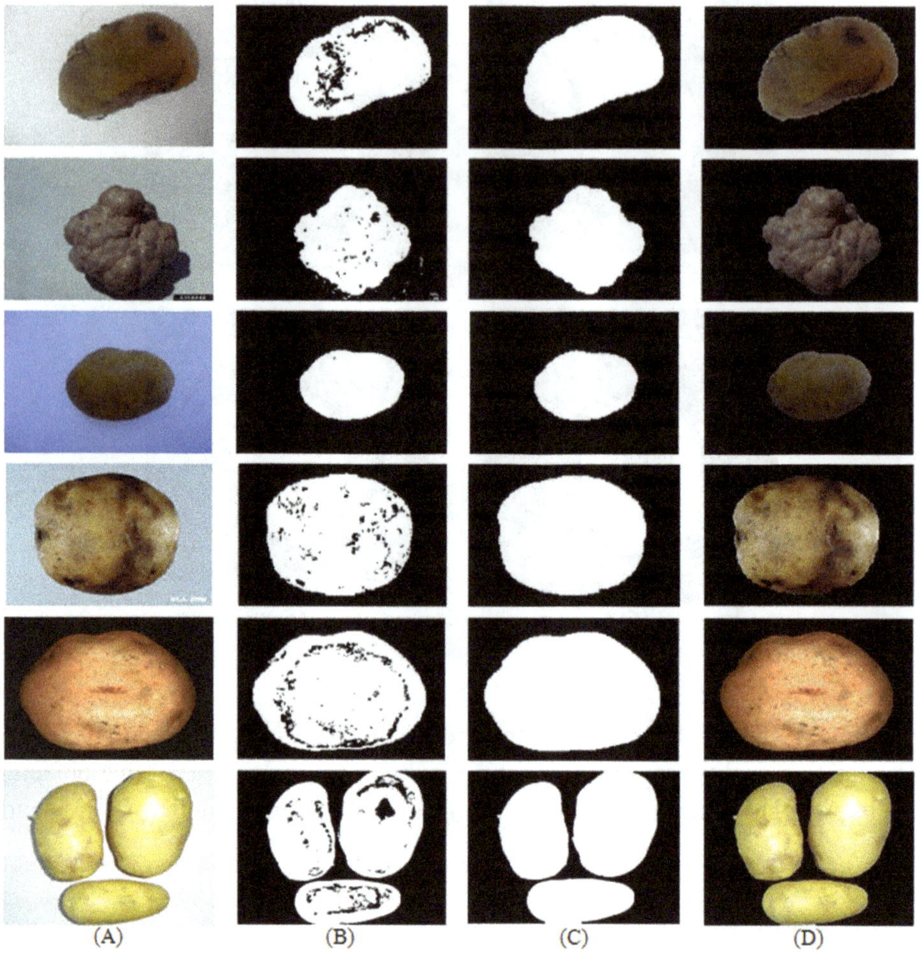

Sugeno type is one the commonly used fuzzy inference method which is employed in this section as well (Zhao et al., 2018). The Sugeno fuzzy model was proposed by Takagi, Sugeno, and Kang in an effort to formalize a system approach to generating fuzzy rules from an input-output data set. Sugeno fuzzy model is also known as Takagi-Sugeno model. A typical fuzzy rule in a Sugeno fuzzy model has the following format:

IF x is A and y is B THEN $z = f(x, y)$, (23)

Figure 27. Applications of the fuzzy method in industrial and outdoor environments

where A and B are fuzzy sets in the antecedent; $Z=f(x,y)$ is a crisp function in the consequent. Usually, $f(x,y)$ is a polynomial in the input variables x and y, but it can be any other functions that can appropriately describe the output of the system within the fuzzy region specified by the antecedent of the rule. When $f(x,y)$ is a first-order polynomial, we have the first-order Sugeno fuzzy model.

When f is a constant, we then have the zero-order Sugeno fuzzy model, which can be viewed either as a special case of the Mamdani FIS where each rule's consequent is specified by a fuzzy singleton or a special case of Tsukamoto's fuzzy model where each rule's consequent is specified by a membership function of a step function centered at the constant. Moreover, a zero-order Sugeno fuzzy model is functionally equivalent to a radial basis function network under certain minor constraints. Figure 28 depicts an example of Sugeno fuzzy model.

In this part, a different method for achieving a good choice of color space is proposed. The intensity varies according to environmental conditions but simply discarding luminance information affects the model's accuracy. In this research, ((1/2) H, S, (2/3) I) is chosen by trial and error.

Here we introduce a novel method to achieve the most effective color space. An (αH,βS,γI) color space is considered, where α, β, and γ are the key parameters they should be selected to make an optimum color space. To materialize a precise system, genetic algorithm (GA) is applied to set α, β and γ values.

GA is the most extended group of an evolutionary technique known, which rely on the use of a selection, crossover and mutation operators. Here, GA characteristics are as follows:

Initial population = 150
Crossover coefficients= 0.67
Mutation coefficient = 0.005
Migration coefficient = 0.35

α, β and γ are chromosomes of the GA, whose fitness function compares the whole detected target pixels in the sample images with the actual number of these pixels, and attempts to minimize the difference. In this way, optimum values are obtained to accurate segmentation.

Figure 28. An example of Sugeno fuzzy system

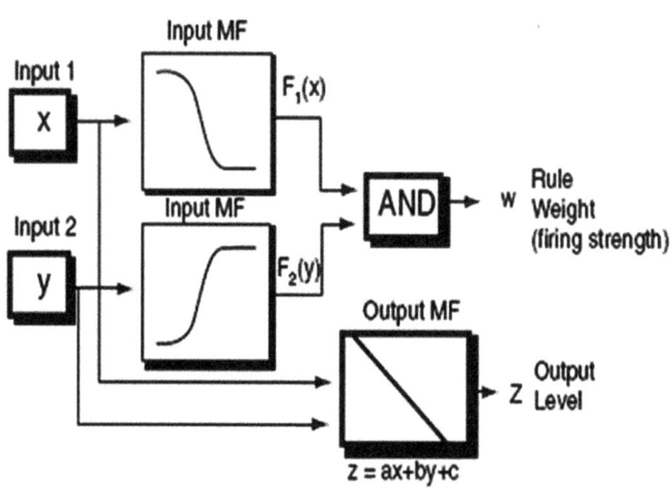

To segment an arbitrary image, pixels with special color through the image should be searched. This color can be any color like red, green, blue, yellow, etc. or the color of a particular object such as vegetables, fruits or even human body skin.

Sugeno fuzzy system used is a 1-input, 1-output system applying the Euclidean distance between the color of each pixel to the average target color sub-space as an input, and the likelihood of being target pixel as an output.

150,000 color vector, contains the pixel of the target color and also any other color, should be provided as a training data set. FIS is designed by neuro-adaptive learning techniques. These techniques provide a method for the fuzzy modeling produce to learn information about a data set, in order to compute the membership function parameters that best allow the associated fuzzy inference system to track the given input/output data. This learning method works similarly to that of neural networks.

Here, first subtractive clustering is implemented on data (Yan & Gelan, 2018).

After deciding about the number of MFs with this method a hybrid learning algorithm to identify parameters of Sugeno-type fuzzy inference system is used. It applies a combination of the least-square method and the backpropagation gradient descent method for training FIS membership function parameters to emulate a given training dataset. The outputs are linear and the weighted average method is used for defuzzification (De & Sil, 2015).

When the sample image is used as an input for designed FIS, an output image is a grey-level image. The intensity of each pixel shows the probability of being a target color vector. We need a binary image to make mask and segment region of interest. So, the value as the threshold should be selected. After investigating various results and with trial and error, 87% is chosen as the best threshold value. It means that the pixels with the %87 likelihood or more are regarded as target pixels.

The binary image is formed by setting target pixels to 1 and all other pixels to 0. After this, morphological processing, consist of opening followed by closing (Razmjooy, Ramezani, & Ghadimi, 2017), is accomplished to acquire separated and connected regions.

To show the efficiency of this method, we design two systems for potato detection. Moreover, proposed algorithms are applied to a wide range of applications to segment any region of interest (ROI) with a special color.

Recognizing different kinds of vegetables and fruits is a recurrent task in supermarkets, where the cashier must be able to point out not only the species of a particular fruit (i.e., banana, apple, pear) but also its variety (i.e., Golden Delicious, Jonagold, Fuji), which will determine its price. The use of barcodes has mostly ended this problem for packaged products (Fernandcz, Xian, & Tian, 2017).

Besides, vegetable recognizer system is applicable in the industry to recognize defected products and quality control. Generally, a computer vision algorithm for a defect detection system includes two main stages; in the first stage, a proper segmentation algorithm should be applied on the input image to separate purpose objects from the background and the second stage consists a proper defect detection algorithm which is used on the purpose objects. The quality of the segmentation algorithm plays an important role in improving the output of defect detection stage.

Utilizing the proposed method, a system for segmentation potato color is proposed.

Designing steps are similar to that detailed in the previous section. The differences just lie in a number of obtained clusters, which are three and called: (Potato, Rather-Potato, Not-potato), and obviously optimum color space. Here (0.91H, 0.68S, 0.5I) is obtained as the optimum color space. The input membership function is shown in Figure 29.

To evaluate the performance of the proposed color image segmentation, some experiments are carried out to detect human body skin and potato color detection, in two databases.

To segment potato color, the proposed algorithm is tried on over more than totally 250 images. These images are from different databases with different backgrounds, including USDA, CFIA, and an obtained potatoes image database from Ardabil (Iran's northern west area).

Figure 29. Input Membership Function for potato segmentation system

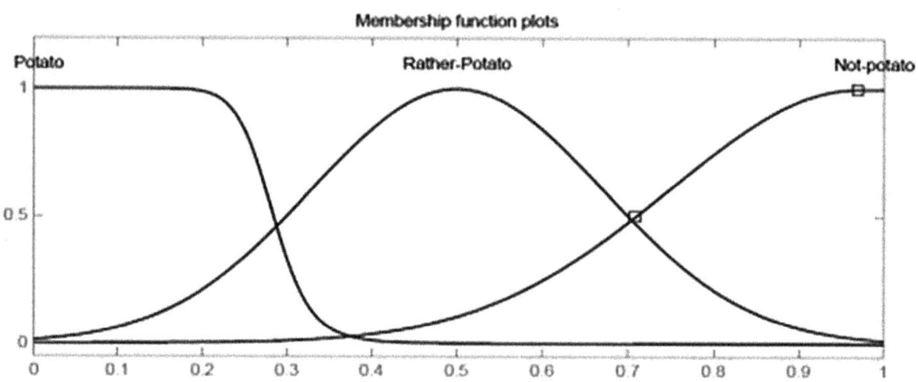

Table 6. Obtained Detection Rates (%)

Color Space	*(H,S,I)*	*((1/2) H, S, (2/3) I)*	*Optimized Color space*
Potato Segmentation	91.6	95.2	98

To better understanding of color space effect, each algorithm is applied in three different color spaces including (H, S, I), ((1/2) H, S, (2/3) I), and obtained optimum color space. Results are summarized in Table 6. As it is observable, the optimized color space improves the output performance compare to other spaces. So, it can be concluded that the designed system, especially in optimized color space, is applicable as a reliable technique to segment any color through the input color image. Some examples of segmented skin and potato regions are depicted in Figure 30.

Figure 30. Obtained images from:(A) skin color segmentation and (B) potato color segmentation, algorithms

(A) (B)

CONCLUSION

Automatic quality inspection of potato tubers based on machine vision has a lot of benefits like enhancing the speed of the process, enhancing the system accuracy, etc.

Image segmentation is a significant part of machine vision which has a large effect on the total automatic system. In this study, we discuss and evaluate main image segmentation techniques used for the purpose of potatoes image segmentation. The analysis illustrated that there is no perfect image segmentation technique and each technique has its special advantages and disadvantages.

Therefore, selecting a definite method for potatoes segmentation is not possible; i.e. using a definite method for all type of images cannot perform well.

REFERENCES

Ahmadvand, G., Mondani, F., & Golzardi, F. (2009). Effect of crop plant density on critical period of weed competition in potato. *Scientia Horticulturae*, *121*(3), 249–254. doi:10.1016/j.scienta.2009.02.008

Boyat, A. K., & Joshi, B. K. (2015). *A review paper: noise models in digital image processing*. arXiv preprint arXiv:1505.03489

De, I., & Sil, J. (2015). *A Fuzzy Regression Analysis Based No Reference Image Quality Metric. In Advances in Intelligent Informatics* (pp. 87–95). Springer.

Dogra, A., & Bhalla, P. (2014). Image sharpening by gaussian and butterworth high pass filter. *Biomedical & Pharmacology Journal*, *7*(2), 707–713. doi:10.13005/bpj/545

Fan, F., Cong, W., & Wang, G. (2018). Generalized backpropagation algorithm for training second-order neural networks. *International Journal for Numerical Methods in Biomedical Engineering*, *34*(5), e2956. doi:10.1002/cnm.2956 PMID:29277960

Fernandcz, W. P., Xian, Y., & Tian, Y. (2017). *Image-Based Barcode Detection and Recognition to Assist Visually Impaired Persons.* Paper presented at the 2017 IEEE 7th Annual International Conference on CYBER Technology in Automation, Control, and Intelligent Systems (CYBER). 10.1109/CYBER.2017.8446388

Garces, E., & Reinhard, E. (2018). *Light-Field Surface Color Segmentation with an Application to Intrinsic Decomposition.* Paper presented at the 2018 IEEE Winter Conference on Applications of Computer Vision (WACV). 10.1109/WACV.2018.00166

García-Lamont, F., Cervantes, J., López-Chau, A., & Ruiz, S. (2018). *Contrast Enhancement of RGB Color Images by Histogram Equalization of Color Vectors' Intensities.* Paper presented at the International Conference on Intelligent Computing. 10.1007/978-3-319-95957-3_47

Hau, C. C. (2015). *Handbook of pattern recognition and computer vision.* World Scientific.

Jain, L. C., Kandel, A., & Teodorescu, H.-N. L. (2017). *Fuzzy and neuro-fuzzy systems in medicine.* CRC Press.

Khmag, A., Ramli, A. R., Al-Haddad, S., & Kamarudin, N. (2018). Natural image noise level estimation based on local statistics for blind noise reduction. *The Visual Computer, 34*(4), 575–587. doi:10.100700371-017-1362-0

Kyi, T. M., & Zin, K. C. M. (2018). *Color Segmentation Based on Human Perception Using Fuzzy Logic.* Paper presented at the International Conference on Big Data Analysis and Deep Learning Applications.

Moallem, P., & Razmjooy, N. (2012). A multi layer perceptron neural network trained by invasive weed optimization for potato color image segmentation. *Trends in Applied Sciences Research, 7*(6), 445–455. doi:10.3923/tasr.2012.445.455

Moallem, P., Razmjooy, N., & Ashourian, M. (2013). Computer vision-based potato defect detection using neural networks and support vector machine. *International Journal of Robotics and Automation, 28*(2), 137–145. doi:10.2316/Journal.206.2013.2.206-3746

Qi, Y.-n., Lü, C.-x., Zhang, J.-n., Li, Y.-s., Zeng, Z., Mao, W.-h., . . . Yang, B.-n. (2018). *Image recognition on raw and processed potato detection: a review.* Paper presented at the Fourth Seminar on Novel Optoelectronic Detection Technology and Application. 10.1117/12.2315467

Razmjooy, N., Mousavi, B. S., & Soleymani, F. (2012). A real-time mathematical computer method for potato inspection using machine vision. *Computers & Mathematics with Applications (Oxford, England), 63*(1), 268–279. doi:10.1016/j.camwa.2011.11.019

Razmjooy, N., Mousavi, B. S., & Soleymani, F. (2013). A hybrid neural network Imperialist Competitive Algorithm for skin color segmentation. *Mathematical and Computer Modelling, 57*(3), 848–856. doi:10.1016/j.mcm.2012.09.013

Razmjooy, N., Ramezani, M., & Ghadimi, N. (2017). Imperialist competitive algorithm-based optimization of neuro-fuzzy system parameters for automatic red-eye removal. *International Journal of Fuzzy Systems, 19*(4), 1144–1156. doi:10.100740815-017-0305-2

Russ, J. C. (2016). *The image processing handbook.* CRC Press.

Salaman, R. N. (2014). *Potato varieties.* Cambridge University Press.

Xu, X., Xu, S., Jin, L., & Song, E. (2011). Characteristic analysis of Otsu threshold and its applications. *Pattern Recognition Letters, 32*(7), 956–961. doi:10.1016/j.patrec.2011.01.021

Yan, W., & Gelan, Y. (2018). A new brain magnetic resonance imaging segmentation algorithm based on subtractive clustering and fuzzy C-means clustering. *Journal of Medical Imaging and Health Informatics, 8*(3), 602–608. doi:10.1166/jmihi.2018.2309

Yeager, M., Gregory, B., Key, C., & Todd, M. (2018). On using robust Mahalanobis distance estimations for feature discrimination in a damage detection scenario. *Structural Health Monitoring*, 1475921717748878.

Zhao, F., Liu, H., Fan, J., Chen, C. W., Lan, R., & Li, N. (2018). Intuitionistic fuzzy set approach to multi-objective evolutionary clustering with multiple spatial information for image segmentation. *Neurocomputing*.

ENDNOTE

1 It is important to know that to select a correct colorspace for different datasets, we should apply the technique.

Chapter 2
WT-MO Algorithm:
Automated Hematological Software Based on the Watershed Transform for Blood Cell Count

Ana Carolina Borges Monteiro
Universidade Estadual de Campinas, Brazil

Yuzo Iano
Universidade Estadual de Campinas, Brazil

Reinaldo Padilha França
Universidade Estadual de Campinas, Brazil

Navid Razmjooy
https://orcid.org/0000-0002-4465-7691
Independent Researcher, Belgium

ABSTRACT

Visual examination of blood smears is an essential tool for analysis, prevention, and remediation of several types of maladies. The interest of computer-aided decision has been acknowledged in many medicinal instances (e.g., automatic ways and means are being explored to spot, classify, and measure visual items in hematological cytology [HC]). This chapter proposes an entirely automated blood smear diagnosis system for hemograms, which can lessen the time spent to scrutinize a slide. The present framework relies on morphological operations (MOs) and soft segmentation by means of the watershed transform (WT). Experiments demonstrate the method efficacy to count white blood cells (WBCs) and red blood cells (RBCs). Some considerations about implementations, design advice and possible variants, as well as improvements are discussed. The future of automated medical analysis is contemplated.

DOI: 10.4018/978-1-5225-8027-0.ch002

INTRODUCTION

Hematologic examinations are relevant for diagnosing disorders of the hematologic system, and they help in the identification of many systemic and organ illnesses. Even though in most cases the disease diagnosis cannot be established solely on a Complete Blood Cell (CBC) count, the hemogram may add valuable evidence in the diagnosis, surveying, and elaboration of a prognosis concerning the future disease progression in an individual. This chapter provides physicians, veterinaries and laboratory technicians with the essentials on the design of an automated digital system for the analysis of Blood Cells (BCs), and diseases related to anomalies in the hematological profile. Since automated cell counters have become increasingly popular, the focus of the present system is on anomalies noticeable by automated cell counters such as abnormalities in the amount, volume, and hemoglobin concentration of BCs rather than on abnormal morphology. Nevertheless, if pathological or irregular outcomes are discovered, or abnormal morphology or BC function damage is suspected, a further microscopic blood smear analysis is strongly advised (Gulati et al., 2002) .

Blood analyses entail a blood smear obtained by sliding a few microliters of blood on a glass slide and subsequent use of dyes, which allow the visualization of cellular structures. Only the final portion of the slide is used to perform the counts, because the anterior portions have clustered and/or overlapping cells, preventing a reliable count. This blood smear is used to analyze the morphology and staining of Red Blood Cells (RBCs) as well as White Blood Cells (WBCs) and it can help to investigate the presence of anisocytosis, such as sickle cell anemia (sickle-shaped RBCs), anemia spherocytosis (round RBCs), hereditary elliptosis (elongated RBCs), among others besides aiding in the identification of blood parasites, such as leishmaniosis, malaria, and so on (Melo, & Silveira, 2013; Ciesla, 2012; Xavier et al., 2016).

Given the above, making the BC methodology more accessible to the less favored populations may help in the detection and consequent early treatment of the pathologies.

It is important to note that there are many tools to solve health problems using a high-level programming language and interactive environment for numerical computation, visualization and programming. This work relies on the MATLAB IDE that is a proprietary software but programs developed in it can be executed in Free Open Source packages like Octave and Scilab (Hansen, 2011; Roux, 2016; Wouwer et al., 2014). It can analyze data, develop algorithms, create models and applications, build mathematical tools capable of exploring multiple approaches and search for a faster solution than traditional programming languages such as C/ C++ or Java (Reyes-Aldosoro, 2015).

First, image preprocessing of the digitized smears is applied for suppressing noise, ameliorating the luminance/contrast differences in the blood smear color image and, in some cases, removing unnecessary image items such as platelets. Second, a segmentation process discovers and isolates the important entities in the image. The third stage characterizes the objects previously extracted to be used in the last phase, i.e., the classification stage. Feature selection can lessen the redundant information required to describe important parts of the image. Selected features are fed to the classification process, which makes the class assignment (Diaz, & Manzanera, 2009).

It is important to stress the fact that the framework discussed here can be easily adapted to veterinary use provided the parameters and databases are adjusted to the animal undergoing examinations. The system in (Herman et al., 2018) is a dedicated analyzer for cows. Figure 1 shows a blood smear for a human and a bovine.

Figure 1. Blood smears for a human (a) and a cow (b)

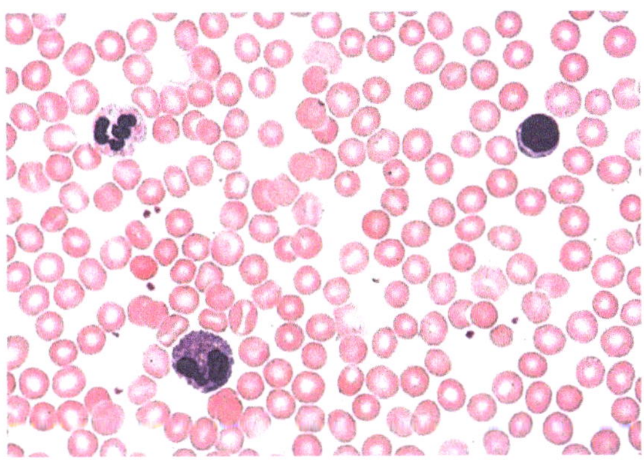

(a) Human blood

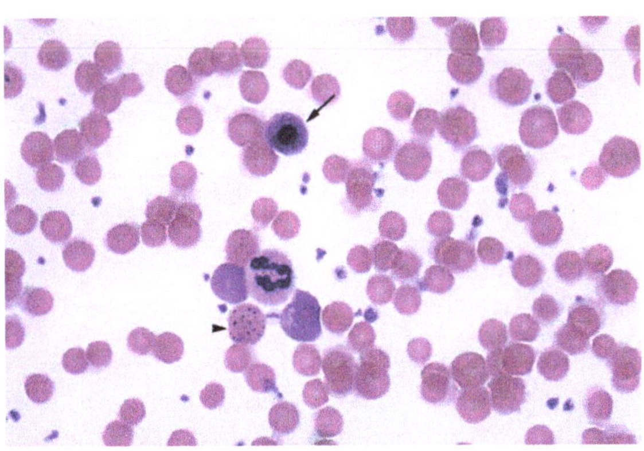

(b) Bovine blood (Cotter, 2018; Eclinpath 2013)

Thus, the present study aims to develop framework such as depicted by Figure 2 that is capable of simultaneously detecting and counting the erythrocytes (RBCs) and leukocytes (WBCs) present in the blood smear images. The methodology must meet the low-cost criteria without loss of quality and reliability. The system modeling, as well as the program, were implemented using an IDE the Matlab® version 2014a.

This chapter talks about the importance of hematological cell counting and analyzing. Later, it discusses a digital image software system (Figure 2 depicts it) for hematological studies based on a variant of the Watershed Transform (WT). Finally, experimental results are shown and discussed.

HEMATOLOGY BACKGROUND

The blood contains approximately 55% plasma and 45% BCs (RBCs, WBCs, and platelets) in human beings. The blood carries gases and nutrients to the tissues, helps the excretion of metabolites, distributes hormones, while transmitting chemical messages to distant organs, regulates and maintains the body temperature besides maintaining the acid and osmotic balances (Junqueira, & Carneiro, 2004; Junqueira, & Carneiro, 2012; Verrastro, 2005). The process of forming BCs within the bone marrow is called hematopoiesis, which highly depends on the presence and stem cell performance. For this, this process relies on a series of actions as well as responses leading to duplication, differentiation and maturation processes that generate and release of mature cells into the bloodstream (Fischbach, & Duynning, 2009).

Figure 2. Image analysis scheme for blood cell identification

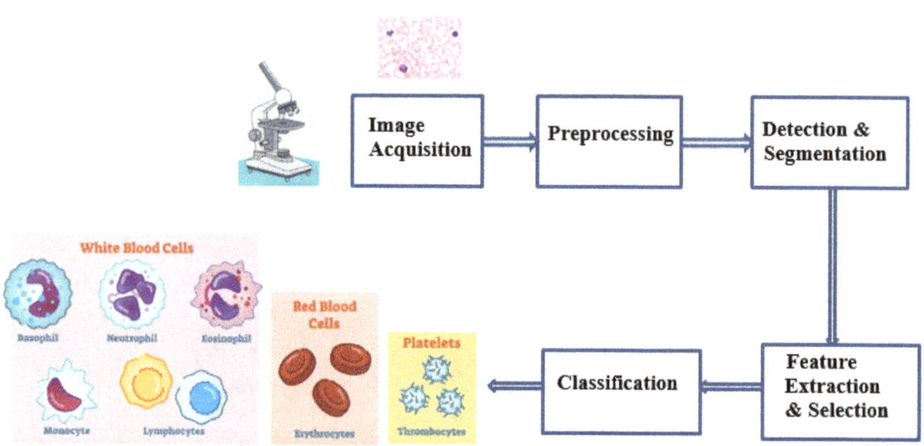

Erythrocytes aka RBCs are morphologically characterized by anucleation, biconcave disk format, formed by a tetram (two chains α and two chains β) responsible for transporting oxygen (Hashimoto, & Bertassoni, 2009; Verrastro, 2005).

Leukocytes or WBCs are the body defense cells (aka make up the innate and the adaptive immune responses). WBCs are classified into 2 distinct classes: leukocytes granulocytes in addition to leukocytes agranulocytes. Granulocytic leukocytes are the defense cells that have granules diffused in the cytoplasm, which have anti-inflammatory and antimicrobial action. Agranulocyte leukocytes do not present granules visible in light microscopy, and have only one lobe (Abbas et al., 2012).

Hemogram

The hemogram is a laboratory exam for the quantification and qualification of BCs (Buttarello, & Plabani, 2008). Regardless of the methodology used to perform the blood count, this test should be performed from the collection of 4 ml of venous blood collected in a tube containing anticoagulant EDTA (Ethylenediamine tetraacetic acid) or heparin, which can be collected with or without fasting of the patient. Samples should avoid the presence of hemolysis (ruptured erythrocytes causing a lower count in the total amount of RBCs), lipemia (presence of fat in the form of triglycerides which makes it difficult to read the results by colorimetric methods), clots (result in lower counts of platelets), antibiotics, anti-inflammatory and antiallergic use hours before the test (which decrease the amount of leukocytes, and may mask infectious or allergic conditions) and avoid smoking 2 hours before examination (change in hemoglobin, hematocrit and hematocrit indices) (Nemer et al., 2010; Godon et al., 2012).

The erythrogram is the hemogram part responsible for the evaluation of the total number of erythrocytes, hemoglobin, hematocrit and erythrocyte indices: (i) Mean Corpuscular Volume (MCV), (ii) Mean Corpuscular Hemoglobin (MCH), and (iii) Mean Corpuscular Hemoglobin Concentration (MCHC). The total erythrocyte count is used for the detection of hematological disorders. Decreases in the number of erythrocytes are called erythropenia, being a strong indication of the presence of anemia or blood loss. On the other hand, erythrocytosis is characterized by increased production of erythrocytes, either by genetic or external factors, being responsible for the appearance of polycythemia (Turgeon, 2004; Melo, & Silveira, 2015).

Anemias can be classified as the genetic origin (sickle cell anemia, thalassemias, anemia spherocytosis, among others) as well as by food deficiency (iron deficiency anemia, megaloblastic anemia, among others). Both types inspire care and medical follow-up, because food anemia causes tiredness, a decrease in work performance, pallor and vertigo. Genetic anemias, in turn, require lifelong care, because they are responsible for pain crises, hospitalizations and the need for periodic blood transfusions. Some types of polycythemia are of genetic origin and, like genetic anemias, require lifelong care, as it is characterized by exacerbated production of erythrocytes, causing damage to the body such as infarction and stroke (Turgeon, 2004; Melo, & Silveira, 2015; Kawthalksr, 2013).

The erythrocyte indices are dependent on the hemoglobin, hematocrit and RBCs values, whose function is to evaluate the staining and the size of the erythrocytes. The size of erythrocytes is evaluated by MCV, which classifies these cells into normocytic, microcytic and macrocytic cells. The normocytic RBCs are those that present normal size, while the microcytic ones have a smaller size and the macrocytes have a larger volume than those expressed by the reference values. Both MCH and CMCH are responsible for the evaluation of erythrocyte staining according to the parameters: normal staining, insufficient staining or exacerbated staining, being classified as normochromic, hypochromic and hyperchromic, respectively (Turgeon, 2004; Melo, & Silveira,2015).

The hemogram also evaluates hemoglobin that is a protein inside the erythrocytes responsible for the red pigmentation of the cells and the transport O_2 and CO_2 to the tissues, through the association of these gases with iron molecules. The hematocrit, in turn, expressed as a percentage the amount of BCs present in the plasma (Provan et al., 2015).

The leukogram evaluates quantitatively the leukocytes granulocytes and agranulocytes, through the the total leukocyte count and differential leukocyte count. The total leukocyte count indicates or excludes the presence of inflammation and/or infections from a variety of sources. The decrease in total leukocyte count is called leukopenia and occurs in cases of cancer patients or in cases of a sedentary lifestyle. In turn, leukocytosis is represented by the increase in the amount of leukocytes, due to the presence of allergies; parasitic, viral or bacterial infections; or leukemias (Ciesla, 2012; Erichsen et al.,2015).

The leukocyte differential count evaluates the amount of each leukocyte: neutrophils, eosinophils, basophils, monocytes, and lymphocytes. Neutrophils are the most abundant BCs, accounting for about 70% of all circulating BCs. The quantitative increase of neutrophils is called neutrophilia, indicating the presence of acute bacterial infections. However, neutrophilia is also present in patients with burns, myocardial infarction, postoperative period, stress, glucocorticoid use and intense physical exercise (Abbas et al., 2012; Murphy et al., 2010).

Neutrophilia is divided into two types: neutrophilia with left deviation and neutrophilia with right deviation. The left-sided deviation is a present condition in cases of acute infections, when large numbers of neutrophils are released into the bloodstream. On the other hand, the deviation to the right, is characterized by the presence of segmented or hyper segmented neutrophils (more than 5 lobes), being indicative of chronic infections. The decrease in the amount of neutrophils is called neutropenia and may indicate, spinal cord infiltration, immunosuppression by the HIV, alcoholism, food deficiency, exposure to radiation or use of cytotoxic drugs (Provan et al., 2015; Abbas et al., 2012; Murphy et al., 2010).

Eosinophils are cells that act primarily on the intestines, the epithelium, and the respiratory system. Thus, eosinophilia is characterized by the numerical increase of these cells, being indicative of intestinal infections by helminths (schistosomiasis, ancylostomiasis, ascariasis, among others), dermatitis, dermatoses, acute myeloid leukemia, chronic myeloid leukemia, pernicious anemia or respiratory allergic processes. However, the decrease (eosinophilia) or the absence of eosinophils in the bloodstream is normal. Thus, eosinopenia usually results from the administration of adrenocorticotrophic hormones or corticosteroids (Murphy et al., 2010; Xavier et al., 2010; Abbas et al., 2012).

Basophils are poorly present cells in the bloodstream. However, they play important roles during allergic reactions through the release of IgE and histamines. Basophilia is a process in which there are more circulating basophils, being indicative of allergic reactions or presence of leukemias. Basopenia, in turn, is not classified as a pathological process. Monocytes are extensively related to phagocytosis processes. In this way, the increase in the production and release of these cells is called monocytosis. This increase is suggestive of tissue repair or indicates the presence of chronic infections. However, monopenia is indicative of decreased production and release of monocytes, and is often associated with the lifestyle of the patient who is exposed to stress. All leukocytes show variations in their amounts according to the age of the patient (Abbas et al., 2012; Murphy et al., 2010).

Lymphocytes are BCs that are active in viral infection processes. The increase in the number of these cells in the bloodstream (lymphocytosis) is directly related to the presence of viral pathologies such as hepatitis, HIV, rubella, among others. The decrease in the amount of lymphocytes (lymphopenia) can be interpreted as chronic HIV, acute stress or the presence of cirrhosis of the liver. It is important to note that lymphocytosis is a normal condition during childhood (Hamdin et al., 2003).

An hemograms can be manual or automated but even with the acquisition of hematological equipment, the manual hemogram is not a dispensed practice and is recommended for the confirmation of hematologic reports of pediatric patients, patients over 75 years of age, cancer patients, patients with suspected leukemia or polycythemia, patients with leukocytosis and patients in severe condition (Failace, & Fernandes, 2015).

Manual Hemogram

The manual hemogram totally depends on a trained human and it has a low cost, but it requires more time and it is less reliable because it depends on the counts and calculations performed by the health professional. Therefore, it is considered a good alternative for small laboratories, where the demand for exams is small and does not justify the cost of hematological equipment and reagents (Failace, & Fernandes, 2015). A standard microscope performance is influenced by its space-bandwidth product, leading to a trade-off between the image resolution and the microscope Field-of-View (FOV). Hence, a conventional microscope needs an objective with an adequate magnification to view and count BCs. Since the FOV is very small, the counting process necessitates mechanical scanning of the specimen, which calls for correct alignment and regulation to bypass any overlap of the regions undergoing scanning. Another drawback is the physical strain on the professionals performing the manual scans by direct observation of the microscope, which may be detrimental to the physician's diagnostic.

Automated Hemogram

The automation of the hemogram implies a greater agility in the accomplishment of the exams and in the release of the reports, however, they are a more expensive methodology when compared to the manual methodology.

The impedance principle and the threshold concept are responsible for the introduction of multi-parameter devices on the market. These devices are capable of performing simultaneous cell counts using separate channels for each type of BC. However, in these techniques the RBCs are undetectable. To solve this problem the erythrocytes began to be counted by means of flow cytometry and hydrodynamic focus, where these cells are counted one by one through an extremely fine capillary. The RBCs are then subjected to a laser beam, where the light scattering is analyzed at different angles of deviation, where at zero degree is indicated the cell size, the ten-degree indication of the internal structure and the 90-degree indication of leukocytes and its characteristics of looseness and the content of granulations (Chabot-Richards, & George, 2015, Failace & Fernandes, 2015; Green, & Wachsmann-Hogiu 2015).

Currently, there are a large number of multi-parameter devices, which use the impedance, conductivity and dispersion techniques of emitted light. These technologies may also be associated with the cytochemical characteristics of cells (such as myeloperoxidase) and the use of reagents that perform the analysis of certain cell types. However, before purchasing a hematology device, it is necessary to take into consideration the following parameters: Automation device vs type of patient attended; number of hemograms day x samples/hour of the apparatus; cost of each blood count; quality control; technical assistance; interfacification, and employee training (Failace & Fernandes, 2015).

Segmentation of individual cells and cell nuclei from microscope images usually involves labeling of nuclei with a marker of DNA. However, this methodology often presents flaws when applied to clusters of nuclei, since the distinction of the nuclear borders becomes difficult to identify in these conditions. Thus, the work in (Soltanzade, 2000).presented a methodology that solved such limitation and allowed the segmentation of whole cells. This segmentation was performed through the identification of each nucleus and/or cell, and the boundaries were expanded until reaching the cell as a whole. This process relied on the use of staining, using curvature gradients and flow techniques. The algorithm developed was tested on computer-generated objects to evaluate its robustness against noise and subsequent application to cells in culture. In all cases presented, the algorithm obtained considerable precision results (Soltanzade, 2000).

The Hough Transform (HT) can deliver precise and efficient curve detection in some images along with the use of grouping techniques aiming at the simultaneous identification of multiple curves in the image has been investigated in (Walsh, & Raftery, 2002). Probabilistic arguments for the development of the stopping conditions of the algorithm were also employed. According to (Song et al., 2005), the HT is a powerful tool for extracting graphic elements of images but its application has been limited to small-sized images because the peak detection and the scan line consumed much longer for large-size images.

A study of the segmentation of cells into solid tissues by (Baggett et al., 2005) tried to understand the cellular and molecular structures, which are the basis of tissue development and function. For this, a software was developed that through segmentation of 2-D microscopic images found the cell membrane in a better state of conservation. The samples were labeled with a cell surface fluorescence marker, so that the membranes became brighter than the other cell structures. The developed algorithm required the user to mark two dots per cell, one in the center and one in the edge (Baggett et al., 2005).

A modified implementations of the HT proposing improvements in the detection of circular objects of low contrast appeared in (Smereka, & Duleba, 2008).

A study by (Wu et al., 2016) used the circular histogram-based Otsu Transform for segmentation of leukocytes. In this way, they stated that for detection of BCs, edge detection by this method is weak because not all boundaries are well determined, it being difficult to get all edge information and to locate cells accurately. On the other hand, (Savkare, & Narote, 2011) carried out a study, whose focus was the detection of erythrocytes infected by the intracellular parasite Plasmodium falciparum, which is the etiologic agent of malaria.

RBCs and leukocytes by differentiating the different morphological characteristics that these cells present with each other. The red cell count was performed through the following steps: conversion to the gray scale; segmentation; noise removal; removal of edge elements and counting by the algorithm. In turn, the leukocyte count came from other steps such as image conversion into a binary image; noise reduction; space filling and removal of cells with poorly defined borders (Arivu, & Sathiya, 2012).

An investigation conducted by (Nasreen et al., 2015) had a digital microscope interfaced to the computer, to obtain the BC images in the RBG system and, later, are converted to the grayscale. Then, the medium filtration was used to remove noise from the capture of the images. In addition, image input has been improved for better segmentation. On the other hand, the image segmentation was performed through gray contours through the graytresh function in Matlab. This function uses the Otsu's method, which chooses a limit to minimize the classes entering the variance of black and white pixels. The BCs detection was performed by the Hough transform.

The Fuzzy Logic (FL) methodology was part of an automatic system to detect and count of RBCs and leukocytes. The counting was performed through the centers of the circles, where the radii were determined for each cell since an RBC is smaller than a WBC. Another way to distinguish BCs is the presence or absence of cell nuclei (Bhagavathi, & Niba, 2016).

WHITE BLOOD CELL COUNT VIA MORPHOLOGICAL OPERATIONS

Morphological Operations (MOs) constitute a broad set of image processing operations based on the shape of binary images. The MOs are applied to the structural element at the input of an image creating an output of the same size, as shown in Figure 3 (Gonzalez et al., 2009).

In this way, the value of the pixels in the image output is based on the comparison of pixels corresponding to the pixels of the image input and neighboring pixels.

Figure 3. Leukocyte count by the WT-MO algorithm

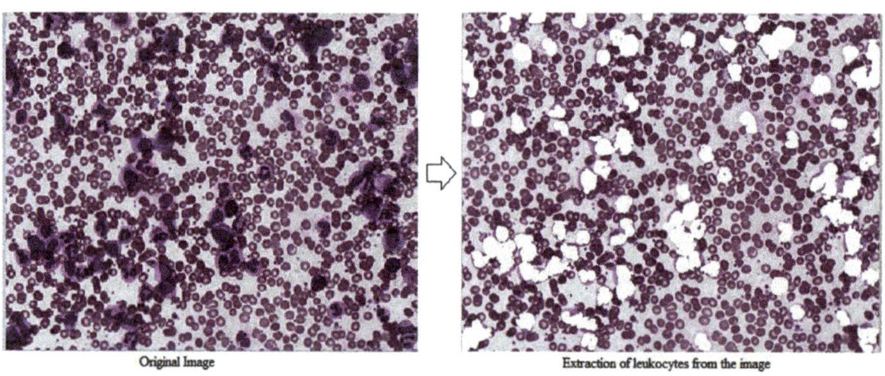

The number of pixels added or removed from objects in an image depends on the size and format of the structural element used to process the image. Morphological operations consist of the following steps: dilatation; erosion; opening; closure and reconstruction (Bhagavathi, & Niba, 2016).

The dilation consists of adding pixels to the edges of objects in an image to the purpose of repairing breaks or possible intruding elements in the image. Erosion, in turn, consists of the removal of pixels from the edges of the image, being used to divide bound objects or remove extrusions. The subsequent equations depict the erosion and dilation operations for an image whose intensity function is $f(x,y)$:

$$\psi_{erosion}(f) = \underset{(\varepsilon,\eta)\subseteq B}{\Lambda} f(x+\varepsilon, y+\eta) = (f \Theta B) \tag{1}$$

$$\psi_{dilation}(f) = \underset{(\varepsilon,\eta)\subseteq B}{\Lambda} f(x-\varepsilon, y-\eta) = (f \oplus B). \tag{2}$$

The aperture is applied for the elimination of protrusions and/or smooth contours (Beucher & Meyer, 1993). Mathematically, it is defined as

$$\psi_{aopn}(f,a) = \underset{t\in\Re}{\Lambda}\{(x,y) \in \cup \{F_S(t) :| F_S(t) |\geq a\}\}, \tag{3}$$

where F_s (t) represents a cross-section of the image intensity f. The F_s $(t)| s = 1; 2; 3$ are granulations of the cross-section $F(t)$ of the image f.

In turn, the closure is used to merge fine and long gaps and gaps, eliminate small holes, and fill spaces in the contour and smooth out the contours. The reconstruction is responsible for extracting relevant information referring to the shapes present in the image (Soltanzade, 2010).

THE RED BLOOD CELL COUNT VIA WATERSHED TRANSFORM

Habitually, image segmentation is employed to identify Regions of Interest (ROIs) or in data annotation (Barnard et al., 2003; Chen, & Wang, 2004). Segmentation fractions an image into sets of spatially related pixels in a homogeneous fashion (Bandyopadhyay, 2011).

The intuitive idea behind the WT image segmentation emanates from the geography: the landscape of topography reliefs accountable for the creation of watersheds. The topography is divided into domain lines aiming the attraction of rainwater over the region. An alternate approach is to envision a landscape undergoing a lake immersion, with holes punctured in local minimums (Vitor, 2012).

Hence, the river basins fill with water beginning at the minimum place, where the water meets barriers. Once the water level is at the uppermost peak in the scenery, the process is terminated. Thus, the landscape is split into areas or basins isolated by dams known as basin hydrography lines (Meyer, 2014).

The WT is used in image segmentation problems with mathematical morphology. The logic through the computational analysis of items in images, consists of considering anything recognizable can be blood cells, tissues, DNA microarray components, or even radioactive markers, among others. Image segmentation can isolate objects in a background image, partitioning that image into disjoint areas with each region homogeneous regarding some property, e.g., gray value and texture (Roerdink, & Meijster, 1996; Roerdink, & Meijster, 2000; Vincente, & Soille, 1991; Bieniek, & Moga, 2000).

Segmenting an image by the WT is basically a two-step process, being (1) to find the markers as well as the chosen segmentation criterion (where the most commonly applied are the contrast and the gradient, even though there are not necessarily rules) and (2) perform a marker-controlled WT with these two elements (Roerdink, & Meijster, 2000; Vincente, & Soille, 1991; Meijster, & Roerdink, 1996; Bieniek, & Moga, 2000).

The WT has been broadly used in CV, specifically, in the segmentation of medical images because of its advantages it adds to this area of science: being an intuitive simple, fast and parallel method, together with an increase semi-linear, sequential and multi-threaded processing found in 32 and 64-bit processors, producing a thorough division of the imagery into separate regions, having the positive differential on the matter if the image contrast is weak, avoiding the requirement for any type of contour junction (Roerdink, & Meijster, 2000; Vincente, & Soille, 1991; Meyer, 1994; Meijster, & Roerdink, 1996; Bieniek, & Moga, 2000).

The algorithm conceived by Vincent and Soille (VS), is founded on the notion of immersion where each local minimum of a grayscale image can be assumed as a surface with a hole and immersed in water. Therefore, the water will gradually fill in different catchment basins of the imagery (aka surface) from the minimum intensity value. The algorithm in this way constructs a dam to evade a situation where water comes from two or more local minima. At the immersion process end, each local minimum is completely delimited by dams consistent with the watersheds of the imagery (or surface) (Roerdink, & Meijster, 2000; Vincente, & Soille, 1991).

The component graphs algorithm is a simple VS algorithm parallel implementation, having the difficulty of segmentation when plateaus happen. Therefore, another tactic was developed first converting an image into a directed value graph with different neighbor values, called the component graph of f. Where in this graph, the WT can be established by a simplified variety of the VS algorithm, where rows with FIFO discipline are not necessary anymore, as no plateaus appear in the graph (Roerdink, & Meijster, 2000; Meijster, & Roerdink, 1996).

A. Watershed Algorithms by Topographical Distance

For these algorithms, their ordered nodes have the shorter topographic distance known and they rely on the shortest path algorithm created by Dijkstra and Moore and minimizing a cost functional between image points employing the distance between the midpoints and the differences between gray levels to determine the regions of the image. These procedures provide a unique solution where a pixel with cost equals two or more regional minima are highlighted as a watershed.

B. Minima Detection

These algorithms rely on the shortest path concept that gets the pixels' treatment order suitable to expedite the application of a traditional raster scanning methodology. Some of these algorithm variants have modifications that permit floods from the local minimum by solving eikonal equations. Its implementation employs an iterative algorithm that integrates the lowest slope of an input image.

C. Distance Transform or Similarity Metric Approach

One tool largely employed to improve the WT for segmentation a similarity metric. When a binary image can be transformed back to a gray-level image, it is proper for WT segmentation via different similarity metrics. Even so, different similarity functionals can yield different effects, so far as Euclidean Distance Transform has a greater likelihood of "Salt and Pepper" noise on the segmentation.

D. Gradient Method

Here, the gradient magnitude is computed using the grayscale image previously to the WT for segmentation. The gradient magnitude image has elevated values on the edges undergoing scrutiny and small values anywhere else. The pixels are now associated with the neighbors.

This method has an excessive segmentation problem (aka oversegmentation), where the topological gradient gives a global analysis of the imagery and in sequence, the undesirable contours owing to the noise can be expressively reduced.

The values can be enriched by the maximum or minimum of the values among each pair of pixels related to the vertices, which can be absolutely separated by this a simple gradient calculation.

E. Marker Controlled Approach

Direct WT application into a gradient image can result in super segmentation owing to noise, performing processes beyond those required in the image, which is a disadvantage. Over-segmentation yields a large number of segmented areas. One approach commonly used to control segmentation relying on markers (Roerdink, & Meijster, 2000).

A marker is a connected element from an image, being used to adjust the gradient image. It can be of two types: internal to object and external to image boundaries. A marker-controlled WT segmentation is a robust and flexible scheme for segmentation of closed-contour objects, with boundaries being expressed as ridges. These values can be optimized by the maximum or minimum of the values between each pair of pixels related to the vertices, being able to perform an absolute separation between them, being a simple form for calculation of gradient (Roerdink, & Meijster, 2000; Bieniek, & Moga, 2000).

There are many research works about still image as well as video segmentation. These studies include motion-based segmentation (Shi, & Malik, 1998; Stiller, 1997), spatial segmentation and motion tracking (Naemura et al., 2000; Coelho, & Estrela, 2012a) moving objects extraction (Fernandes et al., 2014; Meier, & Ngan, 1998), and region growing with spatiotemporal similarity (Choi et al., 1997; Moscheni et al., 1998). Direct clustering segmentation procedures in color space (Deng et al., 1999) thrive on homogeneous color areas but natural scenes contain rich colors and intricate texture. The existence of texture and motion can also portray pathologies.

The algorithms for image segmentation rely on the WT need a way of quantifying the similarity or difference between image pixels. Consequently, the WT needs some gradient info about an image in complex scenes. One example of this is the MPEG-4 standard that needs segmentation to object-based video coding (Beucher, & Meyer, 1993).

Different segmentation techniques to solve these image segmentation problems can be used to lighten tasks formerly done with HC. The combination of morphological segmentation (MO) and the WT has improved sports scenes. This process consisted of coarse segmentation via binary reconstruction grounded on the regions detected thru thresholding the color histogram as well as segmentation by WT. This process can segment sports video frames encompassing brisk movement. That is a key feature in automatic morphological scene segmentation during the broadcast (Andrade, et al., 2003).

Figure 4. Block diagram for the WT-MO algorithm

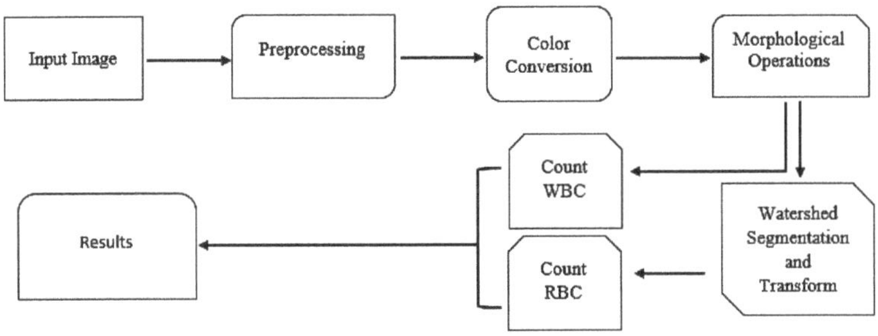

Other studies revealed that the WT works quite well for imagery segmentation. The WT technique computes the dissimilarity of pixels using RGB-D cues with gradient maps, where an RGB-D image is a combination of an RGB image and the related image depth map. The method captures a gradient map to enrich in addition to explain directional variations in image intensities. A distance function or metric measures the similarity between adjacent pixels according to RGB-D values (Jiang, et al., 2015).

Due to the significance and extensive applicability of the WT in image segmentation regardless of the image space dimensionality, and noting the multidisciplinary nature of scientific research. So, the WT application in biomedical engineering can support the development of better blood diagnosis algorithms.

Below, there is the WT-MO algorithm. The WT is actually applied in two versions: an SC WT to obtain a pre-segmentation and WT2. These operations appear inside the block written off as "Watershed Segmentation and Transform" from Figure 4.

Computer Code

```
1. Read f(x, y)
2. Perform the RGB to grayscale conversion
3. Pre-filter with a combination of filters
4. Calculate the gradients
5. Apply the SC WT
6. Create markers
7. Apply MOs
8. Find regional maxima
9. Compute background markers
10. Binarize the image
11. Smooth the image
```

```
12. Compute the Euclidean Distance
13. Apply WT2 to the segment the image
14. If necessary, remove artifacts and return to step 11.
```

WATERSHED TRANSFORM WITH MORPHOLOGICAL OPERATIONS (WT-MO) ALGORITHM

The experiments acquired 25 digital images of blood smear fields containing RBCs and non-pathological leukocytes. These images employed the JPEG, JPG, and PNG formats.

The algorithm developed has as an emphasis on the segmentation process through of the union of WT techniques and morphological operations (the WT-MO methodology). The discovery and erythrocytes counting was done through the WT, while the counted leukocytes result from the MOs process. All this process took into account the distinct morphological characteristics presented between RBCs and WBCs. The WT-MO rationale of the simultaneous counting of RBCs and leukocytes is in Figure 4.

The WT-MO methodology consists of inserting microscopic images of hematoxylin and eosin from stained blood smears in the Matlab environment. After image entry, it goes through preprocessing and color conversion processes, which are accountable for solving small problems like high contrast and low brightness. These problems occur at the image capture, where there may be the little opening of the light condenser in the microscope, which interferes directly with the image brightness. After this step it detects and counts RBCs and WBCs.

Erythrocytes Counts

The segmentation process fractions an image into clusters of spatially related pixels homogeneously. Image segmentation does pixel labeling, where all pixels from the same homogeneous section receive the same label as in Figure 4 (Mogra, & Srivastava, 2014). The definition of homogeneity of an image region corresponds to a particular segmentation concept because each image presents different patterns of pixels (Gonzalez et al., 2009).

For the labeling of the RBCs, the pixel is presumed as a variable directly related to the morphological individualities of the cells under analysis. In this case, the labeling considers the size and staining of the erythrocytes. As the pixels are grouped, the WT-MO algorithm counts and labels each cell in the frame. Labeling consists of inserting a number on top of each cell according to the counting order established by the algorithm as in Figure 5.

Figure 5. Erythrocytes labeled by the WT-MO algorithm

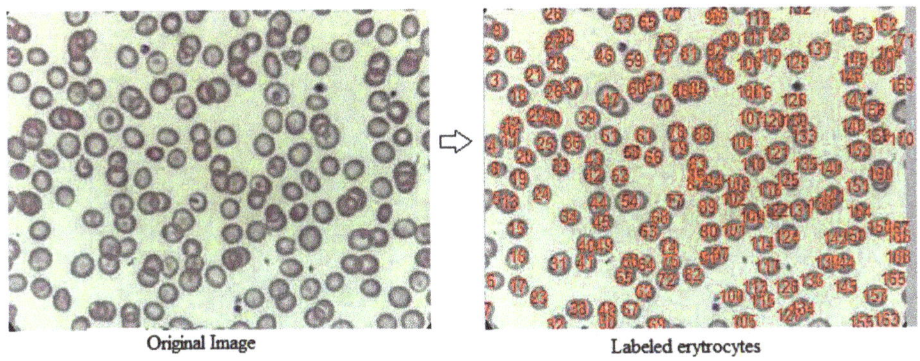

Experiments

1. Cell Counting and Accuracy Determination of the WT-MO Algorithm

The analyses were conducted from 25 microscopy images containing a total of 5458 erythrocytes and 181 leukocytes served as benchmarks. Thus, the average number of erythrocytes and leukocytes per blood smear field was 218 and 7, respectively. These images were submitted to two types of counts: one by the manual methodology, performed by a healthcare expert and another by the WT-MO algorithm. The outcomes obtained in both counts were compared to each other as in Figures 6 and 7.

By relating the results from the manual the WT-MO algorithm counts, the accuracy values were obtained. The high accuracy of the methodology was verified from the values of 96% and 92% in the counts of RBCs and leukocytes respectively.

As important as developing a methodology for detection and counting of BCs that shows high accuracy, is to make the knowhow available to less-favored populations, so the algorithm was submitted to simulations on hardware platforms other than Dual Core processor with 2GB of RAM: an Intel Core i3 processor using 4GB of RAM and an Intel Core i5 processor with 8GB RAM.

2. Processing Time for the WT-MO Algorithm

The applicability of the WT-MO methodology was analyzed by means of two analysis parameters: processing time and execution time. For these evaluations, the 25 images of blood smear microscopy fields were entered in the simulation environment, where, through the MATLAB command line, the tic toc and cputime commands

Figure 6. Comparison of the erythrocyte counts by the manual method with the automated counting via the Watershed Transform

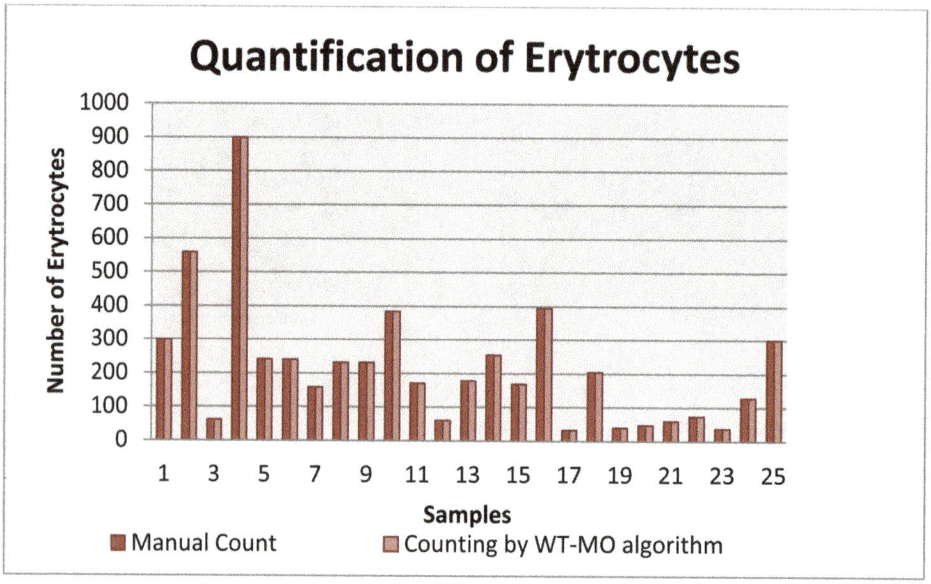

Figure 7. Comparison of the leukocyte counts by the manual method with the automated counting using morphological operations process

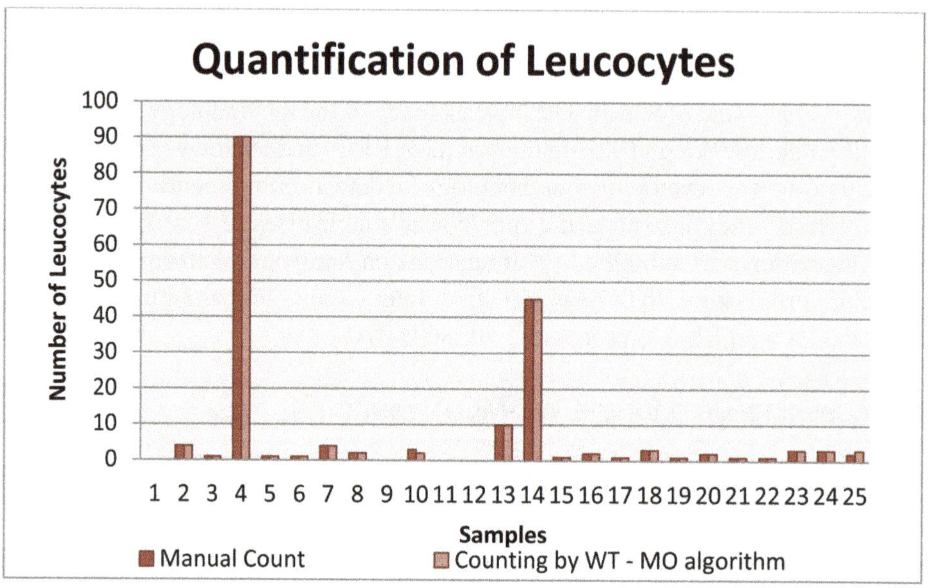

were inserted, which are responsible for measuring in seconds the execution and the processing time, respectively, for each of the blood images (Moore, 2012).

The tic toc command measures the time in seconds, which the WT-MO algorithm needs for its execution with the blood smear images. This function starts the timer to measure the WT-MO algorithm performance in each of the simulations. In this way, this function records the internal time of execution of the Tic command - and displaying the elapsed time with the Toc function. Then, the time elapsed from the stopwatch timer initiated by the Tic function, and together the internal time of the Toc command execution is displayed, displaying the elapsed time since the last call by the Tic that had no way out, in seconds (Moore, 2012). Figures 8, 9 and 10 depict the runtime.

The processing time is obtained with the cputime function, which measures the computational time of the WT-MO algorithm in each of the submitted images. The cputime command gives the total CPU time (in seconds) in the MATLAB IDE from the time the algorithm started. The processing times for the WT-MO algorithm are expressed in the Figures 11, 12 and 13. The processing times found previously are in Table 1.

Figure 8. Runtime in a Dual Core processor

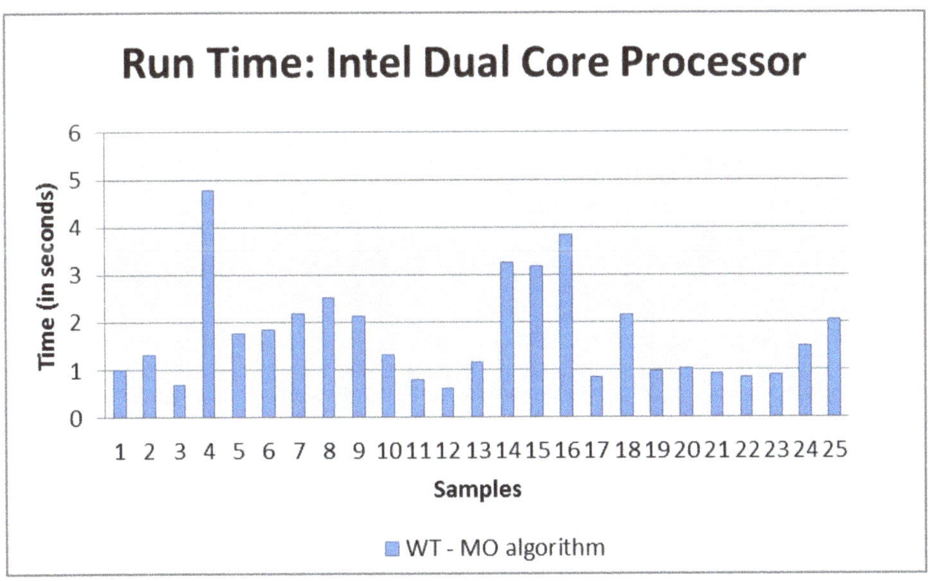

Figure 9. Runtime in Intel Core i3 processor

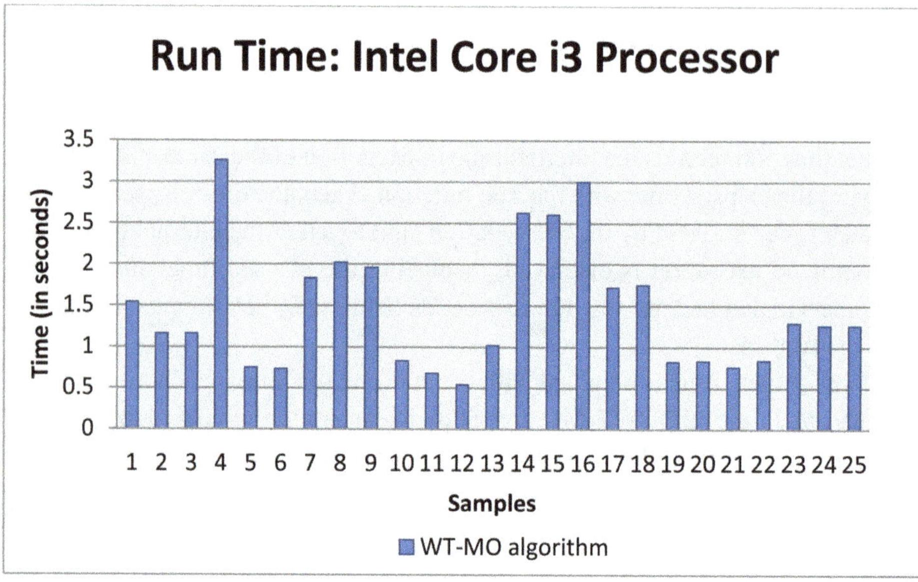

Figure 10. Runtime in Intel Core i5 processor

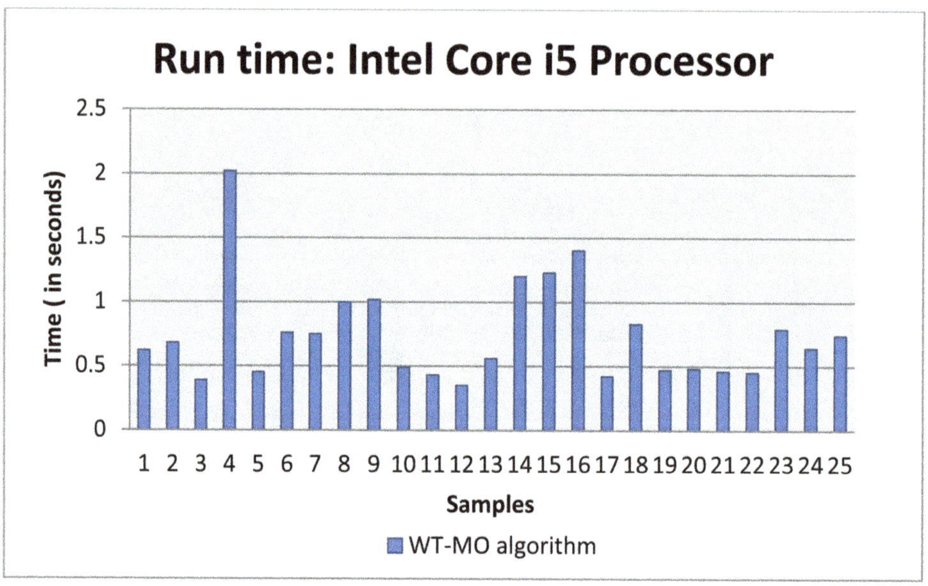

Figure 11. Execution times for the Dual Core processor

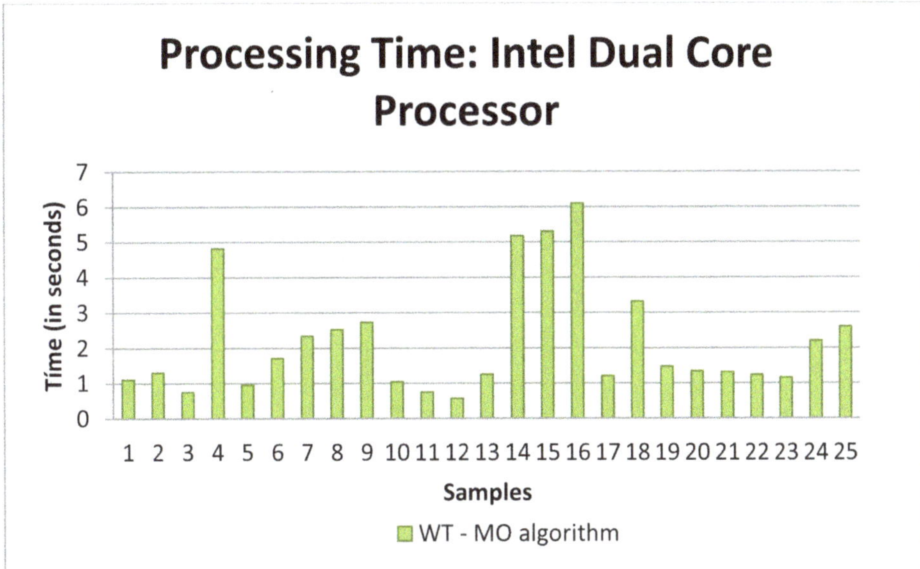

Figure 12. Execution times for the Intel Core i3 processor

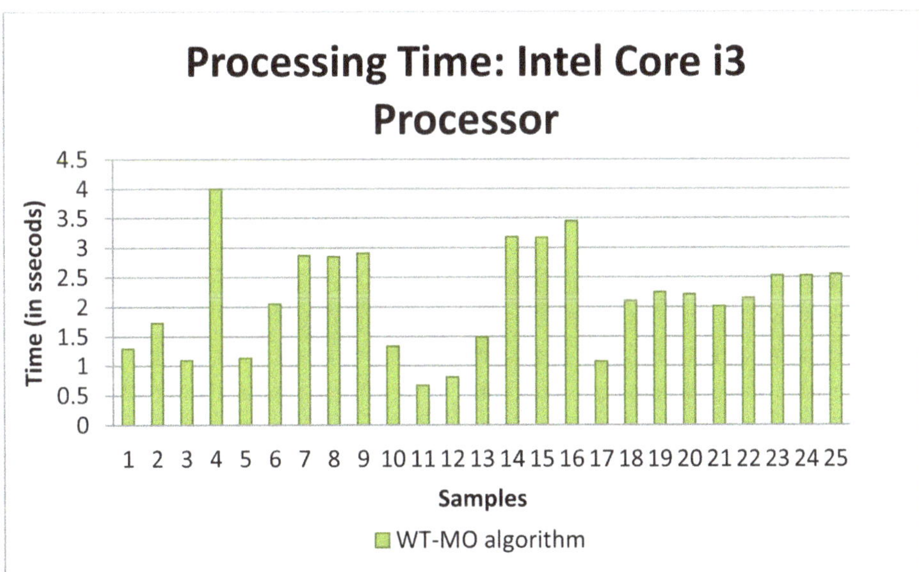

Figure 13. Execution times for the Intel Core i5 processor

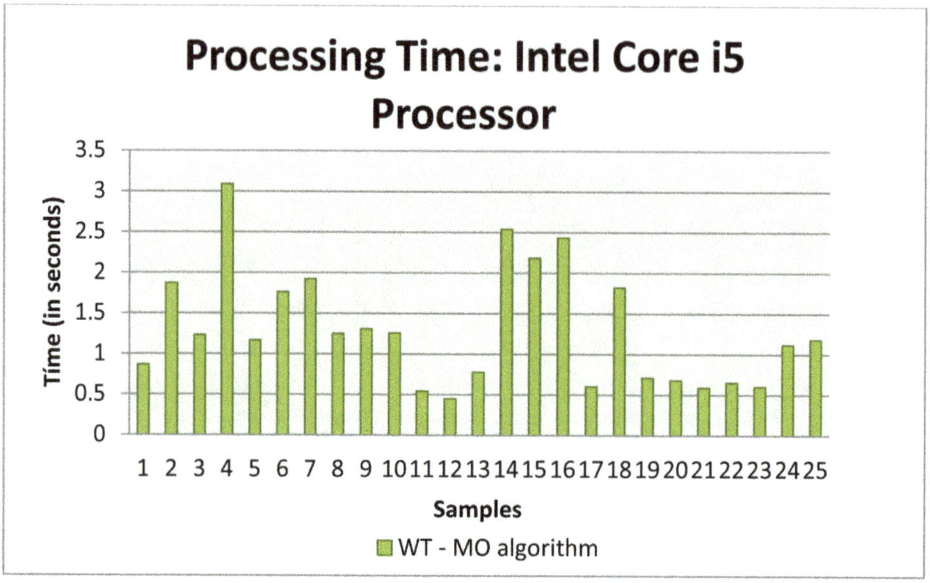

Table 1. Average runtime and processing time

	Average	
Hardware	**Runtime**	**Processing Time**
Intel Dual Core Processor	1.74''	2.17''
Intel Core i3 Processor	1.44''	2.13''
Intl Core i5 Processor	0.74"	1.30''

DISCUSSION

The accuracy of a methodology is an extremely important variable in the medical and laboratory areas, because of the greater the accuracy of a medical methodology, the greater the safety transmitted to the patient and the healthcare professional. Considering the values of 96% and 92% accuracy in the RBC and WBC counts, respectively, it is demonstrated that the WT-MO algorithm is a reliable tool. Medical reports with low reliability can easily exclude patients from medical treatment or lead the individual to incorrect medical treatment. When this occurs, serious damage can happen to both patients and professionals.

These obtained values of accuracy can be compared with other research found in the literature, whose objectives are to identify and count BCs. In this context, it can be seen that some research works presented values of accuracy referring to 91%, 75% and 91, 87% and 90%, respectively (Arivu, & Sathiya, 2012; Mahmood et al; Sahastrabuddhe et al,). In this way, the WT-MO algorithm presents a superior accuracy of 5.21 to 21.88% and 4.30 to 6.25% correspondingly.

Another relevant parameter to facilitate routine medical laboratory practices is the time to process the sample and, consequently, the time consumed until the release of the medical report. Delayed results can negatively impact patients' lives. The blood count can detect the first signs of leukemia in the body, such as the production and disordered release of WBCs in the bloodstream, and can easily reach or exceed 100,000 leukocytes / mm3 of blood. Thus, if the release of the award is due to a delayed process, that patient will be impaired, with delays in referral for complementary exams and correct treatment.

In this context, the WT-MO algorithm presents excellent results of average processing time and execution time, varying from 0.74 - 2.17 seconds. Other works in the literature show average processing time values of 4 seconds to 6 minutes. Thus, the methodology of counting and simultaneous detection of red cells and leukocytes shows an improvement of 440 to 16490% approximately.

Another central parameter to evaluate the WT-MO approach is its applicability. The simulations performed on the hardware platforms Dual Core processor, with 2GB of RAM, an Intel Core i3 processor, with 4GB of RAM in addition to an Intel Core i5 processor using 8GB of RAM clearly demonstrate that the WT-MO does not involve neither a high processing time nor prohibited investments, and, consequently, release users from the purchase of reagents specific for a device. Given the ease of performing hematological exams from simple daily tools such as a simple personal computer or notebook, it is confirmed that the WT-MO cost and the associated need of specific reagents are drastically reduced, which brings in thousands of people to the right to quality health, enabling the early detection of a wide range of diseases, particularly infections.

SOLUTIONS AND RECOMMENDATIONS

Noise Modeling

Multiple steps are required from the specimen images to meaning. The sample can be registered using assorted imaging modalities, types of microscopes, magnifications, viewing angles, and markers. These possibilities clearly cause information loss, which be circumvented with many different image modalities from the same item. Likewise, the intensity values are not easily related to actual material concentrations, since the microscope has an imperfect impulse-response function, alias the Point Spread

Function (PSF). The excitation light intensity may also be unknown. Furthermore, light can suffer absorption and scattering through the sample, which demands some assumptions on the objects within the image related to noise or blur. Image deconvolution in addition to contrast enhancement help to mitigate these problems.

Without a doubt, errors happen at every single phase of the image processing pipeline and are propagated to the subsequent processing stages, including uncertainties in the blood smear itself, such as unknown labeling densities among other factors. Additional image acquisition uncertainties are introduced by light scattering, optical aberrations/noise from the photon-detection procedure and the camera electronics. Image processing can also have flaws, may strengthen noise, there are false discoveries, quantification errors, and neglected detections. Humans also make mistakes when obtaining, understanding and quantifying images. Lastly, the subsequent information understanding is subject to uncertainties, since quite often the model encompasses inexactitudes. If possible, these uncertainties should be acknowledged and enumerated to limit their effects. Otherwise, it is unmanageable to select whether a perceived difference between two investigated images owing to the biological specimen characteristics, or the result from analysis artifacts.

Overlapping Cells

The overlapped BCs can be diagnosed mistakenly as an abnormal cells, which calls for better classification/cluster strategies to distinguish the overlapped BCs from the non-overlapped BCs.

The number of BCs in a particular region is a significant feature for physicians and hematologists to provide appropriate diagnosis. For counting BCs, the overlapped

Figure 14. Situations encompassing overlapping and atypical blood cells

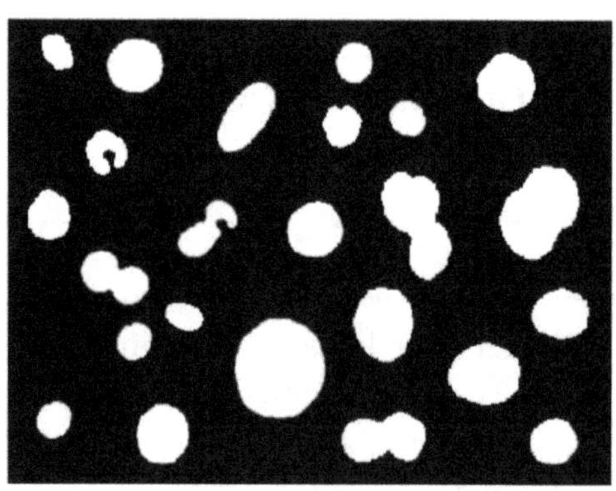

cells should be identified and split into the real number of BCs, which can be hard. Ordinary counting procedures will count the overlapped BCs as one object. When the WT is applied to images containing overlapped BCs to segment it to estimate the overlapped cells number. Figure 15 shows an image with overlapped cells. The WT can be improved by means of more advanced variants that use soft computing techniques like the ones in (Moallem et al., 2014; Mousavi et al., 2011; Bahadure et al., 2016; Patak & Sejwar, 2017; Romero-Záliz, & Reinoso-Gordo, 2017).

FUTURE RESEARCH DIRECTIONS

Deep Learning

Deep Learning (DL) paradigms are end-to-end unsupervised methods that obtain features taking advantage of hefty volumes of training data according to multi-layered Neural Network (NN) architectures. Histopathology imageries involve data complexity and their densities are ideally appropriate for data mining via DL approaches as they experimented with Deep NN (DNN) architectures to learn intricate features from the information. Additionally, this rationale requires less understanding about the domain complexity particulars.

Figure 15. Situations involving overlapping BCs that can be further resolved using heuristics (Abushmmala, & Abushmmala, 2014)

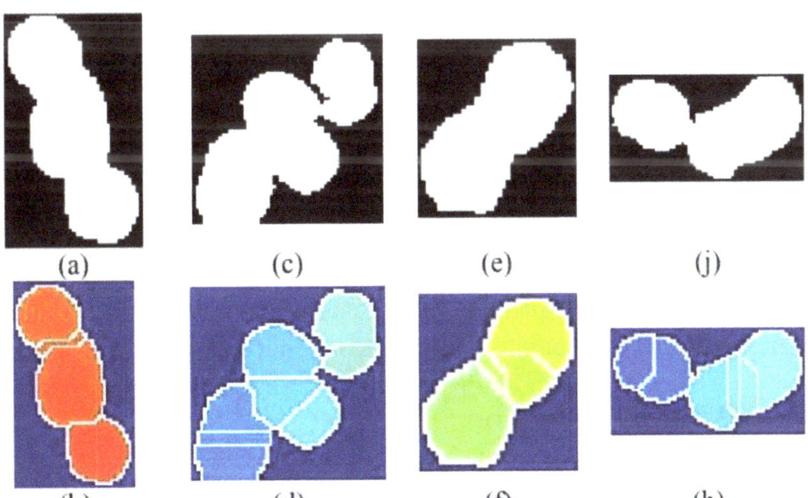

The use of DL to automatically distinguish and classify distinct nuclei in colon cancer images has been probed by (Madabhushi, & Lee 2016; Sirinukunwattana et al., 2016) where inflammatory, epithelial, and fibroblast nuclei were detected and marked with different colored dots, respectively.

An area of ample concern has been the usage of DL methodologies for identifying and counting mitoses appearing in cancer pathology imageries, which are strenuous and time-consuming tasks for pathologists. Several mitosis detection challenges from routine stained tissue images have been tackled using a DL framework (Madabhushi, & Lee 2016; Veta et al., 2015). Nevertheless, DL approaches have received criticism for (i) reliance on huge training data volumes; and (ii) the lack of instinctual knowledge concomitant with the DL generated image features. These difficulties have been addressed by a convergent scheme recommended by (Wang et al., 2014). Their work combines domain inspired crafted features with DL features to identify mitoses with an integrated tactic that produced superior recognition accuracy compared to DL or handcrafted (man-made) feature based stratagems alone.

New DL approaches appear almost everyday and designers should learn from other research areas too for novel DL architectures (Hemanth, & Estrela, 2018).

Content-Based Image Retrieval (CBIR)

Content-Based Image Retrieval (CBIR) entails an image query where the visual information is categorized by a group of features for instance color, dimension, silhouette and texture organized in a feature vector. The image search process looks for the best matches for the image query, which means the recovered (candidate) images satisfy a matching criteria. Examples of metrics to infer the similarity between images are the Euclidean distance (Herrmann, & Estrela, 2016), Mahalanobis distance (Herrmann, & Estrela, 2016), combinations of norms (Estrela et al. 2005), Bregman divergence (Coelho et al., 2012b) and statistical moments to name a few. CBIR can handle multi-dimensional imagery retrieval, multimodality healthcare data, and recovery of unusual data sets. Considering the principles employed by the CBIR, this methodology can spot BCs in both healthy and pathological conditions. Therefore, the WT-MO methodology can be an initial step for developing telemedicine tools in hematology, which allows patients in the region of difficult access to qualified medical professionals to elaborate a diagnosis in a shorter time.

Motion and Shape Detection

Body scanning help to explore phenomena that happen in time-lapses in video, aka motion, to follow moving items over time while identifying their trajectories. This motion tracking requires determining object correspondences across time points and is usually king done after object detection or segmentation (Coelho, & Estrela 2012c; Sbalzarini, 2016). Given identified or segmented objects in a video frame, tracking explains if a detection in a frame matches to an occurrence in the next frame so that the two match up to the same objective at different times (Bar-Shalom, & Blair, 2000). An affluence of tracking methods exist in biological imaging, for both particle (Chenouard et al., 2014) and cell tracking (Maska et al., 2014). Most of them are now libraries compatible with various standard software packages. The moving objects trajectories contain evidence about dynamics, motion types (Wieser et al., 2008; Ruprecht et al., 2011), and displacement patterns (Sbalzarini, 2016; Helmuth et al., 2007). This has, e.g., been employed to analyze virus movement on and inside diseased host cells (Ewers et al., 2005; Helmuth et al., 2007; Yamauchi et al., 2011) and to scrutinize the movement of molecules in plasma membranes (Wieser, & Schutz, 2008). For spatially extended entities, their outlines deformations can be tracked to find the point on a shape equivalent to a point on the later outline. Answers grounded on mechanical equivalents (Machacek, & Danuser, 2006) and level-set approaches (Shi, & Karl, 2005; Machacek, & Danuser, 2006) can be fruitfully applied.

Level-set procedures (Sethian, 1999) can track high-resolution shapes of polarizing and drifting fibroblasts in phase-contrast videos (Ambuhl et al., 2012), and to fragment and follow the paths of fluorescent cancer and Chinese Hamster Ovary (CHO) cells (Dzyubachyk et al., 2010), which allows quantifying cells, their shapes, and temporal dynamics.

CONCLUSION

Visual examination of blood spreads is a fundamental instrument for investigation, prevention, and remediation of various ailments. The enthusiasm prompted by the computer-aided decision-making has been recognized in numerous therapeutic examples, e.g., automatic frameworks are being considered to identify, group and measure items in hematological cytology. This research effort recommends a comprehensive automatic blood smear finding framework for hemograms, which can diminish the time consumed to perform slide examination. The present system is called WT-MO and it hinges on Morphological Operations (MOs) and delicate division by methods for the Watershed Transform (WT). Investigations exhibit the strategy adequacy to check WBCs in addition to RBCs. A few contemplations about usage, deployment and conceivable variations just as upgrades are examined. The fate of automatic tissue examination is mulled over.

Of course, semantic visual image processing has a lot to contribute to tissue analysis (Mousavi et. al., 2012) and the new automation efforts such as the Health 4.0 (Estrela et. al., 2019).

REFERENCES

Abbas, A. K., Lichtman, A. H., & Pillai, S. (2012). *Cellular & Molecular Immunology*. Elsevier.

Abushmmala, F. F., & Abushmmala, F. F. (2014). Processing overlapped cells using k-means and watershed. *International Journal of Intelligent Information Systems, 3*(1), 8–12. doi:10.11648/j.ijiis.20140301.12

Ambuhl, M. E., Brepsant, C., Meister, J. J., Verkhovsky, A. B., & Sbalzarini, I. F. (2012). High-resolution cell outline segmentation and tracking from phase-contrast microscopy images. *Journal of Microscopy, 245*(2), 161–170. doi:10.1111/j.1365-2818.2011.03558.x PMID:21999192

Andrade, E. L., Khan, E., Woods, J. C., & Ghanbari, M. (2003). Segmentation and tracking using region adjacency graphs, picture trees and prior information. In *Proc. IEEE Visual Information Engineering (VIE'03)* (pp. 45–48). IEEE. 10.1049/cp:20030483

Arivu, S. K., & Sathiya, M. (2012). Analyzing blood cell images to differentiate WBC and counting of linear & non-linear overlapping RBC based on morphological features. *Elixir Comp. Sci. & Engg, 48*, 9410–9413.

Baggett, D., Nakaya, M., Mcauliffe, M., Yamaguchi, T. P., & Lockett, S. (2005). Whole cell segmentation in solid tissue sections. *Cytometry. Part A, 67*(2), 137–143. doi:10.1002/cyto.a.20162 PMID:16163696

Bahadure, N. B., Ray, A. K., & Thethi, H. P. (2016). Performance analysis of image segmentation using watershed algorithm, fuzzy C-means of clustering algorithm and Simulink design. *Proc. 2016 3rd International Conference on Computing for Sustainable Global Development (INDIACom)*, 1160-1164.

Bandyopadhyay, S. K. (2011). Method for blood cell segmentation. *Journal of Global Research in Computer Science, 2*(4).

Bar-Shalom, Y., & Blair, W. D. (Eds.). (2000). *Multitarget/multisensor tracking: Applications and advances* (Vol. 3). Artech.

Barnard, K., Duygulu, P., Freitas, N., Forsyth, D., Blei, D., & Jordan, M. I. (2003). Matching words and pictures. *JMLR*, *3*, 1107–1135.

Beucher, S., & Meyer, F. (1993). The morphological approach to segmentation: The watershed transformation. In E. Dougherty (Ed.), *Mathematical Morphology in Image Processing* (pp. 433–481). Marcel Dekker.

Bhagavathi, S. L., & Niba, T. S. (2016). An automatic system for detecting and counting RBC and WBC using fuzzy logic. *J. of Eng. and Appl. Sc.*, *11*(11), 2006–2016.

Bieniek, A., & Moga, A. (2000). An efficient watershed algorithm based on connected components. *Pattern Recognition*, *6*(33), 907–916. doi:10.1016/S0031-3203(99)00154-5

Buttarello, M., & Plebani, M. (2008). Automated blood cell counts: State-of-the-art. *American Journal of Clinical Pathology*, *130*(1), 104–116. doi:10.1309/EK3C7CTDKNVPXVTN PMID:18550479

Chen, Y., & Wang, J. Z. (2004). Image categorization by learning and reasoning with regions. *JMLR*, *5*, 913–939.

Chenouard, N., Smal, I., de Chaumont, F., Maska, M., Sbalzarini, I. F., Gong, Y., ... Meijering, E. (2014). Objective comparison of particle tracking methods. *Nature Methods*, *11*(3), 281–289. doi:10.1038/nmeth.2808 PMID:24441936

Choi, J. G., Lee, S. W., & Kim, S. D. (1997). Spatio-temporal video segmentation using a joint similarity measure. *IEEE Transactions on Circuits and Systems for Video Technology*, *7*(2), 279–286. doi:10.1109/76.564107

Ciesla, B. (2012). *Hematology in practice* (2nd ed.). Davis Company.

Coelho, A. M., & Estrela, V. V. (2012c). Data-driven motion estimation with spatial adaptation. *Intl. J. of Image Proc*, *6*(1), 53–67. http://www.cscjournals.org/csc/manuscript/Journals/IJIP/volume6/Issue1/IJIP-513.pdf

Coelho, A. M., Estrela, V. V., do Carmo, F. P., & Fernandes, S. R. (2012b). Error concealment by means of motion refinement and regularized Bregman divergence. In H. Yin, J. A. F. Costa, & G. Barreto (Eds.), Lecture Notes in Computer Science: Vol. 7435. *Proc. Intelligent Data Engineering and Automated Learning - IDEAL 2012. IDEAL 2012*. Berlin: Springer. doi:10.1007/978-3-642-32639-4_78

Coelho, A.M., & Estrela, V. V. (2012a). EM-based mixture models applied to video event detection. In *Principal Component Analysis - Engineering Applications*. Intech Open. doi:10.5772/38129

Cotter, S. M. (2018). *Merck and the Merck veterinary manual.* Retrieved October 13, 2018, from https://www.merckvetmanual.com/circulatory-system/hematopoietic-system-introduction/white-blood-cells

Deng, Y., Kenney, C., Moore, M. S., & Manjunath, B. S. (1999). Peer group filtering and perceptual color image quantization. *Proc. of IEEE Intl. Symposium on Circuits and Systems, 4*, 21-24.

Díaz, G., & Manzanera, A. (2009). Automatic analysis of microscopic images in hematological cytology applications. In F. A. González & E. Romero (Eds.), Biomedical image analysis and machine learning technologies: Applications and techniques. Academic Press. doi:10.4018/978-1-60566-956-4.ch008

Dzyubachyk, O., van Cappellen, W. A., Essers, J., Niessen, W. J., & Meijering, E. (2010). Advanced level-set-based cell tracking in time-lapse fluorescence microscopy. *IEEE Transactions on Medical Imaging, 29*(3), 852–867. doi:10.1109/TMI.2009.2038693 PMID:20199920

Eclinpath. (2013). *Atlas of the Cornell University College of Veterinary Medicine.* Retrieved October 23, 2018, from http://www.eclinpath.com/ngg_tag/bovine/nggallery/page/2

Erichsen, E. S., Viana, L. G., Faria, R. M. D., & Santos, S. M. E. (2015). *Medicina laboratorial para o clínico.* Belo Horizonte, Brazil: Coopmed.

Estrela, V. V., Franz, M. O., Lopes, R. T., & de Araújo, G. P. (2005). Adaptive mixed norm optical flow estimation. Proc. SPIE 5960, Visual Communications and Image Processing 2005, 59603W. doi:10.1117/12.632674

Estrela, V. V., Monteiro, A. C. B., França, R. P., Iano, Y., Khelassi, A., & Razmjooy, N. (2019). Health 4.0 as an application of Industry 4.0 in healthcare services and management. *Medical Technologies Journal, 2*(4), 262-76. Retrieved from http://ichsmt.org/Journals/ojs/index.php/MTJ/article/view/205

Ewers, H., Smith, A. E., Sbalzarini, I. F., Lilie, H., Koumoutsakos, P., & Helenius, A. (2005). Single-particle tracking of murine polyoma virus-like particles on live cells and artificial membranes. *Proceedings of the National Academy of Sciences of the United States of America, 102*(42), 15,110–15,115. doi:10.1073/pnas.0504407102 PMID:16219700

Failace, R., & Fernandes, F. (2015). *Hemograma. Manual de interpretação* (6th ed.). Porto Alegre, Brazil: Artemed.

Fernandes, S. R., Estrela, V. V., Magalhaes, H. A., & Saotome, O. (2014). On improving sub-pixel accuracy by means of B-spline. *Proc. of the 2014 IEEE Int'l Conf. on Im. Syst. and Techn. (IST 2014)*, Santorini, Greece. 10.1109/IST.2014.6958448

Fischbach, F., & Duynning, M. B. A. (2009). *Manual of laboratory and diagnostic tests*. Lippincott Williams e Wilkins.

Godon, A., Genevieve, F., Marteau-Tessier, A., & Zandecki, M. (2012). Anomalies et erreurs de détermination de l'hémogramme avec les automates d'hématologie cellulaire Partie 3. Hémoglobine, hématies, indices érythrocytaires, réticulocytes. *Annales de Biologie Clinique*, *70*(2), 155–168. doi:10.1684/abc.2012.0685 PMID:22484526

Gulati, G. L., Alomari, M., Kocher, W., & Schwarting, R. (2002). Criteria for blood smear review. *Laboratory Medicine*, *33*(5), 374–377. doi:10.1309/TVJW-ENEK-G1U3-WE0A PMID:23301216

Hamdin, R. I., Lux, S. E., & Stossel, T. P. (2003). *Blood: Principles and practice of hematology*. Lippincott Williams e Wilkins.

Hansen, J. S. (2011). *GNU Octave. Beginner's guide*. Packt Publishing.

Hashimoto, Y., Da Silva, P. H., & Bertassoni, H. (2009). *Hematologia laboratorial*. Rio de Janeiro: Revinter.

Helmuth, J. A., Burckhardt, C. J., Koumoutsakos, P., Greber, U. F., & Sbalzarini, I. F. (2007). A novel supervised trajectory segmentation algorithm identifies distinct types of human adenovirus motion in host cells. *Journal of Structural Biology*, *159*(3), 347–358. doi:10.1016/j.jsb.2007.04.003 PMID:17532228

Herman, N., Trumel, C., Geffré, A., Braun, J.-P., Thibault, M., Schelcher, F., & Bourgès-Abella, N. (2018). Hematology reference intervals for adult cows in France using the Sysmex XT-2000iV analyzer. *Journal of Veterinary Diagnostic Investigation*, *30*(5), 678–687. doi:10.1177/1040638718790310 PMID:30027829

Herrmann, A. E., & Estrela, V. V. (2016). Content-based image retrieval (CBIR) in remote clinical diagnosis and healthcare. In M. Cruz-Cunha, I. Miranda, R. Martinho, & R. Rijo (Eds.), *Encyclopedia of E-Health and Telemedicine* (pp. 495–520). Hershey, PA: IGI Global. doi:10.4018/978-1-4666-9978-6.ch039

Irshad, H., Veillard, A., Roux, L., & Racoceanu, D. (2014). Methods for nuclei detection, segmentation, and classification in digital histopathology: A review–current status and future potential. *IEEE Reviews in Biomedical Engineering*, *7*, 97–114. doi:10.1109/RBME.2013.2295804 PMID:24802905

Jiang, L., Lu, H., My, V. D., Koch, A., & Zell, A. (2015). Superpixel segmentation based gradient maps on RGB-D dataset. *Proc. 2015 IEEE Int'l Conference on Robotics and Biomimetics (ROBIO)*. 10.1109/ROBIO.2015.7418960

Jing, J., Huang, M., Li, P., & Ning, X. (2018). Automatic measurement of yarn hairiness based on the improved MRMRF segmentation algorithm. *Journal of the Textile Institute*, *109*(6), 740–749. doi:10.1080/00405000.2017.1368106

Junqueira, L. C., & Carneiro, J. (2004). *Histologia básica* (10th ed.). São Paulo, Guanabara.

Junqueira, L. C., & Carneiro, J. (2012). *Biologia celular e molecular*. São Paulo: Guanabara Koogan.

Kawthalksr, S. M. (2013). Essentials of haematology (2nd ed.). New Delhi: Jaypee.

Machacek, M., & Danuser, G. (2006). Morphodynamic profiling of protrusion phenotypes. *Biophysical Journal*, *90*(4), 1439–1452. doi:10.1529/biophysj.105.070383 PMID:16326902

Maska, M., Ulman, V., Svoboda, D., Matula, P., Matula, P., Ederra, C., ... Ortiz-de-Solorzano, C. (2014). A benchmark for comparison of cell tracking algorithms. *Bioinformatics (Oxford, England)*, *30*(11), 1609–1617. doi:10.1093/bioinformatics/btu080 PMID:24526711

Meier, T., & Ngan, K. N. (1998). Automatic segmentation of moving objects for video object plane generation. *IEEE Transactions on Circuits and Systems for Video Technology*, *8*(5), 525–538. doi:10.1109/76.718500

Melo, M., & Silveira, C. M. (2015). *Laboratório de hematologia: Teorias, técnicas e atlas*. Rubio.

Meyer, F. (2014). Watersheds on weighted graphs. *Pattern Recognition Letters*, *47*(1), 72–79. doi:10.1016/j.patrec.2014.02.018

Moallem, P., Razmjooy, N., & Mousavi, B.S. (2014). Robust potato color image segmentation using adaptive fuzzy inference system. *Iranian Journal of Fuzzy Systems*.

Mogra, M. & Srivastava, V. (2014). A comprehensive review of analysis of counting blood cells using different image processing algorithms. *Int'l J. of Eng. Sc. Invention*, *3*(6).

Moore, H. (2012). *Matlab for Engineers*. Boston: Pearson.

Moscheni, F., Bhattacharjee, S., & Kunt, M. (1998). Spatio-temporal segmentation based on region merging. *IEEE Transactions on Pattern Analysis and Machine Intelligence*, *20*(9), 897–915. doi:10.1109/34.713358

Mousavi, B. S., Sargolzaei, P., Razmjooy, N., Hosseinabadi, V., & Soleymani, F. (2011). Digital Image Segmentation Using Rule-Base Classifier. *American Journal of Scientific Researches*, *35*, 17–23.

Mousavi, B. S., Soleymani, F., & Razmjooy, N. (2012). Semantic image classification by genetic algorithm using optimised fuzzy system based on Zernike moments. *Signal, Image and Video Processing*. doi:10.100711760-012-0311-7

Murphy, K., Travers, P., & Walport, M. (2010). *Imunibiologia* (7th ed.). Porto Alegre: Artmed, Brazil.

Naemura, N., Fukuda, A., Mizutani, Y., Izumi, Y., Tanaka, Y., & Enami, K. (2000). Morphological segmentation of sports scenes using color information. *IEEE Transactions on Broadcasting*, *46*(3), 181–188. doi:10.1109/11.892154

Nasreen, N., Kumar, C., & Nabeel, A. (2015). *Counting of RBC using circular Hough transform with median filtering*. Bonfring.

Nemer, A. S. A., Das Neves, F. J., & Ferreira, J. E. S. (2010). *Manual de solicitação de exames laboratoriais*. Revinter.

Pathak, S., & Sejwar, V. (2017). Optimized noisy image segmentation using genetic algorithm. *Proc. 2017 International Conference on Intelligent Computing and Control Systems (ICICCS)*, 1311-1316. 10.1109/ICCONS.2017.8250681

Provan, D., Baglin, T., Dokal, I., & de Vos, J. (2015). *Oxford handbook of clinical haematology* (4th ed.). Oxford University Press. doi:10.1093/med/9780199683307.001.0001

Reyes-Aldosoro, C. C. (2015). *Biomedical image analysis recipes in Matlab: For life and engineers*. London: Wiley Blackweel. doi:10.1002/9781118657546

Roerdink, J. B. T. M., & Meijster, A. (1996). Computation of watersheds based on parallel graph algorithms. *Proc of Math. Morphology and Its Application to Image and Signal Processing*.

Roerdink, J. B. T. M., & Meijster, A. (2000). The watershed transform: Definitions, algorithms and parallelization strategies. *Fundamenta Informaticae*, *41*(1-2), 187–228.

Roland, L., Drillich, M., & Iwersen, M. (2014). Hematology as a diagnostic tool in bovine medicine. *Journal of Veterinary Diagnostic Investigation, 26*(5), 592–598. doi:10.1177/1040638714546490 PMID:25121728

Romero-Záliz, R., & Reinoso-Gordo, J.F. (2017). *An Updated Review on Watershed Algorithms.* Academic Press.

Roux, P. (2016). Scilab: I. Fundamentals, from theory to practice. Paris, France: Academic Press.

Ruprecht, V., Axmann, M., Wieser, S., & Schutz, G. J. (2011). What can we learn from single molecule trajectories? *Current Protein & Peptide Science, 12*(8), 714–724. doi:10.2174/138920311798841753 PMID:22044145

Savkare, S.S., & Narote, S.P. (2011). Automatic classification of normal and infected blood cells for parasitemia detection. *Int'l J. of Comp. Science and Network Security, 11*(2).

Sbalzarini, I. F. (2016). Seeing Is believing: Quantifying Is convincing: computational image analysis in biology. *Advances in Anatomy, Embryology, and Cell Biology, 219*, 1–39. doi:10.1007/978-3-319-28549-8_1 PMID:27207361

Sethian, J. A. (1999). *Level Set Methods and Fast Marching Methods.* Cambridge, UK: Cambridge University Press.

Shi, J., & Malik, J. (1998). Motion segmentation and tracking using normalized cuts. *Proc. of IEEE 6th Intl. Conf. on Computer Vision*, 1154-1160.

Shi, Y., & Karl, W. (2005). Real-time tracking using level sets. *Proc. IEEE Conf. CVPR*, 2, 34–41. doi: 10.1109/CVPR.2005.294

Smereka, M., & Duleba, I. (2008). Circular object detection using a modified Hough transform. *International Journal of Applied Mathematics and Computer Science, 18*(1), 85–91. doi:10.2478/v10006-008-0008-9

Soltanzade, R. (2010). Classification of three types of red blood cells in peripheral blood smear based on morphology. *Proceedings of the 2010 ICSP*. 10.1109/ICOSP.2010.5655754

Song, J., & Lyu, R. M. (2005). A Hough transform based line recognition method utilizing both parameter space and image space. *Pattern Recognition, 38*(4), 539–552. doi:10.1016/j.patcog.2004.09.003

Stiller, C. (1997). Object-based estimation of dense motion fields. *IEEE Transactions on Image Processing, 6*(2), 234–250. doi:10.1109/83.551695 PMID:18282920

Turgeon, M. L. (2004). *Clinical hematology theory and procedures* (4th ed.). Philadelphia: Lippincott Williams and Wilkins.

Verrastro, T., Lorenzi, T. F., & Neto, S. W. (2005). *Hematologia e hemoterapia: Fundamentos de morfologia, patologia e clínica*. Rio de Janeiro: Atheneu.

Vincent, L., & Soille, P. (1991). Watersheds in digital spaces: An efficient algorithm based on immersion simulations. *IEEE Transactions on Pattern Analysis and Machine Intelligence*, *13*(6), 583–598. doi:10.1109/34.87344

Vitor, G. B., Körbes, A., Lotufo, R. A., & Ferreira, J. V. (2012). Analysis of a step based watershed algorithm using CUDA. In *Nature-Inspired Computing Design, Development, and Applications*. Hershey, PA: IGI Global; doi:10.4018/978-1-4666-1574-8.ch018

Walsh, D., & Raftery, A. E. (2002). Accurate and efficient curve detection in images: The importance sampling Hough Transform. *Pattern Recognition*, *35*(7), 1421–1431. doi:10.1016/S0031-3203(01)00114-5

Wieser, S., Axmann, M., & Schutz, G. J. (2008). Versatile analysis of single-molecule tracking data by comprehensive testing against monte carlo simulations. *Biophysical Journal*, *95*(12), 5988–6001. doi:10.1529/biophysj.108.141655 PMID:18805933

Wieser, S., & Schutz, G. J. (2008). Tracking single molecules in the live cell plasma membrane–do's and don't's. *Methods (San Diego, Calif.)*, *46*(2), 131–140. doi:10.1016/j.ymeth.2008.06.010 PMID:18634880

Wouwer, A. V., Saucez, P., & Vilas, C. (2014). *Simulation of ODE/PDE models with MATLAB, Octave and Scilab: Scientific and engineering applications*. Springer.

Wu, Z., Lin, D., & Tang, X. (2016). Deep Markov random field for image modeling. In B. Leibe, J. Matas, N. Sebe, & M. Welling (Eds.), Lecture Notes in Computer Science: Vol. 9912. *Proc. Computer Vision – ECCV 2016. ECCV 2016*. Cham: Springer. doi:10.1007/978-3-319-46484-8_18

Xavier, R. M., Dora, J. M., & Barros, E. (2016). *Laboratório na prática clínica: Consulta rápida* (3rd ed.). Porto Alegre: Artmed.

Yamauchi, Y., Boukari, H., Banerjee, I., Sbalzarini, I.F., Horvath, P., & Helenius, A. (2011). Histone deacetylase 8 is required for centrosome cohesion and influenza A virus entry. *PLoS Pathog.*, *7*(10).

ADDITIONAL READING

Basavanhally, A., Ganesan, S., Feldman, M., Shih, N., Mies, C., Tomaszewski, J., & Madabhushi, A. (2013). Multi-field-of-view framework for distinguishing tumor grade in ER+ breast cancer from entire histopathology slides. *IEEE Transactions on Biomedical Engineering*, *60*(8), 2089–2099. doi:10.1109/TBME.2013.2245129 PMID:23392336

Beck, A. H., Sangoi, A. R., Leung, S., Marinelli, R. J., Nielsen, T. O., van de Vijver, M. J., ... Koller, D. (2011). Systematic analysis of breast cancer morphology uncovers stromal features associated with survival. *Science Translational Medicine*, *3*(108), 108ra113. doi:10.1126citranslmed.3002564 PMID:22072638

Brandes, S., Mokhtari, Z., Essig, F., Hünniger, K., Kurzai, O., & Figge, M. T. (2015). Automated segmentation and tracking of non-rigid objects in time-lapse microscopy videos of polymorphonuclear neutrophils. *Medical Image Analysis*, *20*(1), 34–51. doi:10.1016/j.media.2014.10.002 PMID:25465844

Bychkov, D., Turkki, R., Haglund, C., Linder, N., & Lundin, J. (2016). Deep learning for tissue microarray image-based outcome prediction in patients with colorectal cancer. In: Gurcan, M.N., Madabhushi, A. (Eds.), In: *Proceedings of the SPIE9791, Medical Imaging 2016. Digital Pathology*, San Diego, California, United States, 979115. doi:10.1117/12.2217072

Chakravarty, S., Banerjee, M., & Chandel, S. (2017). Spectral-spatial classification of hyperspectral imagery using support vector and fuzzy-MRF. In Lecture Notes in Computer Science (including subseries Lecture Notes in Artificial Intelligence and Lecture Notes in Bioinformatics) (Vol. 10618 LNCS, pp. 151–161). Springer Verlag. doi:10.1007/978-3-319-69155-8_11

de Jesus, M. A., & Estrela, V. V. (2016). An introduction to data mining applied to health-oriented databases [OJCST]. *Oriental J. of Comp. Sc. and Techn.*, *9*(3), 177–185. doi:10.13005/ojcst/09.03.03

de Jesus, M. A., Estrela, V. V., & Saotome, O. (2016) Super-resolution in a nutshell. In *Proceedings of the Brazilian Technology Symposium 2016 (BTSym 2016)*. ISSN 2447-8326. V. 1

de Jesus, M. A., Estrela, V. V., Saotome, O., & Stutz, D. (2018). Super-resolution via particle swarm optimization variants. In J. Hemanth & V. Balas (Eds.), *Lecture Notes in Computational Vision and Biomechanics* (Vol. 25). Cham: Springer.

Devi, S. S., Roy, A., Sharma, M. D., & Laskar, R. H. (2016). kNN Classification Based Erythrocyte Separation in Microscopic Images of Thin Blood Smear. 2016 2nd International Conference on Computational Intelligence and Networks (CINE), 69-72.

Dimililer, K. (2013). Backpropagation neural network implementation for medical image compression. *Journal of Applied Mathematics*, *2013*, 1–8. doi:10.1155/2013/453098

Elvira, C., Chainais, P., & Dobigeon, N. (2017). Bayesian antisparse coding. *IEEE Transactions on Signal Processing*, *65*(7), 1660–1672. doi:10.1109/TSP.2016.2645543

Estrela, V. V., & Galatsanos, N. P. (2000). Spatially adaptive regularized pel-recursive motion estimation based on the EM algorithm. In B. Vasudev, T. R. Hsing, A. G. Tescher, & R. L. Stevenson (Eds.), *In Proc. Image and Video Communications and Processing 2000* (Vol. 3974, pp. 372–383). San Jose, CA, USA: SPIE; doi:10.1117/12.382969

Estrela, V. V., Hemanth, J., Saotome, O., Grata, E. G. H., & Izario, D. R. F. (2019). Emergency response cyber-physical system for flood prevention with sustainable electronics. In: Y. Iano, R. Arthur, O. Saotome, V. Vieira Estrela, & H. Loschi (eds), In *Proceedings of the 3rd Brazilian Technology Symposium. BTSym 2017*. Springer, Cham. 10.1007/978-3-319-93112-8_33

Estrela, V. V., Rivera, L. A., & Bassani, M. H. (2004). Pel-recursive motion estimation using the expectation-maximization technique and spatial adaptation. In *Proc. 2004*. WSCG.

Estrela, V. V., Saotome, O., Loschi, H. J., Hemanth, J., Farfan, W. S., Aroma, J., ... Grata, E. G. H. (2018). Emergency response cyber-physical framework for landslide avoidance with sustainable electronics. *Technologies*, *6*(2), 42. doi:10.3390/technologies6020042

Gopakumar, G. P., Swetha, M., Siva, G. S., & Subrahmanyam, G. R. (2018). Convolutional neural network-based malaria diagnosis from focus stack of blood smear images acquired using custom-built slide scanner. *Journal of Biophotonics*, *11*(3), 3. doi:10.1002/jbio.201700003 PMID:28851134

Helmuth, J. A., Burckhardt, C. J., Greber, U. F., & Sbalzarini, I. F. (2009). Shape reconstruction of subcellular structures from live cell fluorescence microscopy images. *Journal of Structural Biology*, *167*(1), 1–10. doi:10.1016/j.jsb.2009.03.017 PMID:19358891

Helmuth, J. A., Paul, G., & Sbalzarini, I. F. (2010). Beyond co-localization: Inferring spatial interactions between subcellular structures from microscopy images. *BMC Bioinformatics*, *11*(1), 372. doi:10.1186/1471-2105-11-372 PMID:20609242

Hussain, F., & Jeong, J. (2015). Exploiting deep neural networks for digital image compression, In *Proc. 2nd World Symp. on Web Appl. and Networking (WSWAN)* (pp. 1-6), Sousse. 10.1109/WSWAN.2015.7210294

Kamilaris, A., & Prenafeta-Boldú, F. X. (2018, April 1). Deep learning in agriculture: A survey. *Comp. and Electronics in Agriculture*, Elsevier B.V. doi:10.1016/j.compag.2018.02.016

Khalilpour, M., & Razmjooy, N. (2013). A Hybrid Method for Gesture Recognition. *Journal of World's Electrical Engineering and Technology*, *2*(3).

Kit, C., Tomari, R., Zakaria, W. N., Othman, N. B., Safuan, S. N., Yi, J., & Sheng, N. T. (2017). *Mobile based Automated Complete Blood Count (Auto-CBC)*. Analysis System from Blood Smeared Image.

Krishnamachari, S., & Chellappa, R. (1997). Multiresolution Gauss-Markov random field models for texture segmentation. *IEEE Transactions on Image Processing*, *6*(2), 251–267. doi:10.1109/83.551696 PMID:18282921

Liu, Z., Li, X., Luo, P., Loy, C. C., & Tang, X. (2018). Deep learning Markov random field for semantic segmentation. *IEEE Transactions on Pattern Analysis and Machine Intelligence*, *40*(8), 1814–1828. doi:10.1109/TPAMI.2017.2737535 PMID:28796610

Luo, Y., & Ward, R. K. (2003). Removing the blocking artifacts of block-based DCT compressed images. *IEEE Transactions on Image Processing*, *12*(7). PMID:18237958

Mariño, I. P., Zaikin, A., & Míguez, J. (2017). A comparison of Monte Carlo-based Bayesian parameter estimation methods for stochastic models of genetic networks. *PLoS One*, *12*(8), e0182015. doi:10.1371/journal.pone.0182015 PMID:28797087

Meijering, E. H., Dzyubachyk, O., & Smal, I. (2012). Methods for cell and particle tracking. *Methods in Enzymology*, *504*, 183–200. doi:10.1016/B978-0-12-391857-4.00009-4 PMID:22264535

Olivo-Marin, J. C. (2002). Extraction of spots in biological images using multiscale products. *Pattern Recognition*, *35*(9), 1989–1996. doi:10.1016/S0031-3203(01)00127-3

Orlov, N., Shamir, L., Macura, T., Johnston, J., Eckley, D. M., & Goldberg, I. G. (2008). WND-CHARM: Multi-purpose image classification using compound image transforms. *Pattern Recognition Letters*, *29*(11), 1684–1693. doi:10.1016/j.patrec.2008.04.013 PMID:18958301

Orlov, N. V., Chen, W. W., Eckley, D. M., Macura, T. J., Shamir, L., Jaffe, E. S., & Goldberg, I. G. (2010). Automatic classification of lymphoma images with transform-based global features. *IEEE Transactions on Information Technology in Biomedicine*, *14*(4), 1003–1013. doi:10.1109/TITB.2010.2050695 PMID:20659835

Pan, C., Fang, Y., Yan, X., & Zheng, C. (2006). Robust Segmentation for Low Quality Cell Images from Blood and Bone Marrow. *International Journal of Control, Automation, and Systems*, *4*, 637–644.

Pan, C., Yan, X., & Zheng, C. (2006). Recognition of Blood and Bone Marrow Cells using Kernel-based Image Retrieval. *International Journal of Computer Science and Network Security*, *6*, 29–35.

Pan, C., Zheng, C., & Wang, H. (2003). Robust color image segmentation based on mean shift and marker-controlled watershed algorithm. In: *Proceedings of the International Conference on Machine Learning and Cybernetics*.

Paul, G., Cardinale, J., & Sbalzarini, I. F. (2011). An alternating split Bregman algorithm for multi-region segmentation. In: *Proc. 45th IEEE Asilomar Conf. Signals, Systems, and Computers, IEEE*, Asilomar, CA, USA, pp 426–430 10.1109/ACSSC.2011.6190034

Paul, G., Cardinale, J., & Sbalzarini, I. F. (2013). Coupling image restoration and segmentation: A generalized linear model/Bregman perspective. *International Journal of Computer Vision*, *104*(1), 69–93. doi:10.100711263-013-0615-2

Peng, H. (2008). Bioimage informatics: A new area of engineering biology. *Bioinformatics (Oxford, England)*, *24*(17), 1827–1836. doi:10.1093/bioinformatics/btn346 PMID:18603566

Peng, H., Ruan, Z., Long, F., Simpson, J. H., & Myers, E. W. (2010). V3D enables real-time 3D visualization and quantitative analysis of large-scale biological image data sets. *Nature Biotechnology*, *28*(4), 348–353. doi:10.1038/nbt.1612 PMID:20231818

Putzu, L., Caocci, G., & Ruberto, C. D. (2014). Leucocyte classification for leukaemia detection using image processing techniques. *Artificial Intelligence in Medicine*, *62*(3), 179–191. doi:10.1016/j.artmed.2014.09.002 PMID:25241903

Razmjooy, N., & Mohsen Shahrezaei, M. (2018). Image watermarking based on DWT-SVD, In: Proc. of the 2nd International Conference on Combinatorics, Cryptography and Computation (I4C2017), Science and Technology University, Iran.

Razmjooy, N., Naghibzadeh, S. S., & Musavi, B. S. (2014). Digital red eye correction in digital photos. *International Journal of Computers and Applications*, *95*(9).

Seger, W. (2017). Challenges to establish new medical technologies, that are not yet contractually stipulated with the insurers, into the daily medical routine. *Medical Technologies Journal*, *1*(4), 76–79. doi:10.26415/2572-004X-vol1iss4p76-79

Sharma, V., Rathore, A., & Vyas, G. (2016). Detection of sickle cell anaemia and thalassaemia causing abnormalities in thin smear of human blood sample using image processing. *2016 International Conference on Inventive Computation Technologies (ICICT)*, 3, 1-5. 10.1109/INVENTIVE.2016.7830136

Wang, C., Komodakis, N., & Paragios, N. (2013). Markov Random field modeling, inference & learning in computer vision & image understanding: A survey. *Computer Vision and Image Understanding*, *117*(11), 1610–1627. doi:10.1016/j.cviu.2013.07.004

KEY TERMS AND DEFINITIONS

Average Red Blood Count (RBC): This important parameter counts the number of erythrocytes present in the bloodstream ideally between 4.0 to 6.0 x 10^6 mm^3 of blood. These values are variable according to the age and sex of the patient. Changes in the mean value of red blood cells may indicate genetic anemias, deficiency anemia, leukemias, polycythemias, or blood loss.

Average White Blood Count (WBC): This important parameter counts the number of leukocytes present in the bloodstream ideally between 3000 and 10000 cells per mm^3 of blood. The WBC values depend on the age and sex of the patient. More than 10,000 leukocytes per mm3 strongly indicates infections and inflammation, whether acute or chronic.

Blood Smear: It consists of the deposition of a few microliters of blood in a glass slide. It is from the blood smear that the qualitative and quantitative evaluations of blood cells are performed.

Complete Blood Count: Is the result of the counting of blood cells by means of counting of Neubauer chamber with the posterior submission of the values found in consecrated mathematical formulas in the medical literature.

Content-Based Image Retrieval (CBIR): Is a technique that consists of the application of computational vision and pattern recognition methodologies to solve image retrieval problems in large databases. Visual features such as coloring, texture, and so on are used for the creation of a feature descriptor.

Flow Cytometry: Is a technique performed by means of an apparatus called a cytometry whose objective is to analyze particles and/or cells suspended in a liquid medium and submitted to a flow, which simulates the passage of blood from a blood vessel. In this way, it is possible to perform the quantification, analysis, and classification of cells and/or particles.

Mean Corpuscular Hemoglobin (MCH): Is a parameter of evaluation of the erythrocyte indices, being responsible for the evaluation of the staining of the erythrocytes. Thus, this analysis is of great importance for the classification of anemia. It is the average amount of hemoglobin present within each erythrocyte, being represented in picograms (pg).

Mean Corpuscular Hemoglobin Concentration (MCHC): Is a parameter of evaluation of the erythrocyte indices, being responsible for the evaluation of the staining of the erythrocytes. Thus, this analysis is of great importance for the classification of anemia. Is the mean hemoglobin concentration in 100 mL erythrocytes being plotted in gram per deciliter (g/dL). CHCM differs from HCM by the fact that the concentration takes cell volume into account and HCM only relates total hemoglobin to the number of erythrocytes.

Watershed Transform: Is an image processing technique that had its principle based on observations of nature. It presents great applicability in the detection and labeling of objects present in an image.

Chapter 3
Identification of Good One From the Damaged Crops/Fruits Using Decision-Level Information Matching

Muthukumar Arunachalam
Kalasalingam University, India

Meena Arunachalam
KLN College of Engineering, India

ABSTRACT

Identification of useful items that can be picked up from the damaged crops or batch of fruits/vegetables is a challenging task nowadays. Humans may fail to identify them correctly with their naked eyes due to strain. Image processing techniques can help to maximize the amount of the good agro-items easily by comparing the existing goods to templates. This chapter introduces an effective recognition method to spot good agro-items by extracting the local features using Gabor filter for orientation information. Another local information of that fruit/vegetable is extracted by speeded up robust features (SURF) algorithm. The extracted features are matched with their templates which results in the decision of individual feature extraction method. Finally, both local information is fused at decision level individually with AND operation (i.e., both algorithms will give correct decision to identify the good agro-item).

DOI: 10.4018/978-1-5225-8027-0.ch003

INTRODUCTION

Agriculture is one of the old industries in the world. The growth of agriculture decides the development of the whole country. In this modern world, agriculture should be developed along with technical growth. In every field, the technically qualified skill experts will be needed to reduce the investment cost and also to increase the output from the field in order to meet the increasing demand. If all this is done properly, then that industry will get good benefits and profits (Moallem et al., 2013; 2014).

Fruit production industry needs some technical growth in order to meet increasing consumption of them by all the people in the world (Razmjooy et al., 2012; 2013). One important thing to be concerned in this industry is to provide a good quality of fruits for all persons. So the separation or identification of good fruit/crop from the defected one is the most challenging task or problem. In addition, some other difficulties accumulated with this problem are workers cost, investment cost, and proper delivery fruits at the time. Automatic identifying the good crop from defect one is a good solution for the above all problems.

The main purpose of this chapter is to acquire a good quality fruit/crop through image processing methods. The image processing method can inspect the fruits better than human perception in a very fast manner. Pham and Lee (2015) proposed the automatic detection of food fruits from the plant through k-means clustering. The small regions of fruit were obtained through k-means clustering. These regions are merged through Region Adjacent Graph (RAG) method with thresholding values. Finally, RAG is matched to find the good fruit. A novel approach of detecting the fruit using Convolutional Neural Network (CNN) was proposed by Sa et al. (2016). The CNN detects the objects in a very fast manner. CNN is a type of supervised learning done by updating the weights of features through convolution and polling.

The traditional Gabor filter has been used for time and frequency analysis. This filter was modified by Yang et al. (2003) by splitting the single kernel function into two kernel function to analyze x and y spatial domains as bandpass and low pass filter. The image was enhanced due to this modified Gabor filter. The novel dimensionality reduction method was obtained by combining the Gabor feature with Principal Component Analysis (PCA) by Pan and Ruan (2008). The method was applied for palmprint recognition by extracting the features at various orientations in order to recognize more efficiently than the existing one. Zhang et al. (2011) have achieved the low error rate for recognizing the Finger Knuckle Print (FKP) biometric. In that paper, they fused two methods namely local and global information. The local information was attained from the Gabor filter and global one was obtained by Discrete Fourier Transform phase value named as bandlimited phase-only correlation. The matching was done fusion values of stored value with query value.

Using the Gabor feature, Chinese character recognition was proposed by Wang et al. (2005). In this paper, they extracted the features using Gabor and PCA method. The recognition rate was achieved by the method is 98%. Liu and Wechsler (2002) introduced a method of combining the Gabor and Fisher classifier for face recognition. The Gabor vector was derived from the Gabor wavelet and linear discriminant model. From the above literature survey, this chapter concluded that the Gabor filter is the main tool used for extracting the local feature information from an image.

The rotation invariant feature using four types of directional statistics obtained from both the modulus and phase parts of Gabor-filtered images was analyzed in Kim and So (2018). The identification of Pig images is done by extracting the Gabor features which were convolved with the local structural features using the Local Binary Pattern (LBP) identification was proposed Huang et.al (2018).

The speeded Up Robust Features (SURF) is another method for extracting the local feature information an image was proposed by Bay et al. (2008). This paper explains the detailed description detector and descriptor. This algorithm is a scalable and rotation invariant detector which used for object recognition effectively. Valgren and Lilienthal (2010) were investigated the two local feature descriptors namely SIFT and SURF algorithm. The authors used these methods for extracting the information in outdoor environments. Similarly, this method was also adopted for efficient tracking and object recognition which was done by Ta et al. (2009). The FKP biometric recognition was also done by Le-Qing (2011). From the above papers, this chapter finalized that, the SURF is another algorithm used for extracting the local feature descriptors from an image. Until now, most of the above methods are utilized for pattern recognition in order to authenticate a person. This chapter can investigate that how these methods are implemented in identifying the good fruits and results has been analyzed. Suh et.al (2018) had proposed a method to classify the sugar beet and potatoes was achieved based on Bag-of-Visual-Words (BoVW) model using Speeded Up Robust Feature (SURF) features or Scale-Invariant Feature Transform (SIFT) with crop row information in the form of the Out-of-Row Regional Index (ORRI).

The rest of this chapter is arranged as follows. The next section describes the structure of the proposed work. Thereafter section gives the details of the Gabor and SURF feature extraction of fruit. The Experimental results and the analysis are given in pre-last section. Finally, the conclusion was explained in the last section.

Structure of Proposed Work

The automatic separation of fruits from the defected one is useful for agriculture industry because it reduces time consumption and labor cost. In crop automation, several 100-kilogram crops are passing through a conveyer built, from which the good fruit should be identified and separated out from defected ones. In that situation, the sensor senses a fruit and take an image from conveyer built. Good fruit image is already stored in the database. So sensed image is compared with the stored one which results in good or defected one. This is a general workflow of fruit detection. This chapter proposes a new model of combining Gabor and SURF features which are shown in Figure 1.

The databases utilized for this chapter were obtained from the Horea94/Fruit-Images-Dataset (2018). This chapter contains two phases, i.e., registration and identification phase which is shown in Figure 1. The preprocessing is first step done in this chapter which is common for both phases. Conversion of image color and size was done in preprocessing. In registration phase, a template image of a good fruit/crop has been taken and it was preprocessed first. From the preprocessed image, the Gabor features are extracted using a Gabor filter. The key points are extracted from the preprocessed image using SURF simultaneously. The extracted features are stored in the database separately. This process is known as the registration process.

The next phase of this chapter is the identification phase. The same process of extracting the features from query fruit/crop using Gabor and SURF is first to step done in the identification phase also. The Gabor features are extracted from various positions and orientations. The SURF features are extracted in a parallel way to the Gabor features. Interested points and descriptors of SURF features are identified and pointed in the query fruit/crop using Hessian matrix and Haar wavelets respectively. The query image features are extracted using Gabor and SURF are matched with the stored features simultaneously. From that feature matching of individual algorithm displays that fruit is good or bad. From this matching, if both processes say good i.e. features matched (Yes), then the final result will be Good fruit. Otherwise, if

Figure 1. Workflow: Identification of good fruit

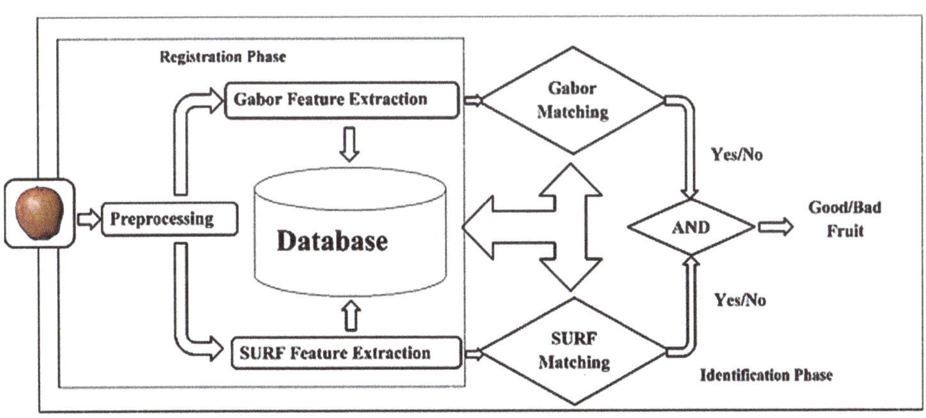

any one of the processes says bad i.e. features not matched (No), then the result will defect fruit. This decision-making process is done by AND operation. This type of fusion is named as decision level fusion.

GABOR FILTER: A REVIEW

The name *Gabor* is titled after Dennis Gabor for the invention in the field of texture analysis, edge detection, feature extraction, stereo disparity estimation etc., which helps in analyzing the distinct frequency content information around the localized region of the image. Here this section describes the Gabor filter and Gabor feature extraction concepts in a detailed. Gabor filter is considered as one of the complex exponentials that are modulated by Gaussian function in the case of the spatial domain. Basically, the Gabor filter is represented by two components:

1. Complex Sinusoidal (c(x,y))
2. Gaussian Function (g(x,y))

This Complex Sinusoidal function is represented as

$$c\left(x,y\right)=e^{-j2\prod F}\left(x\cos\left(\theta\right)+y\sin\left(\theta\right)\right)+p \tag{1}$$

where, F is considered as the magnitude of the sinusoidal function and P denotes the phase of that function and similarly, the Gaussian function also defines as:

$$g\left(x,y\right)=\frac{1}{\sqrt{2\prod\sigma}}e^{-\frac{1}{2}}\left[\frac{x^2}{\sigma x^2}+\frac{y^2}{\sigma y^2}\right] \tag{2}$$

In order to rotate the Gaussian function in the direction of Complex sinusoidal Gaussian function as:

$$g\left(x,y\right)=\frac{1}{\sqrt{2\prod\sigma}}e^{-\frac{1}{2}}\left(\frac{x^t}{\sigma x^2}+\frac{y^t}{\sigma y^2}\right) \tag{3}$$

where (x, y) represents the pixel position in the spatial domain, λ denotes the wavelength in pixels, θ shows the orientation of a Gabor filter and finally Sx, Sy describes the standard deviation along the *x* and *y* directions. The parameters of x and y are given by:

$$x' = \left(x\cos\theta + y\sin\left(\theta\right)\right)^2$$

$$y' = \left(-x\sin\theta + y\cos\left(\theta\right)\right)^2 \tag{4}$$

Therefore the Gabor filter is denoted as:

$$Gabor(x,y) = \left(c\left(x,y\right) * g\left(x,y\right)\right) \tag{5}$$

$$Gabor\left(x,y\right) = e^{-j2\prod F}\left(x\cos\left(\theta\right) + y\sin\left(\theta\right)\right) + p * \frac{1}{\sqrt{2\prod\sigma}} e^{-\frac{1}{2}\left(\frac{x^t}{\sigma x^2} + \frac{y^t}{\sigma y^2}\right)} \tag{6}$$

However, the amplitude and phase of Gabor filter bank denote reasonable cues about the specific pattern which are present in particular images. Amplitude confines the Directional Frequency content and phase includes information about the location of edges and image details. Basically, Gabor filters with various frequencies and orientations are very impressive in conquering the local information present in the particular image. This Gabor filter defines various parameters that are to be calculated as "wavelength", "orientation", "Phasc offsct", "Aspcct ratio", "Bandwidth" which helps in displaying the intensity map of the output window. Likewise, the impulse response of Gabor filter is also created by multiplication of Gaussian envelope function with complex oscillation. It basically helps in extending the space (time) to two dimensions by creating the filters which are selective for orientation. Likewise, the phase difference of Gabor filters is linear in nature. The impulse response of Gabor filter g(x) is denoted as:

$$g_{mn}(x) = \frac{1}{2\prod a_n b_n} e^{-\frac{1}{2}x^T A_{mn}{}^x e^{jk_0 T} mn^x} \tag{7}$$

Here the matrix A represents as bandwidth and orientation selectivity of the filter:

$$A_{mn} = \begin{bmatrix} \cos\varphi_m & -\sin\varphi_m \\ \sin\varphi_m & \cos\varphi_m \end{bmatrix} \begin{bmatrix} an^{-2} & 0 \\ 0 & bn^{-2} \end{bmatrix} \begin{bmatrix} \cos\varphi_m & -\sin\varphi_m \\ \sin\varphi_m & \cos\varphi_m \end{bmatrix} \tag{8}$$

Modulation frequency vector k_0 in the same direction is denoted by

$$k_{om} = k_{om} \begin{bmatrix} \cos\varphi_m \\ \sin\varphi_m \end{bmatrix} \tag{9}$$

Importance of Gabor filters has attained reasonable attention based on the human visual system. Basically, they are orientation selectivity filters and used in many Real-time applications like texture segmentation, optical character recognition, object recognition, image coding, fractal dimension management etc.,

Gabor Feature Extraction: A Review

The Gabor features are mainly calculated by the convolution procedure of input Image with Gabor Filter bank:

$$G_{U,V}\left(x,y\right) = I\left(x,y\right) * \psi\left(x,y\right) \tag{10}$$

These Gabor features are further tangible based on the filtering operation of Gabor with the size u and orientation. This convolution operation is accomplished separately for real and imaginary part. Originally the Gabor feature representation is defined as $\left|o\left(x,y\right)\right|_{m,n}$.

$$\mathrm{Re}\left(O\left(x,y\right)\right)_{m,n} = I\left(x,y\right) * \mathrm{Re}\left(\psi\left(x,y,\lambda_m,\theta_n\right)\right) \tag{11}$$

$$\mathrm{Im}\left(O\left(x,y\right)\right)_{m,n} = I\left(x,y\right) * \mathrm{Im}\left(\psi\left(x,y,\lambda_m,\theta_n\right)\right) \tag{12}$$

Therefore amplitude calculation involves the process of both Real and Imaginary parts:

$$O\left(x,y\right)_{m,n} = \left(\left(\mathrm{Re}\left(O\left(x,y\right)\right)_{m,n}\right)^2 + \left(\mathrm{Im}\left(O\left(x,y\right)\right)m,n\right)^2\right)^{1/2} \tag{13}$$

These Gabor Features are Multi-dimensional and have high redundant features which help in reducing the dimension using this Gabor filter. These filtered features are defined as a Feature vector in order to classify the required input. This classification task helps to determine the class of test data based on a supervised learning algorithm.

Figure 2 comprises of steps for Gabor Feature Extraction for the case of 5 Scales and 8 orientations (with the angle notation as 0^0,30^0, 45^0, 60^0, 120^0, 150^0, 180^0) are projected based on Gabor function Equation for the generation of Gabor Matrices. Finally, feature redundant vector involves the sampling method for the projection of Feature vector values. These Feature vectors are classified to determine the desired classification results for the predetermined input.

Figure 2. Flow diagram representation of Gabor Feature Extraction

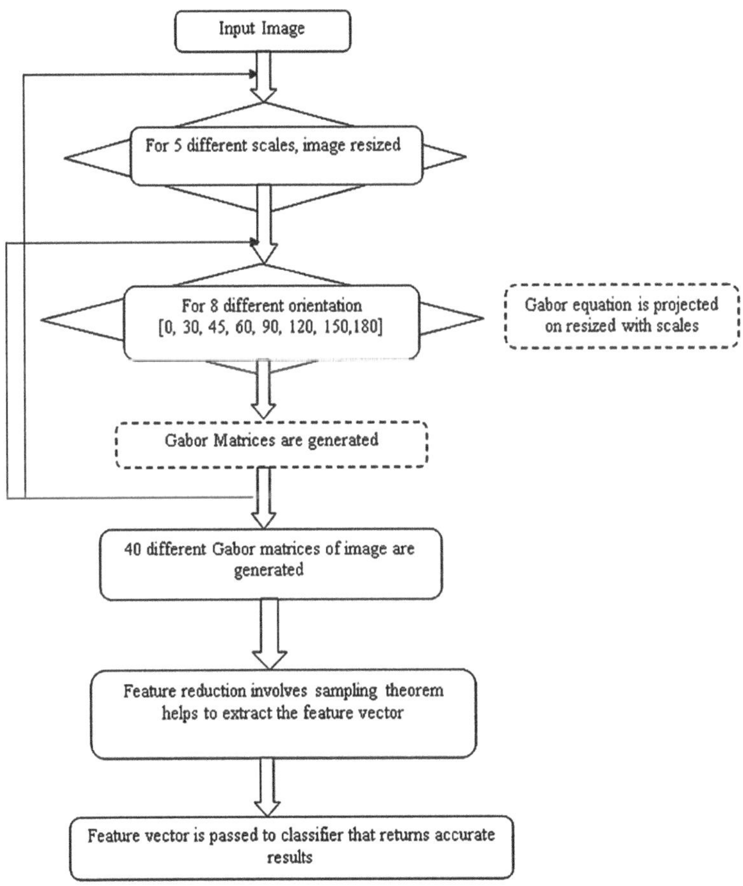

Gabor Wavelet Transform: An Introduction

This Gabor Wavelet is considered to be an essential tool for extracting the local features that are applied to images at specific angles(orientation). Basically, these Gabor filter captures several visual features like spatial localization, Frequency Selectivity, and Phase Relationship

This Gabor Kernel can be determined as

$$\psi\left(x, y, \varpi, \theta\right) = \frac{1}{2\prod\sigma^2} e^{\frac{-\left(x'+y'\right)^2}{2\sigma^2}} e^{i\varpi x'} \tag{14}$$

where (x,y) denotes the pixel position in the spatial domain, ϖ describes the central frequency of sinusoidal Plane Wave, θ explains the orientation of Gabor filter and finally σ defines the standard deviation along X, Y directions:

$$X' = x\cos\theta + \gamma\sin\theta$$

$$Y' = -x\sin\theta + \gamma\cos\theta \tag{15}$$

This Gabor Wavelet algorithm is basically defined as

- Describing the Input Image which represents I(x,y) of Size M*N with Gabor wavelet kernels $\psi_{u,v}\left(x, y, \varpi, \theta\right)$
- Gabor Feature Representation is obtained by convoluting the input with 40 Gabor wavelet Kernels

$$\psi_{u,v}\left(x, y\right) = I\left(x, y\right) * \psi_{u,v}\left(x, y, \varpi, \theta\right) \tag{16}$$

- Concatenate the magnitude of convoluted output images of all 40 features vectors
- Dimension reduction algorithm in size of feature vector

Properties of Gabor Wavelet include (i) Directional Selectivity (ii) Invariance to shifts and rotation (iii)Incentive to illumination variation. Likewise in the case of facial recognition methods, it faces major difficulties like

- High Computational complexity
- High Memory capacity requirement
- Feature vectors dimension are extremely large

SURF: A REVIEW

The SURF algorithm investigates the detailed description of the boundary as well as features of an image with the detector and descriptor. It also changes of the important parameter that describes the features of the image. The basic idea of the SURF algorithm depends on the intensity change of pixels to detect the various points in an image. The detection of points, lines, edges are obtained by intensity changes occur in the image. The changes in the intensity can be identified by first and second derivatives. The difference function is used for the defining the derivatives of a function. The first order derivative of a one-dimensional signal can be expanded using Taylor series which is given by following equations,

$$\frac{\partial f}{\partial x} = f\left(x+1\right) - f\left(x\right) \tag{17}$$

The second order derivative of the one-dimensional signal is given by:

$$\frac{\partial^2 f}{\partial x^2} = f'\left(x+1\right) - f'\left(x\right)$$

$$= f\left(x+2\right) - f\left(x+1\right) - f\left(x+1\right) + f\left(x\right)$$

$$= f\left(x+2\right) - 2f\left(x+1\right) + f\left(x\right)$$

$$\frac{\partial^2 f}{\partial x^2} = f''\left(x\right) = f\left(x+1\right) + f\left(x-1\right) - 2f\left(x\right) \tag{18}$$

Second order derivative will give the fine information of lines, points, and edges. The second order derivative can be achieved by the following equation,

$$\frac{\partial^2 f(x,y)}{\partial x^2} = f(x+1,y) + f(x-1,y) - 2f(x,y) \tag{19}$$

$$\frac{\partial^2 f(x,y)}{\partial y^2} = f(x,y+1) + f(x,y-1) - 2f(x,y) \tag{20}$$

Using the above equations, the intensity transformation is clearly defined. Based on this concept, Bay et al. (2008) proposed a new method named as Speeded Up Robust Features (SURF) to find various isolated points of an image. This method is used for object recognition by extracting and matching the local feature information. This paper explains the detailed description Harris detector and descriptor. This algorithm is a scalable and rotation invariant detector which used for object recognition effectively. Hereafter, the various stages of SURF algorithm explain one by one. The various stages of SURF algorithm are as follows,

- Integral Image
- Detection of points: using Hessian detection, Scale-Space representation
- Descriptors points: using Haar wavelet, orientation assignment (optional)

The overview of the SURF algorithm is shown in Figure 3. The first process of surf algorithm is integral images. In that, a portion of the image has been chosen alone by subtracting the remaining sides of the portion from the total image which is given in the following equation,

$$S(x,y) = \Sigma\Sigma I(x,y) \tag{21}$$

$$S(x,y) = I(x,y) - S(x-1,y-1) + S(x-1,y) + S(x,y-1) \tag{22}$$

The next process of SURF algorithm is the detection of interested point using Hessian matrix. The generation of Hessian matrix is given following derivation.

$$H(x,y) = \sum_{u,v} W_{u,v}(S(x+u,y+v) - S(x,y))^2 \tag{23}$$

Where w represents Gaussian value and s(x,y) is the integral image. The second order derivative of the above equation is given by

Figure 3. Overview of the SURF algorithm

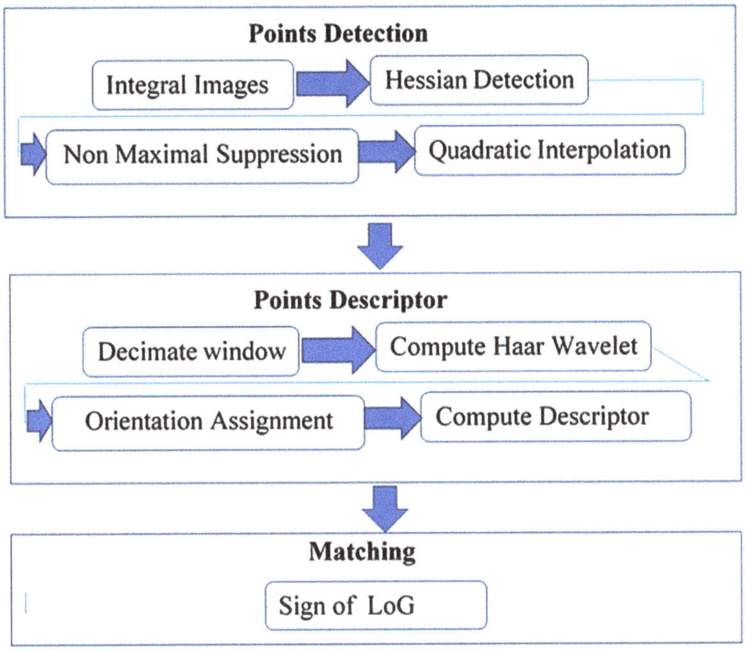

$$H\left(x,y\right)=\sum_{u,v}W_{u,v}(S\left(x,y\right)+uS_{x}\left(x,y\right)+vS_{y}\left(x,y\right)-S(x,y))^{2}$$ 1st Order derivative

$$H\left(x,y\right)=\sum_{u,v}W_{u,v}(u^{2}S_{x}^{2}\left(x,y\right)+v^{2}S_{y}^{2}\left(x,y\right)+2uvS_{x}\left(x,y\right)S_{y}\left(x,y\right))^{2}$$ 2nd Order derivative (24)

The above equation (24) expressed in matrix form as follows,

$$H\left(x,y\right)=\sum_{u,v}W_{u,v}\left[u,v\right]\begin{bmatrix}S_{x}^{2} & S_{x}S_{y}\\ S_{x}S_{y} & S_{y}^{2}\end{bmatrix}\begin{bmatrix}u\\ v\end{bmatrix}$$

$$H\left(x,y\right)=\left[u,v\right]\sum_{u,v}W_{u,v}\begin{bmatrix}S_{x}^{2} & S_{x}S_{y}\\ S_{x}S_{y} & S_{y}^{2}\end{bmatrix}\begin{bmatrix}u\\ v\end{bmatrix} \tag{25}$$

$$H(x,y) = [u,v]\ M\begin{bmatrix} u \\ v \end{bmatrix} \tag{26}$$

where, $M = \sum_{u,v} W_{u,v} \begin{bmatrix} S_x^2 & S_x S_y \\ S_x S_y & S_y^2 \end{bmatrix}$ and $W_{u,v} = \exp-(u^2 + v^2)/2\sigma^2$

Finally, Hessian matrix of equation 25 can be rewritten as:

$$H(x,y,\sigma) = \begin{pmatrix} \dfrac{\partial^2 g}{\partial x^2}(x,y,\sigma) & \dfrac{\partial^2 g}{\partial x \partial y}(x,y,\sigma) \\ \dfrac{\partial^2 g}{\partial x \partial y}(x,y,\sigma) & \dfrac{\partial^2 g}{\partial y^2}(x,y,\sigma) \end{pmatrix} \tag{27}$$

where g is the convolution of Gaussian second order derivative and integral image s(x,y).

The Determinant of Hessian (DoH) is calculated as,

$$DoH(x,y,\sigma) = g_{xx}(x,y,\sigma).g_{yy}(x,y,\sigma) - (0.9)\,g_{xy}(x,y,\sigma)^2 \tag{28}$$

The local maxima of this filter response occur in regions where both $g_{xx}\ and\ g_{yy}$ are strongly positive, and where g_{xy} is strongly negative. Therefore, these extreme occur in regions in the image with large intensity gradient variations in multiple directions, as well as at saddle points. Visually, this means that blob-like structures refer to corners and speckles.

The next step of the SURF algorithm is Descriptors points' identification. The interesting point is identified from the Hessian matrix detection. The window has been placed around the interesting points. Each window is further divided into four sub-windows.

The Haar wavelet is computed for both horizontal and vertical directions which provide spatial data. The decomposition is done by decimating the given image into two levels as lower value components and higher value components using downsampling method through Haar transform. The Haar transform is expressed as:

$$h_k(z) = h_{pq}(z) = \frac{1}{\sqrt{N}} \begin{cases} 2^{\frac{p}{2}} & (q-1)/2^p \le z \le (q-0.5)/2^p \\ -2^{\frac{p}{2}} & (q-0.5)/2^p \le z \le (q)/2^p \\ 0 & \text{otherwise} \end{cases} \tag{29}$$

For example, N=4, k, q, and p consider the values which are shown in Table 1,

When calculating Haar wavelet value, the approximated value and level of decomposition are considered. In this paper, the level of decomposition as 2 is considered which implies J=2, where $N=M=2^J$, so j=0, 1, 2…, J-1. Thereafter, the polarity of intensity changes is obtained by summation of vertical and horizontal values of all 64 descriptors. Finally, the difference between the bright portion from the black background and black portion from the bright background can be obtained by the sign of the Laplacian is expressed as below,

$$sgn\left\{g_{xx}(x,y,\sigma) + g_{yy}(x,y,\sigma)\right\} = \begin{cases} +1 & \textit{bright portion from black backgrounthe d} \\ -1 & \textit{black portion from bright backgrouthe nd} \end{cases} \tag{30}$$

Table 1. Haar Table

K	P	Q
0	0	0
1	0	1
2	1	1
3	1	2

PROPOSED WORK: METHODOLOGY

Gabor Feature Extraction

In this chapter, fruit/crop is considered as an image. The raw fruit/crop image is represented in a color image. So the first step of this chapter is to convert the color image to grayscale image for doing further processing. The texture based image analysis was done extensively using Gabor filter. The Gabor filters are the same as a human vision in terms of frequency and orientation representation (Zhang et al. 2011). In the spatial domain, the Gabor filter is represented as:

$$G\left(x,y\right)=\exp\left(-\frac{1}{2}\left(\frac{\left(x\cos\Theta+y\sin\Theta\right)^{2}}{\sigma_{x}^{2}}+\frac{\left(-x\sin\Theta+y\cos\Theta\right)^{2}}{\sigma_{y}^{2}}\right)\right)\exp(2\pi fi\left(x\cos\Theta+y\sin\Theta\right) \tag{31}$$

where, f is the frequency of Sine function, Θ= Gabor filter orientations, and σ= standard deviation of the Gaussian function.

The Gabor features are extracted using equation 1 with 6 x 5 various orientations by convolving with the input image. It is an operator which obtain the local information of the image in all regions. The Gabor features are mentioned in single row matrix as:

$$S=\left\{s_{1},s_{2},s_{3}\ldots s_{n,}\right\} \tag{32}$$

SURF Feature Extraction

The detailed explanation of the SURF algorithm is to be discussed in this section (Bay et al. 2008). This algorithm consists of the following steps:

1. Integral image construction
2. Interest points based on the Hessian matrix
3. Descriptors based on Haar wavelets

The integral image construction is a process of adding all pixels of the input image by construction small rectangular regions. This construction of the integral image is given in equation 3.

$$I\left(x\right)=\Sigma_{i=0}^{x}\Sigma_{j=0}^{y}I\left(i,j\right) \tag{33}$$

From this image, the interest points are derived from the Hessian matrix. This matrix was constructed by performing the convolution of Gaussian second order partial derivative with the integral image of scaling level 'σ'. The Hessian matrix is given in equation 4.

$$H\left(x,\sigma\right)=\begin{pmatrix}\frac{\partial^2 f}{\partial x^2} & \frac{\partial^2 f}{\partial x\partial y}\\ \frac{\partial^2 f}{\partial x\partial y} & \frac{\partial^2 f}{\partial y^2}\end{pmatrix} \tag{34}$$

The SURF descriptors are constructed using Haar wavelets. Using the image is subdivided into small 16 sub-regions. Splitting of regions is performed by filters into low frequency and high-frequency regions. The regions are sampled with Haar transform.

Fruits/Crop Registration and Identification Phase

This chapter consists of two-phase namely registration and identification phase is shown in Figure 1. The process of identifying good fruit/crop is consolidated in the following steps:

- The image is preprocessed for image color and size conversion is the first step of this chapter
- The next step is extracting the features using Gabor and SURF algorithm
- Features are stored in the database
- The first two steps are repeated with query image in the identification phase
- The features of Gabor and SURF are matched separately with data stored image with query image data
- If the matched result yes is obtained from both algorithm, the final results will be good fruit using AND operation. Otherwise, if anyone algorithm says no, the result will defect fruit.

Experimental Result and Analysis

Experiments in this chapter are performed using Horea94/Fruit-Images-Dataset (2018). This database consists of two folders namely training (37836 images) and test set (12709 images) with 75 subsets of fruits with different varieties in each folder which listed below. Totally database contains 50590 fruit images. This high-quality dataset contains following fruits images:

Apples, Apricot, Avocado, Avocado ripe, Banana, Cactus fruit, Cantaloupe, Carambula, Cherry, Clementine, Cocos, Dates, Granadilla, Grape, Grapefruit, Guava, Huckleberry, Kiwi, Kaki, Kumquats, Lemon, Lime, Lychee, Mandarine, Mango, Maracuja, Melon Piel de Sapo, Mulberry, Nectarine, Orange, Papaya, Passionfruit, Peach, Pepino, Pear, Physalis, Pineapple, Pitahaya Red, Plum, Pomegranate, Quince, Rambutan, Raspberry, Salak, Strawberry, Tamarillo, Tangelo, Walnut.

The first step of this chapter is the image is preprocessed and resized is shown in Figure 4. The output of individual stages and overall final output shown in Figure 4. The Gabor features are extracted in a row vector. The Gabor features of 1x2048 values are matched between the input and query image. The SURF features of 12 x 128 values are matched between the input and query image. For example, an apple is taken and its features of Gabor and SURF features are matched individually, which results as matched i.e. 'yes'. Finally, the values are clubbed together with AND operation to produce as 'fruit ok' which is shown in Figure 4. Likewise, for this chapter, simulation is performed by 10 images of each subset for all types of fruits which listed above. Consider that stored value (S) and query value (Q) on the database are to be compared. The matched value (M) is the output of AND operation two different values. The M is given by equation 35.

$$M = \Sigma_{i=1}^{n} AND\left(S_{i,}Q_{i}\right) \tag{35}$$

In this chapter to analyze the results, False Rejection Rate (FRR) with GAR and False Acceptance Rate (FAR) is to be considered. The GAR, FAR, FRR are

Figure 4. Outputs at each stage with respect to workflow

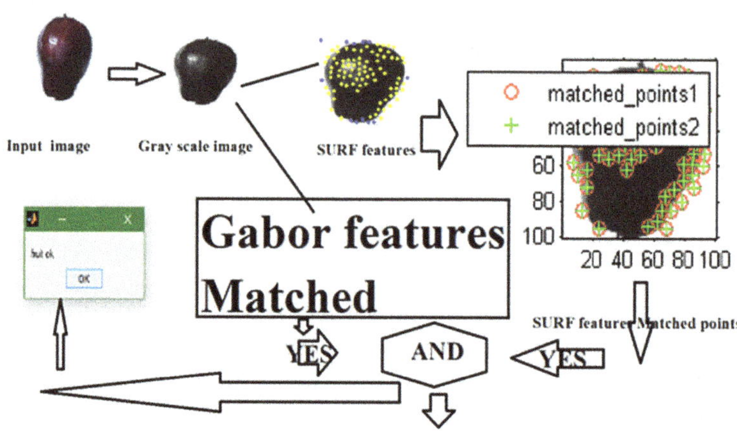

calculated in the equation 36, 37 and 38 respectively. If the number of feature points and key points is very high in fingerprint and FKP, then the GAR is high and the FRR becomes low. A number of tests are conducted to examine the performance of the proposed approach with the above factors. The results attained from the proposed approach are FRR=2.6%, GAR= 97.4%, and FAR=0%

$$GAR=\frac{\text{No. of genuine attempts accepted}}{\text{Total no. of genuine attempts}} \tag{36}$$

$$FRR=\frac{\text{No. of genuine attempts rejected}}{\text{Total no. of genuine attempts}} \tag{37}$$

$$FAR=\frac{\text{No. of impostor attempts accepted}}{\text{Total no. of genuine attempts}} \tag{38}$$

CONCLUSION

In this chapter, the two local feature algorithm namely Gabor and SURF features are investigated to identify a good fruit using decision level fusion. The best performance will be obtained from that algorithm which is shown in the results. Here two methods did individually and they are used fused by decision level fusion. The overall performance is attained from the proposed approach are FRR=2.6%, GAR= 97.4%, and FAR=0%.

ACKNOWLEDGMENT

We acknowledge the Department of Electronics and Communication Engineering of Kalasalingam University (Kalasalingam Academy of Research and Education), Tamil Nadu, India for allowing to use the computational facilities available in Signal Processing and VLSI Design laboratory which was set up with the assistance of the Department of Science and Technology (DST).

REFERENCES

Bay, H., Ess, A., Tuytelaars, T., & Van Gool, L. (2008). Speeded-up robust features (SURF). *Computer Vision and Image Understanding*, *110*(3), 346–359. doi:10.1016/j.cviu.2007.09.014

Fruit Master Database. (2018). *Different Fruits data*. Retrieved from https://github.com/Horea94/Fruit-Images-Dataset

Huang, W., Zhu, W., Ma, C., Guo, Y., & Chen, C. (2018). Identification of group-housed pigs based on Gabor and Local Binary Pattern features. *Biosystems Engineering*, *166*, 90–100. doi:10.1016/j.biosystemseng.2017.11.007

Kim, N. C., & So, H. J. (2018). Directional statistical Gabor features for texture classification. *Pattern Recognition Letters*, *112*, 18–26. doi:10.1016/j.patrec.2018.05.010

Le-Qing, Z. (2011). Finger knuckle print recognition based on SURF algorithm. In *Fuzzy Systems and Knowledge Discovery (FSKD), 2011 Eighth International Conference on* (Vol. 3, pp. 1879-1883). IEEE.

Liu, C., & Wechsler, H. (2002). Gabor feature based classification using the enhanced fisher linear discriminant model for face recognition. *IEEE Transactions on Image Processing*, *11*(4), 467–476. doi:10.1109/TIP.2002.999679 PMID:18244647

Moallem, P., Razmjooy, N., & Ashourian, M. (2013). Computer vision-based potato defect detection using neural networks and support vector machine. *International Journal of Robotics and Automation*, *28*(2), 137–145. doi:10.2316/Journal.206.2013.2.206-3746

Moallem, P., Razmjooy, N., & Mousavi, B. S. (2014). Robust potato color image segmentation using adaptive fuzzy inference system. *Iranian Journal of Fuzzy Systems*, *11*(6), 47–65.

Pan, X., & Ruan, Q. Q. (2008). Palmprint recognition using Gabor feature-based $(2D)^2$ PCA. *Neurocomputing*, *71*(13-15), 3032–3036. doi:10.1016/j.neucom.2007.12.030

Pham, V. H., & Lee, B. R. (2015). An image segmentation approach for fruit defect detection using k-means clustering and graph-based algorithm. *Vietnam Journal of Computer Science*, *2*(1), 25–33. doi:10.100740595-014-0028-3

Razmjooy, N. (2012). A multi layer perceptron neural network trained by invasive weed optimization for potato color image segmentation. *Trends in Applied Sciences Research*, *7*(6), 445–455. doi:10.3923/tasr.2012.445.455

Razmjooy, N., Mousavi, B. S., & Soleymani, F. (2012). A real-time mathematical computer method for potato inspection using machine vision. *Computers & Mathematics with Applications (Oxford, England)*, *63*(1), 268–279. doi:10.1016/j.camwa.2011.11.019

Razmjooy, N., Mousavi, B. S., & Soleymani, F. (2013). A hybrid neural network Imperialist Competitive Algorithm for skin color segmentation. *Mathematical and Computer Modelling*, *57*(3-4), 848–856. doi:10.1016/j.mcm.2012.09.013

Sa, I., Ge, Z., Dayoub, F., Upcroft, B., Perez, T., & McCool, C. (2016). Deep fruits: A fruit detection system using deep neural networks. *Sensors (Basel)*, *16*(8), 1222. doi:10.339016081222 PMID:27527168

Suh, H. K., Hofstee, J. W., IJsselmuiden, J., & van Henten, E. J. (2018). Sugar beet and volunteer potato classification using Bag-of-Visual-Words model, Scale-Invariant Feature Transform, or Speeded Up Robust Feature descriptors and crop row information. *Biosystems Engineering*, *166*, 210–226. doi:10.1016/j.biosystemseng.2017.11.015

Ta, D. N., Chen, W. C., Gelfand, N., & Pulli, K. (2009). Surftrac: Efficient tracking and continuous object recognition using local feature descriptors. In *Computer Vision and Pattern Recognition, 2009. CVPR 2009. IEEE Conference on* (pp. 2937-2944). IEEE.

Valgren, C., & Lilienthal, A. J. (2010). SIFT, SURF & seasons: Appearance-based long-term localization in outdoor environments. *Robotics and Autonomous Systems*, *58*(2), 149–156. doi:10.1016/j.robot.2009.09.010

Wang, X., Ding, X., & Liu, C. (2005). Gabor filters-based feature extraction for character recognition. *Pattern Recognition*, *38*(3), 369–379. doi:10.1016/j.patcog.2004.08.004

Yang, J., Liu, L., Jiang, T., & Fan, Y. (2003). A modified Gabor filter design method for fingerprint image enhancement. *Pattern Recognition Letters*, *24*(12), 1805–1817. doi:10.1016/S0167-8655(03)00005-9

Zhang, L., Zhang, L., Zhang, D., & Zhu, H. (2011). Ensemble of local and global information for finger–knuckle-print recognition. *Pattern Recognition*, *44*(9), 1990–1998. doi:10.1016/j.patcog.2010.06.007

Chapter 4
Soft-Computing-Based Approaches for Plant Leaf Disease Detection:
Machine-Learning-Based Study

Vivek K. Verma
Manipal University Jaipur, India

Tarun Jain
Manipal University Jaipur, India

ABSTRACT

The disease occurrence phenomena in plants are season-based which is dependent on the presence of the pathogen, crops, environmental conditions, and varieties grown. Some plant varieties are particularly subject to outbreaks of diseases; on the other hand, some are opposite to them. Huge numbers of diseases are seen on the plant leaves and stems. Diseases management is a challenging task. Generally, diseases are seen on the leaves or stems of the plant. Image processing is the best way for the detection of plant leaf diseases. Different kinds of diseases occur because of the attack of bacteria, fungi, and viruses. The monitoring of leaf area is an important tool in studying physiological capabilities associated with plant boom. Plant disorder is usually an unusual growth or dysfunction of a plant. Sometimes diseases damage the leaves of plants.

DOI: 10.4018/978-1-5225-8027-0.ch004

INTRODUCTION

Plant diseases cause major production and economic losses in agriculture and forestry. Agriculture in India is one of the biggest sectors of economic development. With increasing population, even though the contribution is continuously falling since independence from 55.1% in 1950 to 14% in 2012, it remained the major employment sector with a marginal difference. Leaf disease detection requires a huge amount of work, knowledge in the plant diseases, and also requires the more processing time. It is crucial to prevent unnecessary waste of financial and other resources, thus achieving healthier production in this changing environment, appropriate and timely disease identification including early prevention has never been more important. India being a tropical region and having three primary seasons- rain, winter, and summer. Most of the crop diseases occur either the Rainy or the Winter Season. Crops in India mostly suffer in the rainy season due to the improper irrigation system. Thus a system needs to be made which can help those farmers in need. We can use Image Processing techniques and Convolution Neural network together for identification of diseases. The basic goal of our project is to help the farmers who want to know what disease their crop has and what is the suitable prevention.

An estimated 15-25 percent of potential crop production is lost due to this menace at a time when India needs not only to raise production but also to ensure the food security and the nutrition for its growing consumption requirements. In 2016 sixth National Agrochemicals Conference was held on the theme of 'Role of Crop Protection Solutions'. Many topics were brought up by various speakers such as Plasticulture, fertigation, seed treatment, crop protection chemicals, biotechnology, precision farming etc. The conference covered topics of relevance to the sector, incl. Facilitating ease of doing business. Making India global manufacturing hub of quality crop protection solutions, Key issues and challenges faced by the industry, the export potential of the sector as also some of the futuristic technologies such as Vertical Farming.

The very important factor is the detection of plant disease and its issue to prevent a serious blow-up. Automatic detection of plant disease is become to a very important topic of the research. Fungi are diagnosed or determine from their morphology. With few exceptions, bacteria exist as a productive cell and increase in numbers via dividing into cells for the duration of a system called binary fission. In popular, a plant turns into diseased whilst it is continuously disturbed by means of a few causal agents that effects in a peculiar physiological method that disrupts the plant's regular shape, boom, feature or different activities. This interference with one or greater of a plant's essential physiological or biochemical systems elicits function pathological conditions or symptoms.

Since the problem has been identified, we need to find a common solution for many types of diseases. There will be some cases where the disease symptoms wouldn't be visible, or the diseased area coverage is large which would make it hard for the farmers to be able to find a cure, hence a sophisticated analysis is necessary. However, most diseases on leaves are visible to the naked eye and so a trained professional could tell what disease the leaf had and how to cure it. This is where the problem starts to occur as sometimes the professionals sent by the government or any organizations are amateur and could have more difficulties determining it than a professional plant pathologist. For this reason, an automated system is to be designed which can detect whether the leaf is diseased or not and if it is then what type of disease is it carrying. This booms the field of computer vision applications for precision agriculture. For this specific application, we explore the field of digital image processing and its common techniques such as feature detection. The feature detection was then used with the aim of detection and classification of plant diseases. For the classification, we used machine learning and convolutional neural networks.

To do the management of different plants actually requires close observation monitoring especially for the management of a disease that can affect complete agricultural production significantly and subsequently the life of after harvesting. The occurrence of the disease in the plants leads to a decrease in quality along with quantity too. This can produce a bad impact on the nations whose economies are often dependent on agriculture. Consequently, the detection of the disease at sooner stage could be very essential to keep away from the loss in terms quantity and finance. Detection of disease with eyes only was the only method taken into the process earlier. But as this requires continuously do monitoring of by some expert which might be prohibitively expensive as well as a time-consuming process for large farms. Automatic detection of the plant may prove benefits in keeping an eye on large farms of crops and thus automatically detect the symptoms of diseases as soon as they appear on plant leaves, so there is a need to have a fast, automatic, less expensive and accurate method to detect disease by identifying disease affected leaf area.

Diseases occur in plants is nowadays becoming an important issue, as its quality and its quantity both are completely depended upon agriculture production (Razmjooy, 2012). Not giving attention to this area in plant disease occurrence which is damaging the crops causes a significant reduction in whole land Production. So, detection and classification should be taken as an important and urgent task. In early time farmers used to identify the diseases by eye observation thus we can say that finding and identifying the disease by only eyes is very strenuous, time-consuming and it is not practical for large agricultural field and this also leads to wrong identification and there is a possibility for them to detect the same part as different disease. To maximize the required accuracy various manual method are used. The disadvantage of these methods is that this kind of methods is strenuous. So what we needed is to work on to get quick result oriented and accurate process to find the present plant diseases.

Several researchers have worked on plant leaf disease detection problem. Plant leaf disease detection has a broad application perspective in agriculture and is especially significant to the environmental science diversity research. Most plant species have their unique leaves which differ from each other by characteristics such as the shape, color, texture and the margin. The biometric features of plants leaf-like venation make this disease detection easy. There are various successful soft computing based object detection techniques available like fuzzy systems (Hosseini et al., 2014), neural networks (Razmjooy et al., 2013), evolutionary computation (Musavi et al., 2013), probabilistic reasoning, and many others. All these soft computing techniques are discussed in this chapter that can be applied over the images of plant leaf and disease if any can be identified to prevent the affecting of all the plants. Deciding on the method for disease detection is often a difficult task because the quality of the results can be different for different input data.

India is essentially an agricultural country. Agriculture is the main and important career for a maximum of the Indian households. In India, agriculture contributes about 16 percentage of overall GDP Over sixty percentage of India's land, agricultural merchandise of extensive economic value consists of rice, wheat, potato, tomato, onion, mangoes, sugar-cane, beans cotton, etc. In India, the rural quarter occupies a primary position inside the trendy monetary device for us as is probably easy from the following:

- **Percentage of Agriculture in Countrywide Earnings:** Agriculture has got a top function in Indian economy. It contributes 16 in keeping with a cent of overall GDP.
- **Agriculture as a Source for a Living:** Agriculture sector, at the gift, presents livelihood to 65 to 70 in line with a cent of the overall population. The agriculture sector provides employment to more than half percent of the country's workforce and is the single private sector occupation at a big level.
- **Crucial Supply of Business Development:** Diverse important industries in India locate their raw material from agriculture area -cotton and jute textile industries, sugar, Vanaspati and so forth are without delay depending on agriculture.
- **Significance in Worldwide Change:** Foreign trading of our country is mainly associated with a sector of agriculture. This has a major role in the economy of the country.

DISEASE OR ILLNESS OF PLANT LEAF

A detriment of the normal condition of a plant that interrupts or modifies its function. The disease occurrence phenomena in plants are season based which is depending on the presence of crops, pathogens, and varieties grown on the other hand some are opposed to them as discussed in the paper Hu, J. et al. (2018).

Huge numbers of disease are seen on the plant leaves and stems and diseases management is a challenging task as discussed in the paper Zhang, S. et al. (2017). Image processing is the best way for the detection of the plant leaf disease. Different kind of disease occurs because of the attack of bacteria, fungi, and viruses. The monitoring of leaf area is an important tool in studying physiological capabilities associated with plant boom. Plant disorder is usually an unusual growth or dysfunction of a plant. Sometimes disease damages the leaves area of the plant.

Diseases occur in plants is nowadays becoming an important issue, as its quality and its quantity both are completely depended upon agriculture production. Not giving attention to this area in plant disease occurrence which is damaging the crops causes a significant reduction in whole land production. So, detection and classification should be taken as an important and urgent task. In early time farmers, used to identify the diseases by eye observation thus we can say that finding and identifying the the disease by only, eyes are very strenuous, time-consuming and it is not practical for large agricultural field and this also leads to wrong identification and there is a possibility for them to detect the same part as different disease.

At present, a new concept immerges of smart farming and has been introduced where the agricultural field conditions are controlled and monitored using the own self-operating systems. The self-recognition of the various diseases is based on the finding of the symptoms of diseases. Thus the information about the occurrence of the present disease could be more quickly and correctly will be given to the farmers, experts, and researchers. Thus monitoring of large field by human eye will be reduced. In the disease present in a plant leaf and to recognize its presence from a picture or an image, the only key is to extract the required characteristic feature of the area which is affected by the disease. Accordingly, the features of the plant leaf vary as per their characteristic. The various features which we have extracted from the taken image are cooler, shape, and texture. Sometimes for disease detection features which we are extracting increases overall cost. So this leads in maximizing the complexity as well as computing time. Hence it is necessary to minimize data of features. Mostly the methods which are adapted for monitoring and management of plant leaf disease are manual. One of the main approaches is with a naked eye observation. But the requirement of this method is continuous keeping an eye over the field by a person with superior knowledge about the plants as well as its corresponding diseases. Moreover, getting appointed such kind person may prove costly. On the second hand seeking advice from any expert also may add the cost. Apart from that, we need an expert at all the time. A laboratory test can be carried out to find diseases present in the plant. However, this method requires first-rate laboratory conditions alongside professional information. The pathogen detection techniques can provide more correct outcomes. But its results in high costing and could be time-consuming.

Image processing & pattern recognition techniques have been developed for the purpose of detection of diseases occurring on plant leaves, stems, lesion and so forth by the researchers (Razmjooy et al., 2013; Moallem et al., 2013, Moallem et al., 2014; Musavi et al., 2014). The sooner disorder appears on the leaf it should be detected. diagnosed and corresponding measures should be taken to avoid loss. Hence an automatic, fast, accurate and less expensive system is required. Many Types of research opted no. of methods for the purpose of detection and identification of disease correctly.

Here we need to get and obtain precise and accuracy to meet this purpose we are using Image processing technique even for leaves having increased or max dimensions. It will require a minimum time period to process the image as discussed in the paper Zhu, X. et. al. (2018).

Table 1. Review of different author work

S. No.	Author	Method Used
1	Camargo, A., & Smith, J. S. (2009)	The RGB image is convened into HSV then ANN is used to classify the present disease
2	Sabrol, H., & Kumar, S. (2015)	In this paper, the classifier is being compared
3	Zhang, S., Wu, X., You, Z., & Zhang, L. (2017).	Quantifying disease affected area of leaf and determine the size and shape of the diseased leaf
4	Jasim, S. S., & Al-Taei, A. A. M. (2018).	The disease is detected using SVM, KNN & ANN
5	Kashyap, Y., Sharma, T., & Ali, S. S. (2017).	HIS, YCBCR, and CIELAB color model are compared to find the accuracy of diseased spot detection using a median filter(image pre-processing) and Otsu method (for classification)
6	DeChant, C. et al. (2017).	3 tier architecture is used to detect the disease of the plant leaf
7	Singh, V., & Misra, A. K. (2017).	Using color co-occurrence Matrix texture features are extracted and then compared with Normal leaf

GENERAL METHODOLOGY

In the complete study till now, we have studied the concept of image processing. We found two main issues which need to be cover in our study for to achieving accuracy and fast detection of leaf diseases.

- **Segmentation:** Whenever we move input images to the system, then there is a requirement for determination of the exact disease present area or region in a particular leaf image that is always been a too long process. So it becomes mandatory for us to execute or go with the process of segmentation of disease affected leaf image and this process of segmentation is done using K-means algorithm of clustering.
- **Detection of Severity:** The leaf images which are processed using other image processing technique none of them identify that how severe the present disease in a particular leaf, so we came across for its implementation to, shows the severity of the present disease.

For applying all the above attention we implemented following techniques defining the quick and correct disease detection for that purpose we offer a software program implementation for performing automatic detection on the disease presence in a plant leaf. The complete system basically includes 4 pre-main stages to start with Plant leaf image are been taken or plant diseased leaf are taken and are given to below method. Every photo is segmented by means of the use of the clustering method. From the clustered images some features are taken or extracted as extracted features are vast

to be taken into consideration for future processing category or purpose, then photos stored in the database are labeled using Neural network So we will say that proposed approach can efficiently come across and classify tested illnesses with a precision among 82% and 92%.

- **Image Acquisition:** In this step, we need to initiate processing of work by taking images by a database which already exists or pre-existing and are a move towards thus Images am selected through any pre-existing database or captured at out of control lights situation.
- **Image Pre-Processing:** When we take leaf image and move it towards the system as input entity then pre-processing of image will be taken to get done some improvement over the pictures data which will use to eliminate distortions which are of no worth, Some functions of pictures are improved, which are required for purpose of further processing. It stores up the color space conversion process, enhancement of the image, and then segmentation

Figure 1. Basic procedure for disease detection

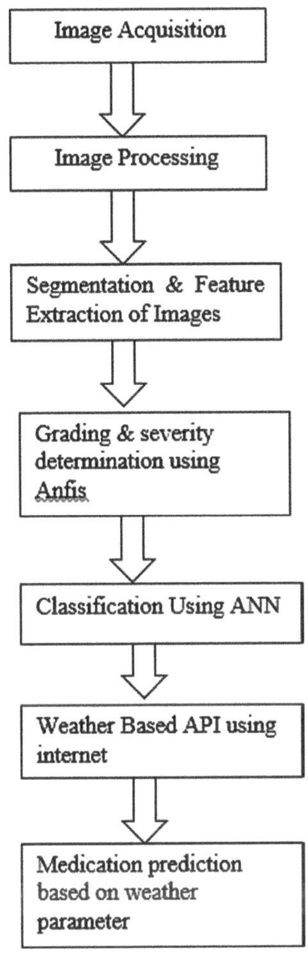

of the image. The transferred image to input as input data is present in the Form of specific color space which is RGB so it will be transformed into a different color space representation which is acceptable and can be used in further processing purpose. The main motto or objective to doing so is to facilitate the specification of colors in some standard accepted way.

- **K-Means Clustering Technique:** K-Means very firstly used by James Mac Queen in 1967 which is considered is a fast and accurate method for the purpose of detection of leaf disease using K-Means. This method is based on the segmentation process. In this clusters is formed which is made up of equal objects and is not similar to items of another cluster. The method clustering is basically sorting of items into different parts such that information in every cluster subset shows some common part according to predefined standard measures. The image of the photo can be classified by its shape, textures and other parameters which can be taken from the input image. We will see into every point or pixel which is having any connection with given data set and will make a bond between them. Now on the basis of closeness of pixels with each other, we will assign them cluster, and to find closeness of pixel Euclidian distance measure is used.

- **Features Extraction Procedure:** We can extract the function of clustered which is anticipated through the machine and contains maximum information. Grey degree Co-incidence Matrices (LLCM) is a statistical method this is used right here for characteristic extraction or texture class. It's been a critical function extraction method inside the domain of texture class that computes the relationship among pixel pairs in the photograph. The texture features can be calculated from the generated GLCMs, for example, these features or characteristic are comparison, correlation, electricity, entropy, and homogeneity.
- **Anfis & Anfis Editor:** Neural networks and Fuzzy good judgment structures arc frequently taken into consideration as part of gentle Computing vicinity. We may one opine the operation of Anfis & can use neural networks and fuzzy systems, which try to execute the operation of the human mind.

Neural networks mainly focus on the structure of the human mind, i.e., at the "hardware" emulating the fundamental functions, whereas fuzzy good judgment structures focus on "software program". emulating fuzzy and symbolic reasoning. Anfis is used as Modeling artificial structures and associated statistics: sample clustering and recognition, characteristic approximation, machine parameter estimation, and various others. So we will say that the mixture of both fuzzy logic structures & NN is parameterized computational nonlinear algorithms to process information in the mathematical form.

- **N/W of Artificial Neuron (ANN):** ANN is basically a method of computational works at the shape and the functions of the biological neural network. Information that flows via the community influences the structure due to the fact neural network adjustments – based totally on that enter and output. ANNs are taken into consideration nonlinear statistical statistics modeling gear where the relationships between inputs and outputs are modeled.

Figure 2. ANN & Fuzzy interface flowchart

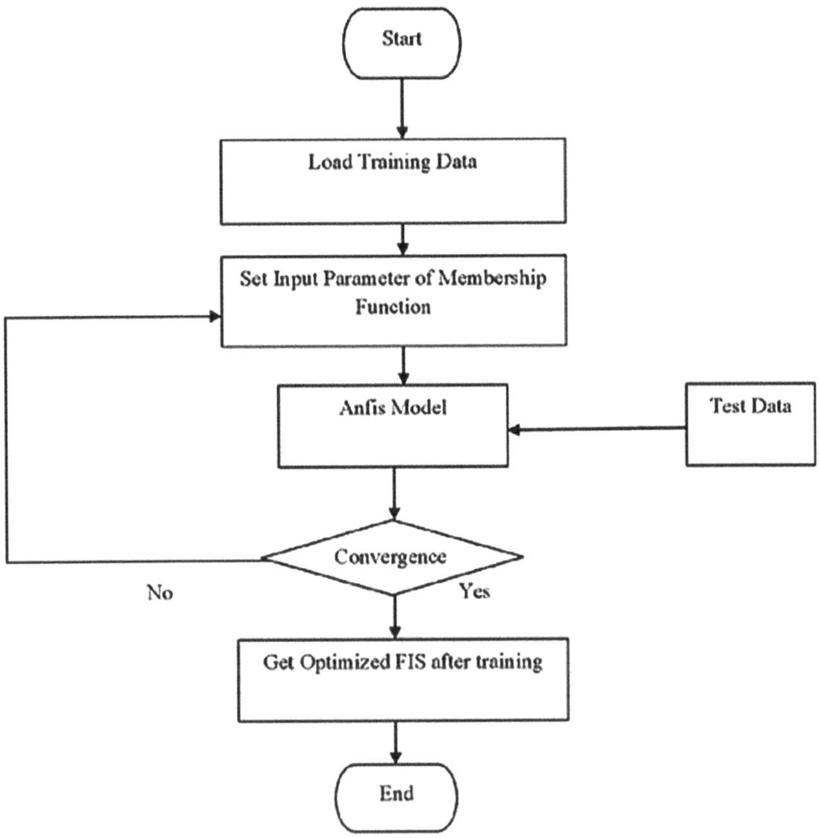

In computer vision, Pattern recognition is a very important field. The aim is to classify patterns based on extracted features from them. This process of pattern recognition is done using ANN.

CONCLUSION

A lot of researchers have worked on a method of plant leaf detection using image processing technique. Some of them have used a different type of technique and method. But a very least number of researchers have taken it to the implementation of various methods. The work of image processing for detection of plat leaf disease detection can never end. This is the race between farmers and these methods. ANN

Figure 3. Flowchart of ANN

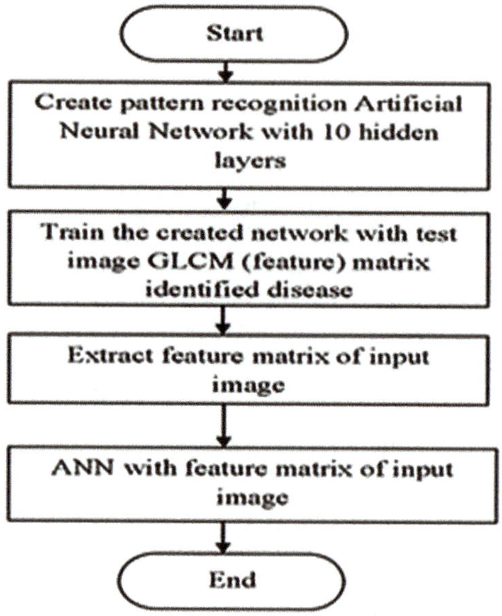

Figure 4. Disease detection & classification

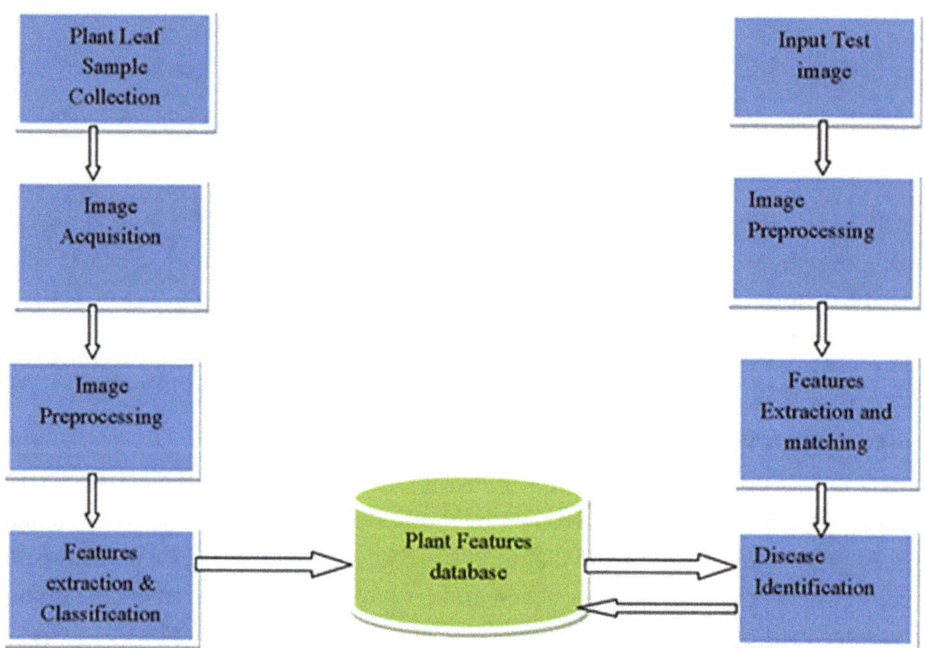

and Anfis for finding disease present in a leaf which all are the best methods as on date but still it is not a destination. As and when new methodologies being evaluated, it is essential to update the tool as per requirement. This chapter aims to identify diseased leaf and detecting its presence in a leaf. It is important to detect disease fast and accurately in a plant leaf.

REFERENCES

Camargo, A., & Smith, J. S. (2009). An image-processing based algorithm to automatically identify plant disease visual symptoms. *Biosystems Engineering*, *102*(1), 9–21. doi:10.1016/j.biosystemseng.2008.09.030

DeChant, C., Wiesner-Hanks, T., Chen, S., Stewart, E. L., Yosinski, J., Gore, M. A., & Lipson, H. (2017). Automated identification of northern leaf blight-infected maize plants from field imagery using deep learning. *Phytopathology*, *107*(11), 1426–1432. doi:10.1094/PHYTO-11-16-0417-R PMID:28653579

Hosseini, H., Tousi, B., & Razmjooy, N. (2014). Application of fuzzy subtractive clustering for optimal transient performance of automatic generation control in restructured power system. *Journal of Intelligent & Fuzzy Systems*, *26*(3), 1155–1166.

Hu, J., Chen, Z., Yang, M., Zhang, R., & Cui, Y. (2018). A Multiscale Fusion Convolutional Neural Network for Plant Leaf Recognition. *IEEE Signal Processing Letters*, *25*(6), 853–857. doi:10.1109/LSP.2018.2809688

Jasim, S. S., & Al-Taei, A. A. M. (2018). A Comparison between SVM and K-NN for classification of Plant Diseases. *Diyala Journal for Pure Science*, *14*(02), 94–105. doi:10.24237/djps.1402.383B

Kashyap, Y., Sharma, T., & Ali, S. S. (2017). Analysis of Plant Diseases with Detection using Image Processing Methods. *International Journal of Computers and Applications*, *166*(7).

Moallem, P., Razmjooy, N., & Ashourian, M. (2013). Computer vision-based potato defect detection using neural networks and support vector machine. *International Journal of Robotics and Automation*, *28*(2), 137–145. doi:10.2316/Journal.206.2013.2.206-3746

Moallem, P., Razmjooy, N., & Mousavi, B. S. (2014). Robust potato color image segmentation using adaptive fuzzy inference system. *Iranian Journal of Fuzzy Systems, 11*(6), 47–65.

Mousavi, B. S., Soleymani, F., & Razmjooy, N. (2013). Color image segmentation using neuro-fuzzy system in a novel optimized color space. *Neural Computing & Applications, 23*(5), 1513–1520. doi:10.100700521-012-1102-3

Mousavi, B. S., Soleymani, F., & Razmjooy, N. (2014). Semantic image classification by genetic algorithm using optimised fuzzy system based on Zernike moments. *Signal, Image and Video Processing, 8*(5), 831–842. doi:10.100711760-012-0311-7

Razmjooy, N. (2012). A multi layer perceptron neural network trained by invasive weed optimization for potato color image segmentation. *Trends in Applied Sciences Research, 7*(6), 445–455. doi:10.3923/tasr.2012.445.455

Razmjooy, N., Mousavi, B. S., & Soleymani, F. (2012). A real-time mathematical computer method for potato inspection using machine vision. *Computers & Mathematics with Applications (Oxford, England), 63*(1), 268–279. doi:10.1016/j.camwa.2011.11.019

Razmjooy, N., Mousavi, B. S., & Soleymani, F. (2013). A hybrid neural network Imperialist Competitive Algorithm for skin color segmentation. *Mathematical and Computer Modelling, 57*(3-4), 848–856. doi:10.1016/j.mcm.2012.09.013

Razmjooy, N., Mousavi, B. S., Soleymani, F., & Khotbesara, M. H. (2013). A computer-aided diagnosis system for malignant melanomas. *Neural Computing & Applications, 23*(7-8), 2059–2071. doi:10.100700521-012-1149-1

Sabrol, H., & Kumar, S. (2015). Recent studies of image and soft computing techniques for plant disease recognition and classification. *International Journal of Computers and Applications, 126*(1).

Singh, V., & Misra, A. K. (2017). Detection of plant leaf diseases using image segmentation and soft computing techniques. *Information Processing in Agriculture, 4*(1), 41–49. doi:10.1016/j.inpa.2016.10.005

Zhang, S., Wang, H., & Huang, W. (2017). Two-stage plant species recognition by local mean clustering and Weighted sparse representation classification. *Cluster Computing, 20*(2), 1517–1525. doi:10.100710586-017-0859-7

Zhang, S., Wu, X., You, Z., & Zhang, L. (2017). Leaf image based cucumber disease recognition using sparse representation classification. *Computers and Electronics in Agriculture*, *134*, 135–141. doi:10.1016/j.compag.2017.01.014

Zhu, X., Zhu, M., & Ren, H. (2018). Method of plant leaf recognition based on improved deep convolutional neural network. *Cognitive Systems Research*, *52*, 223–233. doi:10.1016/j.cogsys.2018.06.008

Chapter 5
Identification of Agricultural Crop Residues Using Non-Destructive Methods

Dimitrios Kateris
Technological Educational Institute of Thessaly, Greece

Ioannis Gravalos
Technological Educational Institute of Thessaly, Greece

Theodoros Gialamas
Technological Educational Institute of Thessaly, Greece

ABSTRACT

Biomass is a bulky and inhomogeneous material, making it difficult to transport and store. In order to solve this problem, it has been found that the most common way to overcome the limitation of the biomass bulk density is to increase it with fine shredding. This chapter investigated the ability to identify specific operation conditions in a prototype biomass shredder by developing and utilizing non-destructive testing and artificial intelligence techniques. In order to demonstrate the performance of proposed methods, three different case studies investigated the different operation conditions from the vibration signals acquired through the ball bearings of the biomass shredder. The results showed that the two classifiers can provide reliable results using as inputs statistical features in time and frequency domain. These statistical features can be used with success for identify different operating condition. The combination of the statistical features with the appropriate classifiers gives a powerful tool for the agricultural biomass shredder condition monitoring.

DOI: 10.4018/978-1-5225-8027-0.ch005

INTRODUCTION

Biomass forms the third accessible energy resource in the world after coal and oil (Bapat, Kulkarni, & Bhandarkar, 1997). The utilization of biomass residues for sustainable heat, power and biofuel production is an important part of future energy concepts (Lesage, Graf, & Westphal, 2010). One of the major challenges of biomass utilization for heat, power and biofuel production is its unfavorable handling properties. Biomass is an inhomogeneous and bulky material, making it both expensive and difficult to transport and store. Furthermore, it is difficult to comminute into small particles and has a relatively low energy density (compared to fossil fuels) and high moisture contents. The typically density range for straws and grasses is 80-100 kg m^{-3} and for wood biomass in the form of chips and sawdust is 150–200 kg m^{-3} (Mitchell, Kiel, Livingston, & Dupont-Roc, 2007; Sokhansanj & Fenton, 2011). This is an important limitation for the use of this material as a source of energy due to storage and transport difficulties (Tumuluru, Wright, Kenny, & Hess, 2010).

Biomass density has a significant effect both on handling and storage. Also, it depends on particle size and shape, specific density, material composition and moisture content (Lam et al., 2008). According to Emami and Tabil, the bulk density of biomass raises while handling, being transported and being stored, something that can be caused by compaction due to vibration, tapping, or normal load (Emami & Tabil, 2008). For Fasina, biomass compaction behavior is very important in terms of capacity sizing and supply logistics (Fasina, 2006). Particle size distribution plays a significant role in flowability and other properties; even a small change in the particle size may cause significant alterations in the resulting flowability (Ganesan, Rosentrater, & Muthukumarappan, 2008). According to this, it has been repeatedly found that the most common way to overcome the limitation of the biomass bulk density is to increase it with the fine shredding.

Although milling is one of the oldest methods of biomass processing, a very little knowledge about this processing based on the mechanical properties of the grinding materials. According to Lopo particle size reduction plays an important role in the utilization of biomass in energy production and animal feedstock (Lopo, 2002). Size reduction is an important pretreatment of biomass for energy conversion. Apart from that, it is also crucial to the densification process. For example, in the production of fuel pellets and briquettes, the biomass has to be shred before being transformed into a denser product. In 2015, Xavier et al. indicated that the size reduction increases the total surface area and the number of contact points in the compaction process making it possible to obtain the agglomerate of density equal to approximately 1200 kg m^{-3}. This affects the chemical structure of biomass losing some of the water under pressure (Xavier, Moset, Wahid, & Møller, 2015).

The reduction of biomass particle size reduction is crucial since biorefinery technologies cannot (yet) digest efficiently the whole stems of grass and wood. Paulrud claims that corn stover particle sizes (0.5 to 3mm) are necessary for corn stover ethanol production. Furthermore, compared to pellets and bales size-reduced biomass for direct combustion produces a more stable flame, have a high burnout, and emit low CO_2 and ash emissions (Paulrud, 2004).

It has been observed that biomass after size reduction has better digestibility in the conversion reactor than being in baled form (Cundiff & Grisso, 2008; Hess, Wright, & Kenney, 2007; Wu et al., 2007). The feedstock should have a particulate form for biorefinery procedures such as hydrolysis, gasification, pyrolysis, fermentation and chemical synthesis. As a result of the last years, a lot of studies have been conducted over the effect on the conversion efficiency of several particle sizes (Kumar, Barrett, Delwiche, & Stroeve, 2009; Wei, Pordesimo, Igathinathane, & Batchelor, 2009).

The knowledge of biomass size reduction equipment's mechanical properties is necessary in order to design it properly and have it optimized. One of such properties is the energy required for size reduction and it is influenced by initial particle size, material properties, moisture content, the feed rate of the material, and machine variables (Mani, Tabil, & Sokhansanj, 2004). Size reduction increases the surface area, the pore size, and the number of contact points for interparticle bonding in compaction operations such as densification (Adapa, Tabil, & Schoenau, 2011).

The most common to use pieces of equipment are tub grinders and hammer mills since they are affordable, easy to operate and produce a wide range of particle sizes. According to Bitra et al. (2009) and Souček et al. (2007) size reduction is an important unit operation for densification and reduction of transportation costs.

According to Drzymala the preprocessing procedure of biomass size reduction raises bulk density and improves flow properties (Drzymala, 1993). Typical size reduction machines for biomass are hammer-, knife-, and disk-mills, and various choppers, chippers, and shredders (Yu, Womac, & Pordesimo, 2003). These machines are usually based on rotary action, with machine elements mounted on a shaft. Hammer-mills have the advantages of simple operation, rugged construction, and wide product particle size range (Lopo, 2002). A high-capacity tub grinder, commonly used with crop and forest biomass residues (Arthur, Kepner, Dobie, Miller, & Parsons, 1982), is essentially a hammer mill with rotary feed hopper.

NON-DESTRUCTIVE TESTING AND ARTIFICIAL INTELLIGENCE TECHNIQUES

Non-destructive Testing (NDT) includes a variety of non-invasive techniques used to evaluate different degrees of the defects in the detected objects. These inspection techniques give the location, size, number, and nature of the defect and determine the technical state (such as the current condition, remaining life, etc.) of the component using sound, light, electricity, magnetism and other characteristics. They may also detect, characterize, or measure the presence of damage mechanisms (e.g. corrosion or cracks). NDT, in contrary to destructive testing, inspects a component in a safe, reliable, and cost-effective manner avoiding any damage to the equipment or the machine operations, whereas in destructive testing the part being tested is damaged or destroyed. In general terms NDT compared to destructive testing offers many advantages, e.g. the testing equipment is often portable and can be performed numerous times on a single component, the component itself can be thoroughly evaluated externally and internally for harmful flaws etc.

The NDT methods are not new techniques. During the last decades with the help of technological development, they have evolved equally in codes, standards, and best practices. The most common NDT methods are: a) Vibration – Acoustic methods (Arumugam, Kumar, Santulli, Sarasini, & Stanley, 2011; Kateris et al., 2014), b) Ultrasonic methods (Oguma, Goto, & Sugiura, 2012; Peng, Zhang, Qiu, & Xue, 2012), c) Radiographic methods (Aidi, Philen, & Case, 2015), d) Electromagnetic methods (Koyama, Hoshikawa, & Kojima, 2013), e) Shearography methods (Hung, Yang, & Huang, 2013), f) Thermographic methods (Kroeger, 2014; Maierhofer et al., 2014) and g) Infrared thermography method (Vavilov, Budadin, & Kulkov, 2015)

In the field of agricultural engineering, non-destructive testing (NDT) methods gain more popularity for their ability to examine the in-situ component properties without damaging the subject structure. The high speed rotating machine elements supported by ball bearings have been widely used in agricultural machinery because it is easy to maintain and the dynamic instability is reduced. In the field of rotor dynamics (rotating machinery dynamics), researchers have proved that many parameters (stiffness distribution of a rotating shaft, disk properties, bearing support stiffness) may influence the dynamic behavior of a rotor-bearing system. NDT methods are used to solve this problem.

Recently research over the automatic interpretation of data from non-destructive testing (NDT) focuses to the fast, accurate and cost-effective assessment of embedded flaws. This is because data from NDT methods come usually in the form of signals or images which often do not present direct information of the structure's condition. Signal processing as a powerful tool has provided many techniques in order to extract the appropriate information on fault detection from NDT signals. This signal processing can be achieved by using Artificial Intelligence techniques.

Artificial Intelligence techniques such as artificial neural networks (ANN) (Samanta, Al-Balushi, & Al-Araimi, 2006; Razmjooy et al., 2012; 2013), and support vector machines (SVM) (Sugumaran & Ramachandran, 2011; Widodo et al., 2009; Moallem et al., 2013) have been successfully applied to automated detection and diagnosis of machines. Some of them were applied to detect the machine faults at an early stage. Also, they are capable of determining the running health of the machine automatically and reliably. Artificial Neural Networks (ANNs) have become widely adopted in ball bearings for fault diagnosis due to their learning ability. Various types of neural networks and learning algorithms have been applied in machinery fault diagnosis including Recurrent Neural Network (Zaremba, Sutskever, & Vinyals, 2015), Flexible Neural Tree (W Bao, Chen, & Wang, 2014; Wenzheng Bao, Wang, & Chen, 2017), Radial Basis Neural Network (Huang & Du, 2008; Razmjooy et al., 2018), Wavelet Neural Network (Pindoriya, Singh, & Singh, 2008; Razmjooy and Ramezani, 2016), Back Propagation trained Feed Forward Networks (BPFF) (Brown, 2011), Radial Basis Function networks (RBF) (Wenzheng Bao, Huang, Yuan, & Huang, 2017), Multi-Layer Perceptrons (MLP) (Li, Zhang, & Zheng, 2009; Lu & Li, 1998), Self-Organized Maps (SOM) (Paya, Esat, & Badi, 1997), Principal Component Analysis (PCA) (Ghate & Dudul, 2010) and Recurrent Neural Networks (RNN) (Moshou et al., 2010). Besides ANN, Support Vector Machine (SVM), k-Nearest Neighbor (kNN), Hidden Markov Model (HMM), and ensemble learning (Parlos AG, 1999; Sun, Chen, & Li, 2007) are also popular supervised algorithms.

In case the recording data are noisy, the most common method to overcome the effects of noise and parameter redundancy, frequently used in real-world training scenarios, is the use of regularizers such as weight decay. The role of regularizers has been analyzed for linear perceptrons and optimal values for regularizers have been calculated. However, the efficacy of regularizers for online learning in multilayer networks is still somewhat unclear (Babak & Ahmed, 2015; Herrmann & Estrela, 2016).

Many studies show that Artificial Intelligence techniques comparing to the traditional statistical methods (linear and non-linear regression) demonstrate better ability in data regression and prediction. In this field, one of the most popular machine learning technique is Artificial Neural Networks (ANNs). ANNs application has been successful in many fields for classification and regression and it may also improve non-destructive test results.

The present work analyses the effects of the different operation parameters on the vibration of the rotor system (shaft with blades) in a biomass shredder. Specifically, the dynamic behavior of a rotor-bearing system was investigated experimentally under different operating conditions: a) different input supply rate (without biomass supply, irregular and continuous input supply rate) b) different types of biomass (reed stalks and cotton stalks) and c) different number of blades on the rotor of the shredder (the rotor of the shredder had all the blades and 10% of the blades are missing - simulating this way broken blades) using non-destructive testing method (vibration signal) and two artificial intelligence techniques. A Multilayer Perceptron neural network with Automatic Relevance Determination (MLP-ARD) and a Least

Squares Support Vector Machine (LS-SVM) are used to distinguish between different operating conditions. It's about a soft computing application to an NDT problem in an alternative domain instead of the spatial domain.

With the aim to demonstrate the performance of the proposed methods, three different case studies are investigated in order to detect the different operating conditions from the real vibration signals acquired through the ball bearings of the biomass shredder. In addition, the effectiveness of the MLP-ARD neural network and of the LSSVM is compared. The results show the effectiveness of the used features from the acquired signals in the diagnosis of the machine condition. The procedure is illustrated using the vibration data of an experimental setup operating in different operating conditions.

BAYESIAN MLP WITH AUTOMATIC RELEVANCE DETERMINATION (MLP-ARD)

For binary classification one commonly opts for a feed-forward multilayer perceptron neural network (MLP) with one hidden layer and output unit (Bishop, 1995). Given a training set $D = \left\{x_i, t_i\right\}_{i=1}^{N}$ with input data $x_i = \left(x_i^{(1)}, \ldots, x_i^{(n)}\right) \in R^n$ and target labels $t_i \in \left\{0,1\right\}$, the MLP classifier then performs the following non-linear function mapping:

$$y(x) = f_2\left(\upsilon\, f_1\left(u\, x\right)\right) \tag{1}$$

where $y(x) \in R$, υ and u are the weight vectors of the hidden and output layer, respectively, that together make up weight vector ω, which needs to be estimated during the training process, while f_1 and f_2 are termed transfer functions and essentially allow the network to perform complex non-linear function mappings. In the hidden layer, we use hyperbolic tangent transfer functions. A logistic transfer function is used in the output layer. The latter allows the MLP's output $y\left(x\right)$ to be interpreted as an estimated probability of the form $p\left(t = 1 \middle| x\right)$, ω is randomly initialized and iteratively adjusted so as to minimize a cross-entropy objective function G defined as (Bishop, 1995):

$$G = -\sum_{i=1}^{N}\left\{t_i \ln\left(y_i\right) + \left(1 - t_i\right)\ln\left(1 - y_i\right)\right\} \tag{2}$$

Overfitting the training data is avoided by adding weight regularization to the objective function as follows:

$$F(\omega) = G + \sum_k \alpha_k Ew_{(k)} \quad (3)$$

where $Ew_{(k)} = \frac{1}{2}\sum_j \omega_j^2$, weights $\omega_j \in R$ and j is running over all weights of weight class $W(k)$.

Automatic relevance determination (ARD) weight regularization now considers $k = 1, \ldots, n+3$ weight classes within weight vector ω, each associated with one weight decay hyperparameter α_k. It associates a single weight regularization hyperparameter α_k with each predictor $x^{(k)} = (K = 1, \ldots, n)$, more specifically, with the set of weights corresponding to the connections from input $x^{(k)}$ to the hidden nodes. Three additional weight decay hyperparameters, i.e. weight classes, are introduced: one associated with the connection from the hidden layer bias neuron to the output neuron, one associated with the connection from the hidden neurons to the output neuron and one associated with the connections from the input layer bias to the hidden neurons (Kateris et al., 2014).

SUPPORT VECTOR MACHINES (SVM)

SVM correspond to a computational intelligence technique, related to the machine learning concept (Vapnik, 1998). In contrast to many classical neural network training algorithms which exhibit many local minima, SVM are always able to find a global minimum and they have a simple geometric interpretation. SVM are frequently used for binary classification tasks in order to estimate a classification function such as:

$$f : x \to \{\pm 1\} \quad (4)$$

Overfitting in SVM is limited according to the statistical theory of learning from small samples (Vapnik, 1998). The procedure that is followed in order to find the maximum margin of separation is formulated as a convex quadratic problem (Vapnik, 1999). The input data are projected into a feature space F using a map such as:

$$\varphi : x \to F \quad (5)$$

In nonlinear problems, one has to one has to supply the kernel functions that define the feature space via:

$$K\left(x_i, x_j\right) = \varphi\left(x_i\right) \cdot \varphi\left(x_j\right) \tag{6}$$

It is not necessary to know the mapping f because it is performed implicitly. SVM can also learn which of the features implied by the kernel are distinctive for the two classes. The selection of an appropriate kernel function may boost the learning process. The Radial Basis Function (RBF) kernel is defined by the following equation:

$$K(x_i, x_j) = exp(-\gamma\left\|x_i - x_j\right\|^2), \quad ^3 > 0. \tag{7}$$

where $(x_i, x_j), i = 1, \ldots, l \ \ x_i \in R^n$ is a training set of instance label pairs, γ is the factor related to the inverse square of the spread of the RBF and $y \in \left\{1, -1\right\}^l$ are the SVM target classes. The RBF kernel has been selected because it offers several advantages. It performs nonlinear mapping of samples so it can handle a nonlinear relation between class labels and attributes in comparison with a linear kernel. On the other hand, the linear kernel can be considered as a special case of RBF (Keerthi & Lin, 2003). The sigmoid kernel shows similar behavior to RBF for certain parameters (Lin & Lin, 2003). Another reason is the number of hyperparameters which are involved in model construction which increases complexity. The polynomial kernel uses more hyperparameters than the RBF kernel. One major point is that the RBF kernel is restricted between 0 and 1 in contrast to the polynomial kernels which can take values up to infinity for higher degree polynomials. However, in certain cases, the RBF kernel might not be suitable. More specifically, for problems where the number of features is large, a more suitable choice is the linear kernel (Hsu, Chang, & Lin, 2010).

Least Squares Support Vector Machines (LSSVM)

LSSVM (Suykens & Vandewalle, 1999) build a similar classification model to the SVM based on the function in Eq. (4) which uses basis functions φ. The following model is taken:

$$f(x) = w^T\varphi(x) + b$$

$$\min_{w,b} J\left(w,b\right) = \frac{\mu}{2} w^T w + \frac{\zeta}{2}\sum_{i=1}^{N} e_i^2 \tag{8}$$

$$s.t. \qquad y_i[w^T\varphi(x_i)+b] = 1-e_i, \qquad i=1,....,N$$

with regularizer $\gamma = \zeta/\mu$

This approach simplifies SVM due to linear system formulation which can be solved efficiently with numerical methods. Figure 1 illustrates the procedure of classification as performed by the LSSVM while the support vectors are shown in circles.

Instead of inequality, it is subject to equality constraints which lead to the formulation of Lagrange multipliers. The support vectors can be found by solving a linear system.

Figure 1. Interpretation of the LSSVM classifier

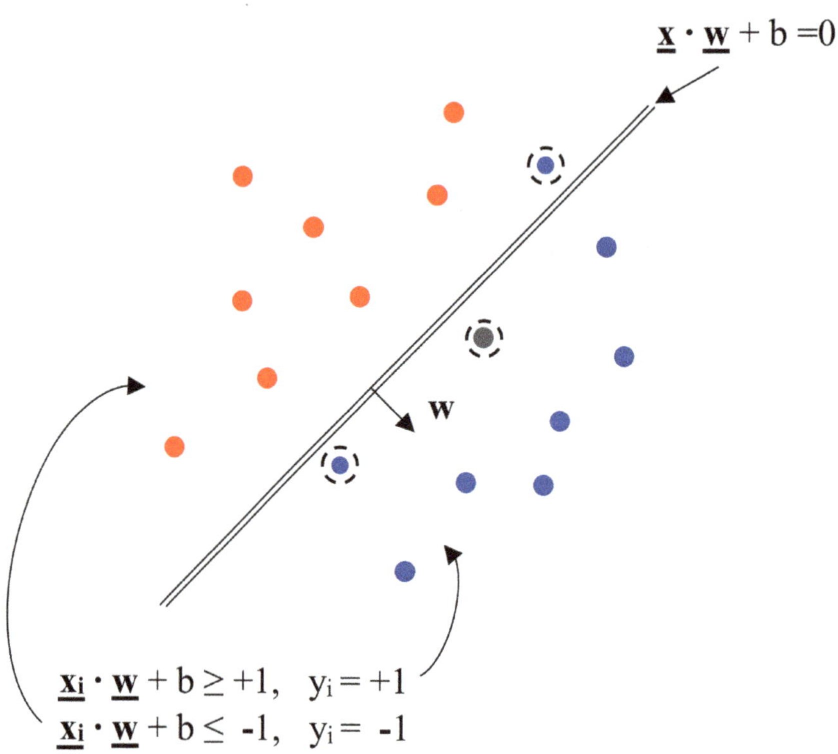

$$\begin{bmatrix} 0 & l_v^T \\ l_v & \Omega + \gamma^{-1} I_N \end{bmatrix} \begin{bmatrix} b \\ \alpha \end{bmatrix} = \begin{bmatrix} 0 \\ Y \end{bmatrix} \tag{9}$$

$$Y = [y_1, \ldots, y_N]^T,\ l_v = [1, \ldots, l]^T,\ e = [e_1, \ldots, e_N]^T,$$

$$\alpha = [\alpha_1, \ldots, \alpha_N]^T, \Omega_{ij} = \varphi(x_i)^T\ \varphi(x_j) = K(x_i, x_j)$$

$$\text{e.g. RBF kernel: } K(x, z) = \exp\left\{-\|x - z\|^2 / \sigma^2\right\}$$

Resulting in the classifier:

$$y(x) = sing\left[\sum_{t=1}^{N} a_i y_i K(x, x_i) + b\right] \tag{10}$$

This approach simplifies SVM due to linear systems formulation can solve efficiently with numerical methods (Moshou, Pantazi, Kateris, & Gravalos, 2014).

EXPERIMENTAL SETUP AND VIBRATION DATA

In order to be carried out all the experimental measurements a prototype biomass shredder was used (Figure 2a). The rotor of the shredder was seated on two identical ball bearings. The rotor was powered by a three-phase AC induction motor at a speed of 1100 rpm with belt transmission. The rated (nameplate) power of the induction motor was 5.5 hp (4.1 kW). The motor was mounted on a frame with rubber dampers in order to insulated from the rest structure. This happens in order to prevent the vibration transfer during motor operation. The blades of the shredder (60 in total) were placed along a shaft in 15 disks. Four (4) blades per disk every 90 degrees. The bearings were mounted on cast iron housings which were in turn fixed to the frame. To measure the vibration of the rotor system on ball bearings, two triaxial vibration acceleration sensors (model 4524B) were used. The sensitivity of the acceleration sensors is 10 mV g^{-1}. The vibration acceleration sensors were mounted on the bearing housings (near the pulley - No.1 (Figure 3a) and on the other side of the rotor - No.2 (Figure 3b)).

Figure 2. a) The prototype biomass shredder and b) The rotor of the shredder after the 6 blades subtraction

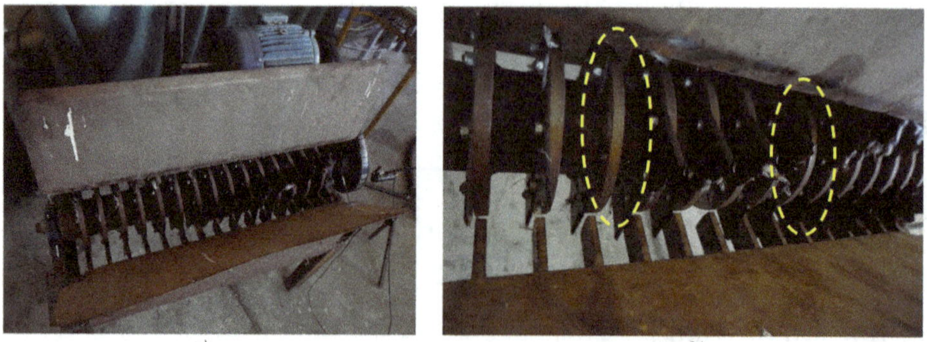

a) b)

Figure 3. The vibration acceleration sensor on the bearing housings a) near the pulley (No.1) and b) on the other side of the rotor (No.2)

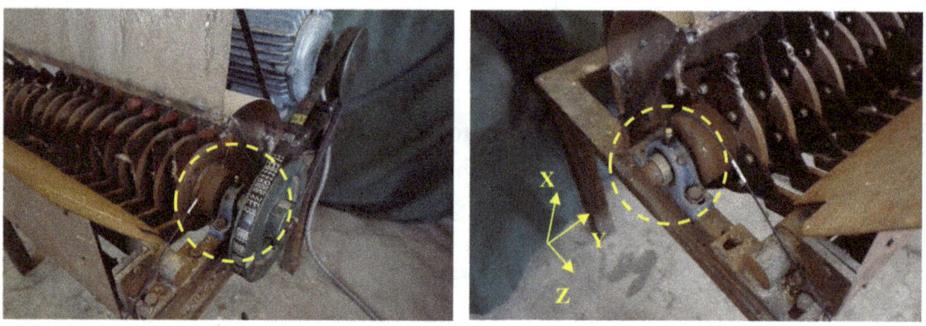

a) b)

In order to record all the vibration signals for each operation condition, a Pulse system type 3560C with 17-channel produced by Brüel & Kjaer Company was used. For measuring the shaft rotation speed, a photo-reflective switch was used to provide a once per rev tacho signal. The 3560C system recorded together vibration and shaft speed signals.

During the experimental process, the 5424B piezoelectric acceleration sensors (Brüel & Kjaer) are used to measure acceleration in three axes (X-Y-Z) (Figure 3b). The sampling frequency was 65,536Hz and the recording duration was 1s. The number of samples collected for each channel was 65,536. In the present work, these time-domain data were preprocessed to extract the features for use as inputs to the classifiers (MLP-ARD and LS-SVM).

SIGNAL PROCESSING AND FEATURE EXTRACTION

The first and maybe the most important step in any classification problem is to be chosen and extracted the appropriate features from the raw signal. The aim of this is to reflect the general changes in the operating conditions. However, some features are closely related to the recognize specific operation condition, others are not.

In this study, ten (10) sets of experimental data each for different operation condition were recorded. In each set was included six (6) vibration signals. Three signals X-Y-Z for each triaxial acceleration sensor. Each signal consisted of 65,536 samples. Figure 4 illustrates indicatively the recording signals for the vertical axis (X) in all the different operating conditions.

After all signals recording, each signal was divided into 131 signal parts with 500 data points each. Each of these signal parts was further processed to Fast Fourier Transform (FFT) in order to transform for time-domain to frequency-domain (Figure 5).

The next step was the feature extraction from the raw signal in time-domain and in frequency-domain. Frequency-domain is another description of a signal. This type of description includes some information that cannot be found in time-domain. The aim of this step was to reflect the general changes in the machine operating conditions. However, though some statistical features are closely related to the specific operating conditions, others are not. In this direction, each of the signal parts was further processed to extract twelve (12) statistical features. The first six (6) features were time-domain features, Mean value (T_1), Standard deviation (T_2), Root mean square (T_3), Peak (T_4), Skewness (T_5), Kurtosis (T_6) and the other six (6) features were frequency-domain features (F_1 – F_6) (Lei, He, & Zi, 2008). Feature F_1 may indicate the vibration energy in the frequency domain. Features F_2-F_4, F_6 may describe the convergence of the spectrum power and finally, F_5 give information about the position change of main frequencies (Lei et al., 2008). All the used features provide statistical information about the nature of data and were found to be reasonably good features for bearing. All the statistical features are shown in Table 1.

After the feature extraction, the two classifiers (MLP-ARD and LS-SVM) were trained and tested in three different case studies. The whole feature points were divided into two parts, training data and testing data. To test the effectiveness of the two classifiers the 72% (94/131) of the feature points have been used for training while the 28% (37/131) have been used in order to test the effectiveness of the two classifiers.

Each case study had different operating condition parameters, as a) the condition of the blade shaft (with all the blades and without 10% of the total blades, simulating in this way possibly broken blades), b) the shredder input supply rate (without supply, with irregular and with continuous supply and c) the type of biomass (reed or cotton

Figure 4. Vibration signals for different operation parameters (a_1) without supply rate and with all the blades, (a_2) without supply rate and without 10% of the blades, (b_1) with irregular reed stalks supply rate and with all the blades, (b_2) with irregular reed stalks supply rate and without 10% of the blades, (c_1) with continuous reed stalks supply rate and with all the blades, (c_2) with continuous reed stalks supply rate and without 10% of the blades, (d_1) with irregular cotton stalks supply rate and with all the blades, (d_2) with irregular cotton stalks supply rate and without 10% of the blades, (e_1) with continuous cotton stalks supply rate and with all the blades and (e_2) with continuous cotton stalks supply rate and without 10% of the blades.

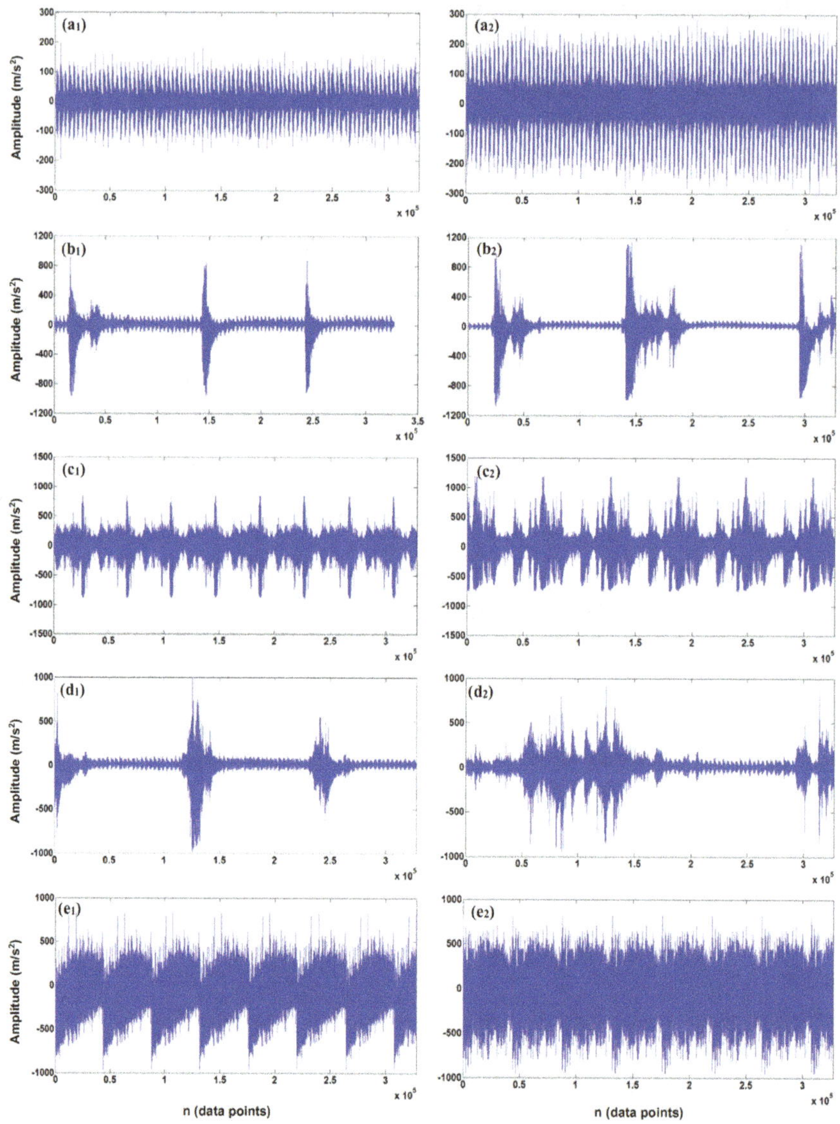

Figure 5. Sequence of spectral vectors and time duration selection

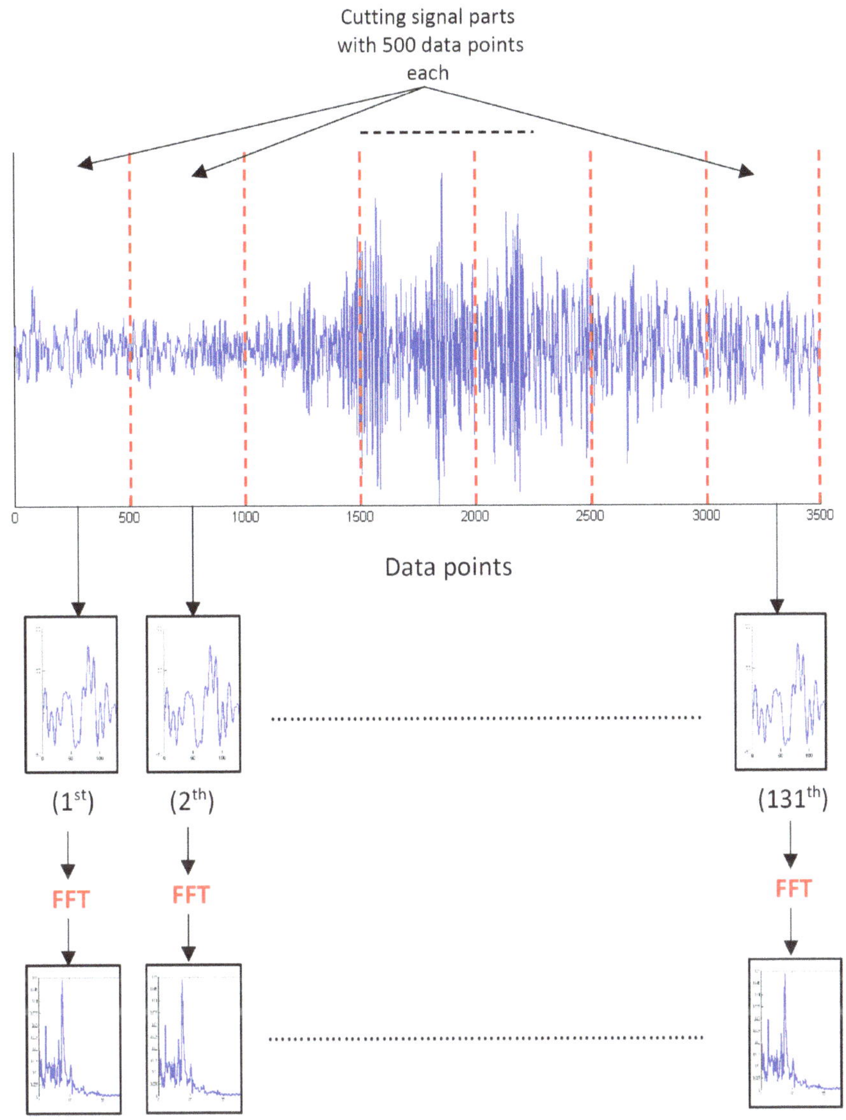

stalks). The aim of these case studies was to investigate the classification accuracy of the MLP-ARD and LS-SVM systems in different operational condition parameters. Each case study operation condition parameters are summarized below in Table 2.

As mentioned, for each case study twelve statistical parameters (features), six time-domain and six frequency-domain were selected. These statistical features were extracted for all the cases studies and were input to the MLP-ARD and LS-SVM

Table 1. Time and Frequency domain features

Time-domain features (T)			Frequency-domain features (F)	
$T_1 = \frac{\sum_{n=1}^{N} x(n)}{N}$	(11)		$F_1 = \frac{\sum_{k=1}^{K} s(k)}{K}$	(17)
$T_2 = \sqrt{\frac{\sum_{n=1}^{N} (x(n) - T_1)^2}{N-1}}$	(12)		$F_2 = \frac{\sum_{k=1}^{K} (s(k) - F_1)^2}{K-1}$	(18)
$T_3 = \sqrt{\frac{\sum_{n=1}^{N} (x(n))^2}{N}}$	(13)		$F_3 = \frac{\sum_{k=1}^{K} (s(k) - F_1)^3}{K\left(\sqrt{F_2}\right)^2}$	(19)
$T_4 = max\lvert x(n)\rvert$	(14)		$F_4 = \frac{\sum_{k=1}^{K} (s(k) - F_1)^4}{KF_2^2}$	(20)
$T_5 = \frac{\sum_{n=1}^{N} (x(n) - T_1)^3}{(N-1)T_2^3}$	(15)		$F_5 = \frac{\sum_{k=1}^{K} f_k s(k)}{\sum_{k=1}^{K} s(k)}$	(21)
$T_6 = \frac{\sum_{n=1}^{N} (x(n) - T_1)^4}{(N-1)T_2^4}$	(16)		$F_6 = \sqrt{\frac{\sum_{k=1}^{K} (f_k - F_5)^2 s(k)}{K}}$	(22)
where $x(n)$ for the time-domain feature is a signal series for $n=1,2,\ldots,N$, N is the number of data points.			where $s(k)$ is a spectrum for $k=1,2,\ldots,K$, K is the number of spectrum lines, f_k is the frequency value of the kth spectrum line.	

in order to be trained. These statistical features were used to distinguish between different operation parameters (number of blades on the shredder rotor, input supply rate and the biomass type). The previous systems were trained in all operating conditions. The MLP-ARD consists of the input layer, one hidden layer, and the output layer. The number of neurons at the input level was equal to the number of selected statistical features. The number of neurons in the hidden layer was 10. The number of output nodes was 3 in all the case studies. In the case of the LS-SVM, the RBF kernel was selected. The method that has been used for obtaining the multiclass LS-SVM classifier was the "one-versus-one" which relies on building a

Table 2. Different operation condition parameters

	The condition of the blade shaft		Shredder input supply rate			Type of biomass	
	with all the blades	without 10% of the blades	without supply rate	with irregular supply rate	with the continuous supply rate	reed stalks	cotton stalks
1st case study	√	–	√	√	√	√	–
	√	–	√	√	√	–	√
2nd case study	–	√	√	√	√	√	–
	–	√	√	√	√	–	√
3rd case study	√	–	√	√	–	√	√
	√	–	√	–	√	√	√
	–	√	√	√	–	√	√
	–	√	√	–	√	√	√

set of one versus one classifier and choosing the class that is selected by the most classifiers. The code of the two classifiers was written in Matlab. Figure 6 illustrates the flowchart of the proposed procedure.

RESULTS AND DISCUSSION

In order to verify the effectiveness of the MLP-ARD classifier, its performance was compared with another well-known classifier; LS-SVM classifier. Three different case studies have tested. Each case study had different operation parameters in order to investigate the efficiency of the classifiers in different operation parameters combinations (Table 2).

The most widely used basic measure of multi-class classifier performance is average accuracy (Sokolova & Lapalme, 2009). Accuracy is the most intuitive performance measure and it is simply a ratio of correctly predicted observation to the total observations. The assumption that if the accuracy is high our model is effective. This applies when the datasets are symmetric, values of false positive and false negatives are almost the same. This measure, qualify the performance of a multi-class classifier or decision process and also are used in to assess the accuracy of a diagnostic procedure. The Eq. (23) gives the average accuracy for multi-class classifiers.

Figure 6. Flowchart of operation condition classification

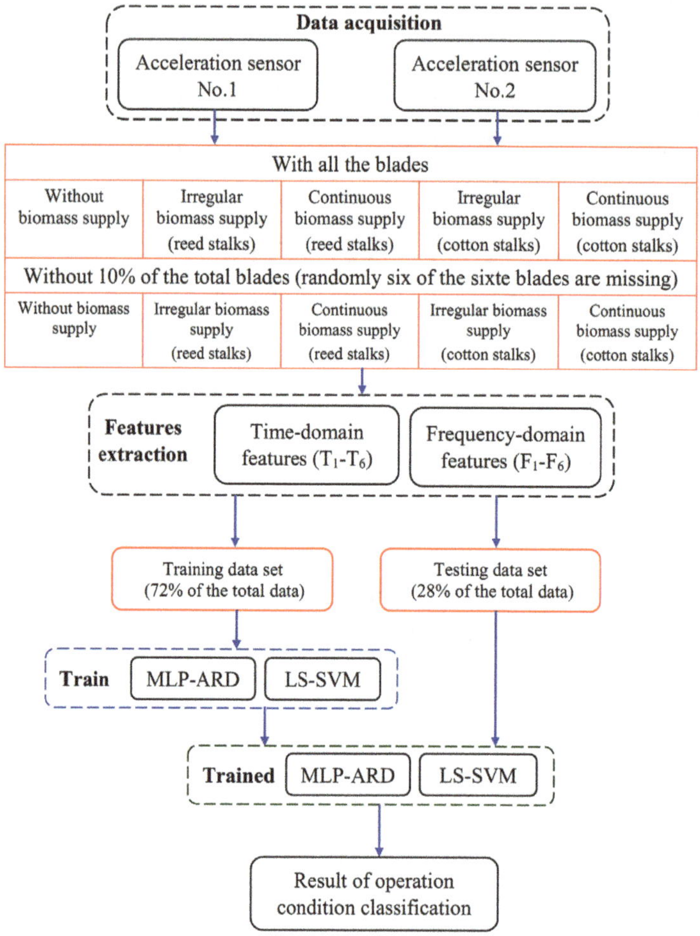

$$Average\ accuracy = \frac{\sum_{i=1}^{l} \frac{TP_i + TN_i}{TP_i + TN_i + FP_i + FN_i}}{l} \tag{23}$$

where, i is the number of the class, l is the number of the classes, True Positives (TP) are the correctly predicted positive values which means that the value of the actual class is yes and the value of predicted class is also yes, True Negatives (TN) are the correctly predicted negative values which means that the value of the actual class is no and value of predicted class is also no, False Positives (FP) - When actual class is no and predicted class is yes and False Negatives (FN) – When actual class is yes but predicted class in no.

A multi-class confusion matrix is a visual tool which summarises the prediction results of the classifier model in machine learning. It can condense the records of the datasets in the form of two criteria based on the real category and classification model. Figure 7 to 14 displays the confusion matrix temples for the three case studies. As is shown, the ordinate axis and horizontal axis represent the actual label and predicted label, respectively.

Case Study 1: The Effect of the Input Supply Rate and the Biomass Type With All the Blades on the Shredder Rotor

As shown in Figure 7, in the first case study with different input supply rate with all the blades on the shredder rotor when the biomass is reed stalks the two classifiers (MLP-ARD and LS-SVM) achieves quite a good classification average accuracy (92.8% and 86.5% respectively). When the shredder operates without supply (empty) the accuracy is 97.3% for the MLP-ARD and 89.2% for the LS-SVM for this operating condition. These percentages are very high. But when the input supply rate becomes irregular the efficiency was significantly reduced (81.1% and 70.3% respectively) and takes the higher accuracy (100%) for a continuous supply. The cause for the degradation in accuracy for the irregular input supply rate is might be the feed rate itself. This gives a first indication that the two classifiers are not as effective in irregular input supply rate conditions.

Also, in Figure 8 it is observed that the MLP-ARD and the LS-SM have given the same results in all input supply rate when the type of biomass was cotton stalks. The classification accuracy in the case of zero and continuous input supply rate was 100%. In case of irregular input supply rate was reduced imperceptible (97.3%) but it still high. This may be due to the biomass type and especially in the cotton stalks density. The average accuracy, in this case, is exactly the same for the two classifiers (99.1%).

Figure 7. Confusion matrix tables for different input supply rate operation conditions with all the blades on the shredder rotor (reed stalks)

MLP-ARD

True Class \ Predicted Class	WS	IISR(R)	CISR(R)	
WS	**36**	1	0	97.3% 2.7%
IISR(R)	3	**30**	4	81.1% 18.9%
CISR(R)	0	0	**37**	100% 0.0%
	92.3% 7.7%	96.8% 3.2%	90.2% 9.8%	**92.8% 7.2%**

LS-SVM

True Class \ Predicted Class	WS	IISR(R)	CISR(R)	
WS	**33**	4	0	89.2% 10.8%
IISR(R)	11	**26**	0	70.3% 29.7%
CISR(R)	0	0	**37**	100% 0.0%
	75.0% 25.0%	86.7% 13.3%	100% 0.0%	**86.5% 13.5%**

WS - Without supply, **IISR (R)** - Irregular input supply rate (reed stalks), **CISR (R)** - Continuous input supply rate (reed stalks)

Figure 8. Confusion matrix tables for different input supply rate operation conditions with all the blades on the shredder rotor (cotton stalks)

MLP-ARD

True Class \ Predicted Class	WS	IISR(C)	CISR(C)	
WS	**37**	0	0	100% 0.0%
IISR(C)	1	**36**	0	97.3% 2.7%
CISR(C)	0	0	**37**	100% 0.0%
	97.4% 2.6%	100% 0.0%	100% 0.0%	**99.1%** **0.9%**

LS-SVM

True Class \ Predicted Class	WS	IISR(C)	CISR(C)	
WS	**37**	0	0	100% 0.0%
IISR(C)	1	**36**	0	97.3% 2.7%
CISR(C)	0	0	**37**	100% 0.0%
	97.4% 2.6%	100% 0.0%	100% 0.0%	**99.1%** **0.9%**

WS - Without supply, **IISR (C)** - Irregular input supply rate (cotton stalks), **CISR (C)** - Continuous input supply rate (cotton stalks)

Case Study 2: The Effect of the Input Supply Rate and the Biomass Type With 10% Fewer Blades on the Shredder Rotor

At the second case study, the blades shaft rotating without 10% of the total blades (6 from 60 blades have been removed, simulating possible breaking blades) in three input supply rate a) without biomass supply, b) with irregular and c) with continuous input reed stalks supply rate. Figure 9 shows that the MLP-ARD classification accuracy was marginally better in case of zero and irregular input supply rate (100% and 91.9% respectively) than the LS-SVM (97.3% and 89.2% respectively). But in case of continuous input supply rate, the classification accuracy between the two classifiers is reversed (LS-SVM 100% and MLP-ARD 86.5%). The average accuracy is similar and in high level (LS-SVM 95.5% and MLP-ARD 92.8%).

The same results with a small exception are presented in case of the biomass type was cotton stalks. As presented in Figure 10 the MLP-ARD classification accuracy was better in case of zero and irregular input supply rate (94.6% and 91.9%

Figure 9. Confusion matrix tables for different input supply rate operation conditions without 10% of the total blades on the shredder rotor (reed stalks)

MLP-ARD

True Class \ Predicted Class	WS	IISR(R)	CISR(R)	
WS	**37**	0	0	100% 0.0%
IISR(R)	2	**34**	1	91.9% 8.1%
CISR(R)	0	5	**32**	86.5% 13.5%
	94.9% 5.1%	87.2% 12.8%	97.0% 3.0%	**92.8%** **7.2%**

LS-SVM

True Class \ Predicted Class	WS	IISR(R)	CISR(R)	
WS	**36**	1	0	97.3% 2.7%
IISR(R)	4	**33**	0	89.2% 10.8%
CISR(R)	0	0	**37**	100% 0.0%
	90.0% 10.0%	97.1% 2.9%	100% 0.0%	**95.5%** **4.5%**

WS - Without supply, **IISR (R)** - Irregular input supply rate (reed stalks), **CISR (R)** - Continuous input supply rate (reed stalks)

Figure 10. Confusion matrix tables for different input supply rate operation conditions without 10% of the total blades on the shredder rotor (cotton stalks)

MLP-ARD

True class \ Predicted Class	WS	IISR(C)	CISR(C)	
WS	**35**	2	0	94.6% 5.4%
IISR(C)	0	**34**	3	91.9% 8.1%
CISR(C)	0	0	**37**	100% 0.0%
	100% 0.0%	94.4% 5.6%	92.5% 7.5%	**95.5% 4.5%**

LS-SVM

True class \ Predicted Class	WS	IISR(C)	CISR(C)	
WS	**33**	4	0	89.2% 10.8%
IISR(C)	0	**32**	5	86.5% 13.5%
CISR(C)	0	0	**37**	100% 0.0%
	100% 0.0%	88.9% 11.1%	88.1% 11.9%	**91.9% 8.1%**

WS - Without supply, **IISR (C)** - Irregular input supply rate (cotton stalks), **CISR (C)** - Continuous input supply rate (cotton stalks)

respectively) than the LS-SVM (89.2% and 86.5% respectively). But in case of continuous input supply rate, the classification accuracy between the two classifiers is the same (100%).

Case Study 3: The Effect of the Biomass Type and the Number of Shredder Blades

In this case study, the results show that the operation condition "without biomass supply" and "irregular input supply rate (cotton stalks)" can be well distinguished by the two classifiers (MLP-ARD - 94.6% and LS-SVM 100%), whereas it is slightly difficult to distinguish the "irregular input supply rate (reed stalks)", (24.3% and 45.9% respectively). This is probably occurred due to an irregular input supply rate operation condition. The weakness of the classifiers in this operation condition significantly reduces the average accuracy of the classifiers (MLP-ARD - 71.2% and LS-SVM - 82%) (Figure 11).

Figure 11. Confusion matrix tables for different biomass type with all the blades on the shredder rotor (irregular biomass supply)

MLP-ARD

True Class \ Predicted Class	WS	IISR(R)	IISR(C)	
WS	**35**	2	0	94.6% 5.4%
IISR(R)	13	**9**	15	24.3% 75.7%
IISR(C)	0	2	**35**	94.6% 5.4%
	72.9% 27.1%	69.2% 30.8%	70.0% 30.0%	**71.2% 28.8%**

LS-SVM

True Class \ Predicted Class	WS	IISR(R)	IISR(C)	
WS	**37**	0	0	100% 0.0%
IISR(R)	9	**17**	11	45.9% 54.1%
IISR(C)	0	0	**37**	100% 0.0%
	80.4% 19.6%	100% 0.0%	77.1% 22.9%	**82.0% 18.0%**

WS - Without supply, **IISR (R)** - Irregular input supply rate (reed stalks), **IISR (C)** - Irregular input supply rate (cotton stalks)

On the other hand, in case of the continuous input supply rate (Figure 12) the results saws that in all the operating conditions the classification accuracy for both classifiers is the highest at 100%.

In Figure 13, with different biomass type without 10% of the total blades on the shredder rotor, the MLP-ARD classification accuracy was better in case of zero and irregular cotton stalks input supply rate (100% and 86.5% respectively) than the LS-SVM (97.3% and 43.2% respectively). But, they don't show any encouraging results in recognizing the irregular reed stalks input supply rate (MLP-ARD- 43.2% and LS-SVM – 24.3%). The weakness of the classifiers in this operation condition significantly reduces the average accuracy of the classifiers (MLP-ARD – 76.6% and LS-SVM - 55%).

Finally, Figure 14 shows the results with different biomass type, without all the blades on the rotor and for continuous biomass supply for both types of classifiers. For both classifiers, test success was 100% in all the operation parameters.

Figure 12. Confusion matrix tables for different biomass type with all the blades on the shredder rotor (continuous biomass supply)

MLP-ARD

True Class \ Predicted Class	WS	CISR(R)	CISR(C)	
WS	**37**	0	0	100% 0.0%
CISR(R)	0	**37**	0	100% 0.0%
CISR(C)	0	0	**37**	100% 0.0%
	100% 0.0%	100% 0.0%	100% 0.0%	**100%** **0.0%**

LS-SVM

True Class \ Predicted Class	WS	CISR(R)	CISR(C)	
WS	**37**	0	0	100% 0.0%
CISR(R)	0	**37**	0	100% 0.0%
CISR(C)	0	0	**37**	100% 0.0%
	100% 0.0%	100% 0.0%	100% 0.0%	**100%** **0.0%**

WS - Without supply, **CISR (R)** - Continuous input supply rate (reed stalks), **CISR (C)** - Continuous input supply rate (cotton stalks)

Figure 13. Confusion matrix tables for different biomass type without 10% of the total blades on the shredder rotor (irregular biomass supply)

MLP-ARD

True Class \ Predicted Class	WS	IISR(R)	IISR(C)	
WS	**37**	0	0	100% 0.0%
IISR(R)	9	16	12	43.2% 56.8%
IISR(C)	0	5	**32**	86.5% 13.5%
	80.4% 19.6%	76.2% 23.8%	72.7% 27.3%	**76.6%** **23.4%**

LS-SVM

True Class \ Predicted Class	WS	IISR(R)	IISR(C)	
WS	**36**	0	1	97.3% 2.7%
IISR(R)	12	**9**	16	24.3% 75.7%
IISR(C)	0	21	**16**	43.2% 56.8%
	75.0% 25.0%	30.0% 70.0%	48.5% 51.5%	**55.0%** **45.0%**

WS - Without supply, **IISR (R)** - Irregular input supply rate (reed stalks), **IISR (C)** - Irregular input supply rate (cotton stalks)

Figure 14. Confusion matrix tables for different biomass type without 10% of the total blades on the shredder rotor (continuous biomass supply)

MLP-ARD

True Class \ Predicted Class	WS	CISR(R)	CISR(C)	
WS	**37**	0	0	100% 0.0%
CISR(R)	0	**37**	0	100% 0.0%
CISR(C)	0	0	**37**	100% 0.0%
	100% 0.0%	100% 0.0%	100% 0.0%	**100% 0.0%**

LS-SVM

True Class \ Predicted Class	WS	CISR(R)	CISR(C)	
WS	**37**	0	0	100% 0.0%
CISR(R)	0	**37**	0	100% 0.0%
CISR(C)	0	0	**37**	100% 0.0%
	100% 0.0%	100% 0.0%	100% 0.0%	**100% 0.0%**

WS - Without supply, **CISR (R)** - Continuous input supply rate (reed stalks), **CISR (C)** - Continuous input supply rate (cotton stalks)

In order to highlight the total efficiency of the two classifiers in all the operating conditions, the average accuracies of all the case studies are presented in Figure 15. The height of each bar represents the average accuracy in any operating conditions. The results show that the two classifiers present occasionally similar results. The two classifiers present high accuracy in all conditions (>80%) except in case of different biomass type – irregular supply (MLP-ARD – 71.2%) and in case of different biomass type – 10% fewer blades and irregular supply (MLP-ARD – 76.6%, LS-SVM – 55%). This implies that the classifiers have difficulties when the operating conditions present variable parameters (irregular input supply and rotor unbalance).

Figure 15. Comparison of the average accuracies

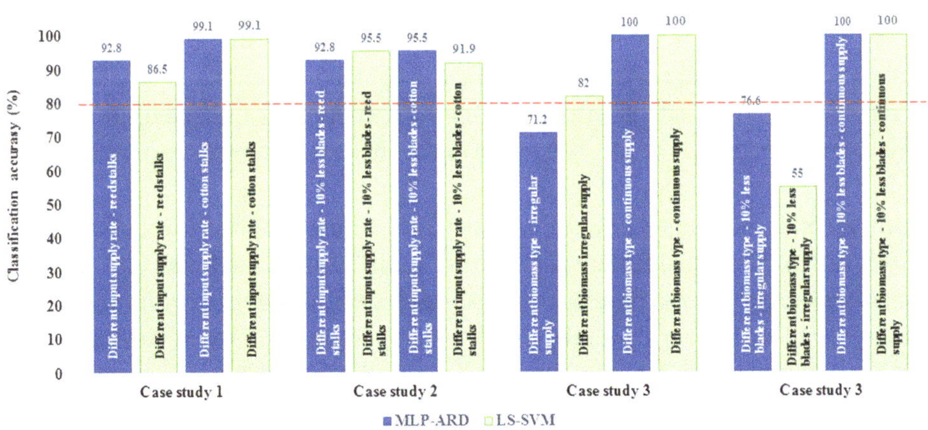

CONCLUSION

In this study, a detailed experimental investigation of the operation condition behavior of a prototype biomass shredder by developing utilizing a Multilayer Perceptron neural network with Automatic Relevance Determination (MLP-ARD) and Least Squares Support Vector Machine (LS-SVM) methods. It's about a soft computing application to an NDT problem in an alternative domain instead of the spatial domain. Specifically, was investigated the ability to identify a specific operation condition (different input supply rate, different types of biomass and a different number of blades on the rotor of the shredder with non- destructive techniques.

The roles of different vibration signals and signal preprocessing techniques were investigated. The performance of MLP-ARD (71.2%-100%) has been found to be imperceptibly better than LS-SVM (55%-100%). More detailed, both classifiers are proven as more efficient in recognizing the zero-biomass supply (83.78% to 100%), the irregular and continuous input supply rate with cotton stalks (92,8%-99,1% and 86,5%-99,1% respectively). Even in the case of the 10% fewer blades, both classifiers are proven efficient (MLP-ARD – 92.8% to 95,5% and LS-SVM – 91.9% to 95.5%). But they don't show any encouraging results in recognizing the irregular input supply rate with reed stalks (MLP-ARD- 24.3% and LS-SVM – 45.9%) and without 10% of the blades (MLP-ARD- 43.2% and LS-SVM – 24.3%).

In order to evaluate the proposed systems, the most appropriate metrics is the Root Mean Squared Error (RMSE), the Mean Absolute Error (MAE), and the percent good classifications in order to determine their consistency for selecting the "better" network for the classification problem. Empirical results of a well-established pattern classification problem demonstrate the necessity of examining all three commonly used metrics.

The present paper is focusing on the basic idea of the methodology, the simple symmetric rotor-bearing system is selected to explain the methodology. Therefore, only six time-domain and six frequency-domain features are selected. In a future work, we believe that it would be interesting to make a comparative discussion of Fourier transform and wavelet transform.

In conclude, the results showed that the two classifiers can provide reliable results using as inputs statistical features in the time domain and in the frequency domain. These statistical features (time or frequency-domain) can be used with success for identifying the different operating condition. The combination of the statistical features with the appropriate classifiers gives a powerful tool for the agricultural biomass shredder condition monitoring.

REFERENCES

Adapa, P., Tabil, L., & Schoenau, G. (2011). Grinding performance and physical properties of non-treated and steam exploded barley, canola, oat and wheat straw. *Biomass and Bioenergy*, *35*(1), 549–561. doi:10.1016/j.biombioe.2010.10.004

Aidi, B., Philen, M. K., & Case, S. W. (2015). Progressive damage assessment of centrally notched composite specimens in fatigue. *Composites. Part A, Applied Science and Manufacturing*, *74*, 47–59. doi:10.1016/j.compositesa.2015.03.022

Arthur, J. F., Kepner, R. A., Dobie, J. B., Miller, G. E., & Parsons, P. S. (1982). Tub grinder performance with crop and forest residues. *Transactions of the ASAE. American Society of Agricultural Engineers*, *25*(6), 1488–1494. doi:10.13031/2013.33751

Arumugam, V., Kumar, C. S., Santulli, C., Sarasini, F., & Stanley, A. J. (2011). A Global Method for the Identification of Failure Modes in Fiberglass Using Acoustic Emission. *Journal of Testing and Evaluation*, *39*(5).

Babak, S., & Ahmed, E. (2015). Large-scale Classification of Fine-Art Paintings: Learning The Right Metric on The Right Feature. *Computer Vision and Pattern Recognition*, *21*.

Bao, W., Chen, Y., & Wang, D. (2014). Prediction of protein structure classes with flexible neural tree. In Bio-Medical Materials and Engineering (Vol. 24, pp. 3797–3806). Academic Press. doi:10.3233/BME-141209

Bao, W., Huang, Z., Yuan, C. A., & Huang, D. S. (2017). Pupylation sites prediction with ensemble classification model. *International Journal of Data Mining and Bioinformatics*, *18*(2), 91. doi:10.1504/IJDMB.2017.086441

Bao, W., Wang, D., & Chen, Y. (2017). Classification of Protein Structure Classes on Flexible Neutral Tree. *IEEE/ACM Transactions on Computational Biology and Bioinformatics*, *14*(5), 1122–1133. doi:10.1109/TCBB.2016.2610967 PMID:28113983

Bapat, D., Kulkarni, S., & Bhandarkar, V. (1997). Design and operating experience on fluidised bed boiler burning biomass fuels with high alkali ash. In *The 15th International Conference on Fluidized bed combustion*. Academic Press.

Bishop, C. M. (1995). Neural networks for pattern recognition. *Journal of the American Statistical Association*, *92*, 482. doi:10.2307/2965437

Bitra, V. S. P., Womac, A. R., Chevanan, N., Miu, P. I., Igathinathane, C., Sokhansanj, S., & Smith, D. R. (2009). Direct mechanical energy measures of hammer mill comminution of switchgrass, wheat straw, and corn stover and analysis of their particle size distributions. *Powder Technology*, *193*(1), 32–45. doi:10.1016/j.powtec.2009.02.010

Brown, G. (2011). Ensemble learning. In *Encyclopedia of Machine Learning*. Springer.

Cundiff, J. S., & Grisso, R. D. (2008). Containerized handling to minimize hauling cost of herbaceous biomass. *Biomass and Bioenergy*, *32*(4), 308–313. doi:10.1016/j.biombioe.2007.10.009

Drzymala, Z. (1993). *Industrial briquetting-fundamentals and methods. Studies in mechanical engineering* (Vol. 13). Warszawa: PWN-Polish Scientific Publishers.

Emami, S., & Tabil, L. G. (2008). Friction and compression characteristics of chickpea flour and components. *Powder Technology*. doi:10.1016/j.powtec.2007.07.012

Fasina, O. O. (2006). Flow and physical properties of switchgrass, peanut hull and poultry litter. *Transactions of the ASABE*, *49*(3), 721–728. doi:10.13031/2013.20470

Ganesan, V., Rosentrater, K. A., & Muthukumarappan, K. (2008). Flowability and handling characteristics of bulk solids and powders - a review with implications for DDGS. *Biosystems Engineering*, *101*(4), 425–435. doi:10.1016/j.biosystemseng.2008.09.008

Ghate, V. N., & Dudul, S. V. (2010). Optimal MLP neural network classifier for fault detection of three phase induction motor. *Expert Systems with Applications*, *37*(4), 3468–3481. doi:10.1016/j.eswa.2009.10.041

Herrmann, A. E., & Estrela, V. V. (2016). Content Based Image Retrieval (CBIR) in Remote Clinical Diagnosis and Healthcare. In Encyclopedia of E-Health and Telemedicine. Academic Press.

Hess, R. J., Wright, C. T., & Kenney, K. L. (2007). Cellulosic biomass feedstocks and logistics for ethanol production. *Biofuels, Bioproducts & Biorefining*, *1*(3), 181–190. doi:10.1002/bbb.26

Hsu, C.-W., Chang, C.-C., & Lin, C.-J. (2010). A Practical Guide to Support Vector Classification. *Technical Report*, 1–16. doi:10.1177/02632760022050997

Huang, D.-S., & Du, J.-X. (2008). A constructive hybrid structure optimization methodology for radial basis probabilistic neural networks. *IEEE Transactions on Neural Networks*, *19*(12), 2099–2115. doi:10.1109/TNN.2008.2004370 PMID:19054734

Hung, Y. Y., Yang, L. X., & Huang, Y. H. (2013). Non-destructive evaluation (NDE) of composites: digital shearography. In V. M. Karbhari (Ed.), *Non-Destructive Evaluation (NDE) of Polymer Matrix Composites* (pp. 84–115). Woodhead Publishing. doi:10.1533/9780857093554.1.84

Kateris, D., Moshou, D., Pantazi, X.-E., Gravalos, I., Sawalhi, N., & Loutridis, S. (2014). A machine learning approach for the condition monitoring of rotating machinery. *Journal of Mechanical Science and Technology*, *28*(1), 61–71. doi:10.100712206-013-1102-y

Keerthi, S. S., & Lin, C. J. (2003). Asymptotic behaviors of support vector machines with gaussian kernel. *Neural Computation*, *15*(7), 1667–1689. doi:10.1162/089976603321891855 PMID:12816571

Koyama, K., Hoshikawa, H., & Kojima, G. (2013). Eddy Current Nondestructive Testing for Carbon Fiber- Reinforced Composites. *Journal of Pressure Vessel Technology*, *135*(4), 041501. doi:10.1115/1.4023253

Kroeger, T. (2014). Thermographic inspection of composites. *Reinforced Plastics*, *58*(4), 42–43. doi:10.1016/S0034-3617(14)70183-3

Kumar, P., Barrett, D. M., Delwiche, M. J., & Stroeve, P. (2009). Methods of pretreatment of lignocellulosic biomass for efficient hydrolysis and biofuel production. *Industrial & Engineering Chemistry Research*, *48*(8), 3713–3729. doi:10.1021/ie801542g

Lam, P. S., Sokhansanj, S., Bi, X., Mani, S., Lim, C. J., Womac, A. R., … Nayaran, S. (2008). Physical characterization of wet and dry wheat straw and switchgrass – bulk and specific density. *ASABE Annual International Meeting, 300*(7), 23. 10.13031/2013.24490

Lei, Y., He, Z., & Zi, Y. (2008). A new approach to intelligent fault diagnosis of rotating machinery. *Expert Systems with Applications*, *35*(4), 1593–1600. doi:10.1016/j.eswa.2007.08.072

Lesage, D., Graf, V. de T., & Westphal, K. (2010). *Global energy governance in a multipolar world.* London: Ashgate.

Li, H., Zhang, Y., & Zheng, H. (2009). Gear fault detection and diagnosis under speed-up condition based on order cepstrum and radial basis function neural network. *Journal of Mechanical Science and Technology*, *23*(10), 2780–2789. doi:10.100712206-009-0730-8

Lin, H.-T., & Lin, C.-J. (2003). *A study on sigmoid kernels for SVMand the training of non-PSD kernels by SMO-type methods*. Academic Press.

Lopo, P. (2002). The right grinding solution for you: roll, horizontal or vertical. *Feed Management*, (53), 6–23.

Lu, Q., & Li, D. (1998). Neural network method for diagnosing faults rolling bearing in electrical machines with frequency signatures. *Qinghua Daxue Xuebao/Journal of Tsinghua University, 38*(4), 94–97.

Maierhofer, C., Myrach, P., Reischel, M., Steinfurth, H., Röllig, M., & Kunert, M. (2014). Characterizing damage in CFRP structures using flash thermography in reflection and transmission configurations. *Composites. Part B, Engineering, 57*, 35–46. doi:10.1016/j.compositesb.2013.09.036

Mani, S., Tabil, L. G., & Sokhansanj, S. (2004). Grinding performance and physical properties of wheat and barley straws, corn stover and switchgrass. *Biomass and Bioenergy, 27*(4), 339–352. doi:10.1016/j.biombioe.2004.03.007

Mitchell, P., Kiel, J., Livingston, B., & Dupont-Roc, G. (2007). *Torrefi ed biomass: A foresighting study into the business case for pellets from torrefi ed biomass as a new solid fuel, in All Energy 2007*. University of Aberdeen, ECN, Doosan Babcock, and ITI Energy.

Moallem, P., Razmjooy, N., & Ashourian, M. (2013). Computer vision-based potato defect detection using neural networks and support vector machine. *International Journal of Robotics and Automation, 28*(2), 137–145. doi:10.2316/Journal.206.2013.2.206-3746

Moshou, D., Kateris, D., Gravalos, I., Loutridis, S., Sawalhi, N., Gialamas, T., ... Tsiropoulos, Z. (2010). Determination of fault topology in mechanical subsystems of agricultural machinery based on feature fusion and neural networks. *Conference Proceeding - 4th International Conference, TAE 2010: Trends in Agricultural Engineering 2010*.

Moshou, D., Pantazi, X.-E., Kateris, D., & Gravalos, I. (2014). Water stress detection based on optical multisensor fusion with a least squares support vector machine classifier. *Biosystems Engineering, 117*, 15–22. doi:10.1016/j.biosystemseng.2013.07.008

Oguma, I., Goto, R., & Sugiura, T. (2012). Ultrasonic inspection of an internal flaw in a ferromagnetic specimen using angle beam EMATs. *Przeglad Elektrotechniczny, 88*(7B), 78–81.

Parlos, A. G. M. S., & A. A. (1999). Adaptive state estimation using dynamic recurrent neural networks in Neural Networks. In *IJCNN '99. International Joint Conference on. Dept. of Nucl,* (pp. 3361–3364). Academic Press.

Paulrud, S. (2004). *Upgraded biofuels d Effects of quality on processing, handling characteristics, combustion, and ash melting*. Swedish University of Agricultural Sciences.

Paya, B. A., Esat, I. I., & Badi, M. N. M. (1997). Artificial neural network based fault diagnostics of rotating machinery using wavelet transforms as a preprocessor. *Mechanical Systems and Signal Processing*, *11*(5), 751–765. doi:10.1006/mssp.1997.0090

Peng, W., Zhang, Y., Qiu, B., & Xue, H. (2012). A Brief Review of the Application and Problems in Ultrasonic Fatigue Testing. *AASRI Procedia*, *2*, 127–133. doi:10.1016/j.aasri.2012.09.024

Pindoriya, N. M., Singh, S. N., & Singh, S. K. (2008). An Adaptive Wavelet Neural Network-Based Energy Price Forecasting in Electricity Markets. *IEEE Transactions on Power Systems*, *23*(3), 1423–1432. doi:10.1109/TPWRS.2008.922251

Razmjooy, N. (2012). A multi layer perceptron neural network trained by invasive weed optimization for potato color image segmentation. *Trends in Applied Sciences Research*, *7*(6), 445–455. doi:10.3923/tasr.2012.445.455

Razmjooy, N. and Ramezani m. (2016). Training Wavelet Neural Networks Using Hybrid Particle Swarm Optimization and Gravitational Search Algorithm for System Identification, International Journal of Mechatronics. *Electrical and Computer Technology*, *6*(21), 2987–2997.

Razmjooy, N., Mousavi, B. S., & Soleymani, F. (2013). A hybrid neural network Imperialist Competitive Algorithm for skin color segmentation. *Mathematical and Computer Modelling*, *57*(3-4), 848–856. doi:10.1016/j.mcm.2012.09.013

Razmjooy, N., Sheykhahmad, F. R., & Ghadimi, N. (2018). A hybrid neural network–world cup optimization algorithm for melanoma detection. *Open Medicine: a Peer-Reviewed, Independent, Open-Access Journal*, *13*(1), 9–16. doi:10.1515/med-2018-0002 PMID:29577090

Samanta, B., Al-Balushi, K. R., & Al-Araimi, S. A. (2006). Artificial neural networks and genetic algorithm for bearing fault detection. *Soft Computing*, *10*(3), 264–271. doi:10.100700500-005-0481-0

Sokhansanj, S., & Fenton, J. (2011). *Cost benefit of biomass supply and preprocessing: BIOCAP (Canada) research integration program synthesis paper*. Retrieved from http://www.biocap.ca/rif/reprt/Sokhansanj_S.pdf

Sokolova, M., & Lapalme, G. (2009). A systematic analysis of performance measures for classification tasks. *Information Processing & Management*, *45*(4), 427–437. doi:10.1016/j.ipm.2009.03.002

Souček, J., Kocánová, V., & Novák, M. (2007). Parametres of energy crop biomass handling. *Research in Agricultural Engineering*, *53*(4), 161–165. doi:10.17221/1952-RAE

Sugumaran, V., & Ramachandran, K. I. (2011). Effect of number of features on classification of roller bearing faults using SVM and PSVM. *Expert Systems with Applications*, *38*(4), 4088–4096. doi:10.1016/j.eswa.2010.09.072

Sun, W., Chen, J., & Li, J. (2007). Decision tree and PCA-based fault diagnosis of rotating machinery. *Mechanical Systems and Signal Processing*, *21*(3), 1300–1317. doi:10.1016/j.ymssp.2006.06.010

Tumuluru, J., Wright, C., Kenny, K., & Hess, R. (2010). A Riview on Biomass Densification Technologies for Energy Application. *Statewide Agricultural Land Use Baseline*, *2015*, 1–59. doi:10.1017/CBO9781107415324.004

Vapnik, V. N. (1998). *Statistical Learning Theory*. John Wiley & Sons. doi:10.2307/1271368

Vapnik, V. N. (1999). *The Nature of Statistical Learning Theory* (Vol. 8). Springer; doi:10.1109/TNN.1997.641482

Vavilov, V. P., Budadin, O. N., & Kulkov, A. A. (2015). Infrared thermographic evaluation of large composite grid parts subjected to axial loading. *Polymer Testing*, *41*, 55–62. doi:10.1016/j.polymertesting.2014.10.010

Wei, L., Pordesimo, L. O., Igathinathane, C., & Batchelor, W. D. (2009). Process engineering evaluation of ethanol production from wood through bioprocessing and chemical catalysis. *Biomass and Bioenergy*, *33*(2), 255–266. doi:10.1016/j.biombioe.2008.05.017

Widodo, A., Kim, E. Y., Son, J.-D., Yang, B.-S., Tan, A. C. C., Gu, D.-S., ... Mathew, J. (2009). Fault diagnosis of low speed bearing based on relevance vector machine and support vector machine. *Expert Systems with Applications*, *36*(3), 7252–7261. doi:10.1016/j.eswa.2008.09.033

Wu, X., Zhao, R., Bean, S. R., Seib, P. A., McLaren, J. S., & Madl, R. L. … Wang, D. (2007). Factors impacting ethanol production from grain sorghum in the dry-grind process. *Cereal Chemistry*. doi:10.1094/CCHEM-84-2-0130

Xavier, C. A. N., Moset, V., Wahid, R., & Møller, H. B. (2015). The efficiency of shredded and briquetted wheat straw in anaerobic co-digestion with dairy cattle manure. *Biosystems Engineering*, *139*, 16–24. doi:10.1016/j.biosystemseng.2015.07.008

Yu, M., Womac, A. R., & Pordesimo, L. O. (2003). Review of Biomass Size Reduction Technology. *2003 ASABE Annual International Meeting*, 1–12. 10.13031/2013.15454

Zaremba, W., Sutskever, I., & Vinyals, O. (2015). Recurrent Neural Network Regularization. *ArXiv Preprint*. Retrieved from http://arxiv.org/abs/1409.2329

ADDITIONAL READING

Hsu, C.-W., Chang, C.-C., & Lin, C.-J. (2010). A Practical Guide to Support Vector Classification. *Technical Report*, 1–16.

Kateris, D., Gravalos, I., Gialamas, Th., Xyradakis, P., & Moshou, D. (2016). A new approach to fault diagnosis in agricultural tractor mechanical gearbox. In *Proceedings of 6th International Conference TAE 2016 (Trends in Agricultural Engineering)* (pp. 290-299). FE CULS Prague, Czech Republic.

Kateris, D., Moshou, D., Gialamas, Th., Gravalos, I., & Xyradakis, P. (2014). Bearing fault diagnosis in mechanical gearbox, based on time and frequency – domain parameters with MLP-ARD. *Journal of Agricultural Machinery Science*, *10*(2), 101–106.

Kateris, D., Moshou, D., Pantazi, X.-E., Gravalos, I., Sawalhi, N., & Loutridis, S. (2014). A machine learning approach for the condition monitoring of rotating machinery. *Journal of Mechanical Science and Technology*, *28*(1), 61–71. doi:10.100712206-013-1102-y

Lei, Y., He, Z., & Zi, Y. (2008). A new approach to intelligent fault diagnosis of rotating machinery. *Expert Systems with Applications*, *35*(4), 1593–1600. doi:10.1016/j.eswa.2007.08.072

Moshou, D., Kateris, D., Sawalhi, N., Gravalos, I., Loutridis, S., Gialamas, T., … Tsiropoulos, Z. (2011). Condition monitoring of Mechanical Subsystems of Agricultural Vehicles Based on Fusion of Vibration Features. In *Proceedings of XXXIV CIOSTA CIGR V Conference*. Vienna, Austria.

Sugumaran, V., & Ramachandran, K. I. (2011). Effect of number of features on classification of roller bearing faults using SVM and PSVM. *Expert Systems with Applications*, *38*(4), 4088–4096. doi:10.1016/j.eswa.2010.09.072

Widodo, A., Kim, E. Y., Son, J.-D., Yang, B.-S., Tan, A. C. C., Gu, D.-S., ... Mathew, J. (2009). Fault diagnosis of low speed bearing based on relevance vector machine and support vector machine. *Expert Systems with Applications*, *36*(3), 7252–7261. doi:10.1016/j.eswa.2008.09.033

KEY TERMS AND DEFINITIONS

Classification: Is a process related to categorization, the process in which ideas and objects are recognized, differentiated, and understood.

Condition Monitoring (CM): Is the process of monitoring a parameter of condition in machinery (vibration, temperature, etc.), in order to identify a significant change. It is a major component of predictive maintenance. The use of condition monitoring allows maintenance to be scheduled, or other actions to be taken to prevent consequential damages and avoid its consequences.

Least Squares Support Vector Machines (LS-SVM): Are least squares versions of support vector machines (SVM), which are a set of related supervised learning methods that analyze data and recognize patterns, and which are used for classification and regression analysis.

Multilayer Perception Neural Network (MLP): Is a class of feedforward artificial neural network. An MLP consists of, at least, three layers of nodes: an input layer, a hidden layer and an output layer. Except for the input nodes, each node is a neuron that uses a nonlinear activation function.

Non-Destructive Testing (NDT): Nondestructive testing or non-destructive testing is a wide group of analysis techniques used in science and technology industry to evaluate the properties of a material, component or system without causing damage.

Vibration: Is a mechanical phenomenon whereby oscillations occur about an equilibrium point. The oscillations may be periodic, such as the motion of a pendulum—or random, such as the movement of a tire on a gravel road.

Chapter 6
Prognosis for Crop Yield Production by Data Mining Techniques in Agriculture

Divya Singh
Amity University, India

Dinesh Sharma
Amity University, India

ABSTRACT

In agriculture, data mining technique is used for extracting information from a large dataset. The techniques for data mining are used in yield prediction for crop at broader spectrum. Agricultural system is very complex and vast therefore to deal with large data situation is a great factor. Different consultancy, industrial production department, organization related to crops is taking keen interest towards crop yield prediction. Here the focus is on the applicability of data mining techniques in agricultural field. The classification and clustering techniques of data mining are used recently in agriculture field. Data mining technology merged with the rapid development of computer science. This chapter focuses on collecting information and overcome the short comes of manual data handling and prediction of yield results of crop production. Data mining is a prominent agricultural research area for analysis of crop yield. These predictions are a very important in solving agricultural problems for crops.

DOI: 10.4018/978-1-5225-8027-0.ch006

INTRODUCTION

Data mining is known as a process which extracts useful information from a vast amount of data. The novel research field of data mining is used in agriculture which is helpful in solving agriculture problems. The yield prediction of agricultural product becomes easier which can benefit farmers to improvise the productivity for specific field and crops.

The comparison of data mining procedures previously used and the latest updated gathered data help in understanding, learning and evaluating how to classify future yield predictions and give out the best out of information. Day by day increase in population worldwide increases the food demands and its production demand. To meet this vision the focus on improving production efficiency the use of crop models and decision tools are in demand in upcoming years. The combined study of agriculture, as well as advanced technology, is upgrading the advancement in the production field of agriculture and benefiting farmers.

All agriculture-related organizations, consultancy, and producers are new technology with the end use of crop yield prediction. This high percentage crop yield depends on a well defined bio-socio-economic system which depends on various factors like water, air and soil environment. This demands a comprehensive model which can facilitate by classical engineering expertise. The crop forecasting is an essential tool for predicting crop yields and production percentage within months before harvesting.

This philosophy based on data's collection from different sources like agro-meteorological (yield), soil (water retention capacity) remotely sensed, agricultural statistics etc. This several conclusions based derives several indices which ultimately gives variables in determining crop yield, for instance, crop water satisfaction, surplus, and excess moisture, average soil moisture, etc. Information technology has played a vital role not only in industry but also in agriculture. The crops are grown in the field and the data collected are analyzed. These data are exact, accurate and small in scale. Data mining techniques benefit not only productivity but also farmer economy.

Data mining is a technique used for agriculture development. Depend on various climatic, ecological, geographical and economical aspects various risk factors for agricultural growth indifferent or a particular zone can be quantified by mathematical, statistical and advanced computing and challenges to extract information from agriculture database is much focused for agricultural progressive growth. The data mining technique used in agriculture is to discover interesting and useful patterns and finding correlation among huge amount data's collected in a large volume in the concerned field. It synchronizes the data archaeology, data pattern analysis, extraction of knowledge, a combination of statistics, probability analysis and data technologies.

It employs the statistical method. Data mining in agriculture aims to predication about the behavior of attributes within the data, classify and segmenting related data in the category and the optimization the use of limited resources etc. The classification and clustering techniques are two fundamental classifications of Data technique and are used in the collection of data and information in the agricultural field, which is used to classify unknown samples. It includes artificial neural networks, support vector machines K- nearest neighbor and K-Means algorithms. These classifications are also methodologies used in data mining.

The crucial role of data mining in agriculture help in determining and understanding various facts related to crop growth i.e. influence of climate on Kharif crop and rabi crops, the total estimate of damage done by pests, crop yield estimation, spatial data mining relating to a pattern related to agriculture, mushroom grading etc.

The different data mining methodologies have different application i.e. neutral networking method focuses on the weather forecast, predication rainfalls and climatic conditions, K-Means method classify soils with a combination of GPS, Wine fermentation problems, yield estimation, and yield production, K- nearest neighbor method measures stimulating daily precipitations and other weather conditions.

DATA MINING TECHNIQUES/METHOD

Data Mining incorporates many techniques like clustering, classification, machine learning, Support Vector Machines, Regression, Association Rules etc. Further, these techniques can be applied to the dataset by different algorithms. Here we focused on two broad classifications i.e. clustering, classification technique.

Data Mining in Agriculture is an upcoming area and have given special attention by data analysts and data mining experts to focus their studies on it. It gives complete information about crop production can help farmers identify crop losses and help in finding preventive measures in the future (Razmjooy et al., 2013).

Other problems based on decision making and managing crop efficiency can be solved using this technique. It can help in predicting hazards and future risk factor in the agriculture sector. Retail industries are using this technology for the collection of huge amount of data on customer shopping trends, sales of the company etc.

This helps the company to analyze both the sales data and data relating to the customer which helps them in making better business decisions. Telecommunication industry also uses data mining which has extended its application from providing telephone services to offer many add-on services like fax, Internet and cellular phone (Ramesh, and Ramar, 2011).

According to recent applicability Data mining techniques are classified into two types: Classification and Clustering techniques

Figure 1. Classification of Data Mining Techniques/Method

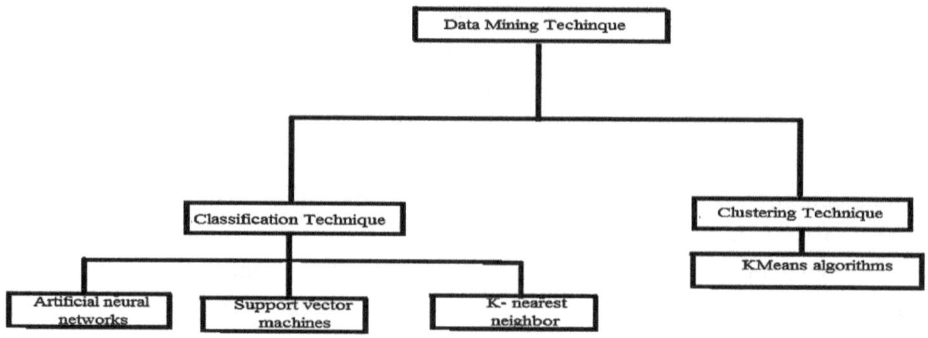

CLASSIFICATION TECHNIQUES

These sets of classifies samples are referred to as training set as it explains about classification process and explains how classification is performed (Devi, et al., 2012). Artificial Neural Networks and Support Vector Machines are such kind of classification techniques. The K- Nearest Neighbor is another classification technique used which uses the training set every time a classification must be performed and it does not have any learning phase. The basic assumption in the K-Nearest Neighbor algorithm is that similar samples should have a similar classification. The parameter K signifies the number of equivalent known samples used for assigning a classification of the unknown sample. It uses the information in training set but it does not cite any rule for classification.

In this technique, the set of classified samples provide information to classify unknown samples. It is of following types:

1. Artificial neural networks (ANN)
2. Support vector machines (SVM)
3. K-nearest neighbor

Artificial Neural Networks (ANN)

Weather prediction is a collection of data based on whether parameter i.e. temperature, precipitation, pressure, wind speed and direction, dew point and rainfall etc. Any changes in such parameters are necessary to be measured for the safety of life and atmosphere conditions (Mousavi et al., 2013; Razmjooy et al., 2013). All the input and output relations are mapped by the neural network technique. Backpropagation algorithm (BPP), collects data relating to the changes which explain gives variation in one parameter which reflects the change in another parameter. It will give a large class of function and very efficient for numerical differentiation. Artificial Neural Network (ANN) provides a methodology for solving nonlinear problems that are difficult to be solved by traditional methods. The Figure 2 explains the ANN working process.

Support Vector Machines (SVM)

SVM are learning a system that uses a hypothesis space of linear function in a high dimensional feature space and trained with a learning algorithm for optimization theory. The basic working function is shown in Figure 3.

It is a supervised machine learning algorithm which can be used for both classification and regression challenges. SVM is majorly applicable in problems classification methods. In this algorithm, we plot each data item as a point in n-dimensional space (where n is a number of features you have) with the value of each feature being the value of a particular coordinate.

Then, we perform classification by finding the hyperplane that differentiates the two classes very well as shown in Figure 4.

Figure 2. Artificial neural networks (ANN)

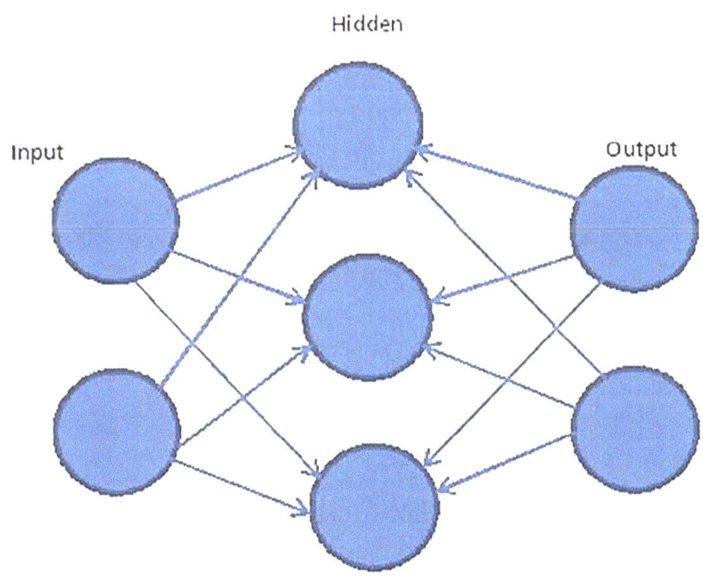

Figure 3. Basic Structure of SVM

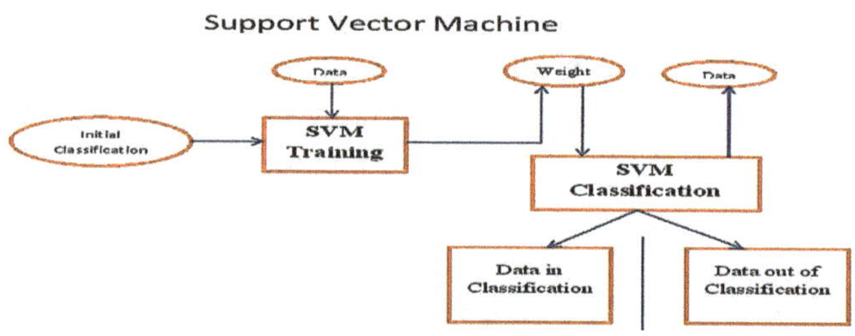

Figure 4. SVM algorithms

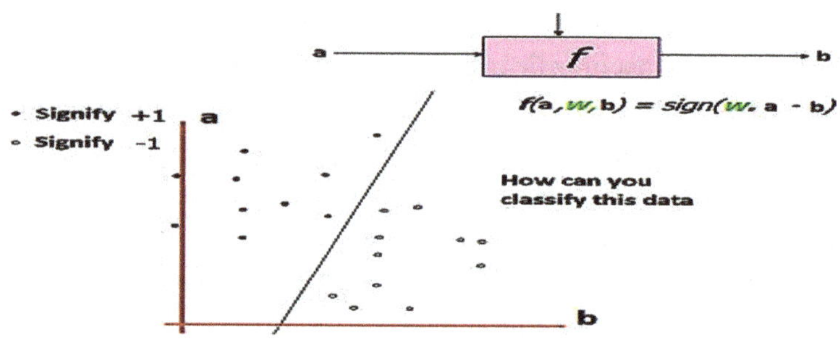

Working of SVM System

The working of SVM is explained below (by an example) in the following steps:

Step 1: Identify the Right Hyper-Plane (Scenario-1)

Three hyper-planes a, b, and c are considered. To classify star and circle identify right hyperplane as shown Figure 5.

Any hyper-plane **a** or **b** or **c** can be selected. Suppose hyper-plane **a** is selected. It has a higher margin compared to **b.** But as it is known that SVM selects only that hyper-plane which classifies the classes accurately prior to maximizing margin. Here, hyper-plane b has a classification error and **a** hyperplane has classified all correctly. Therefore, the right hyper-plane is **a.**

Step 2: How to Classify two Classes?

Figure 6 explains the segregation of the two classes using a straight line is impossible. This can be explained very well from the figure below. It is observed that one of star lies in the territory of other (circle) class as an outlier.

Figure 5. Hyper-planes

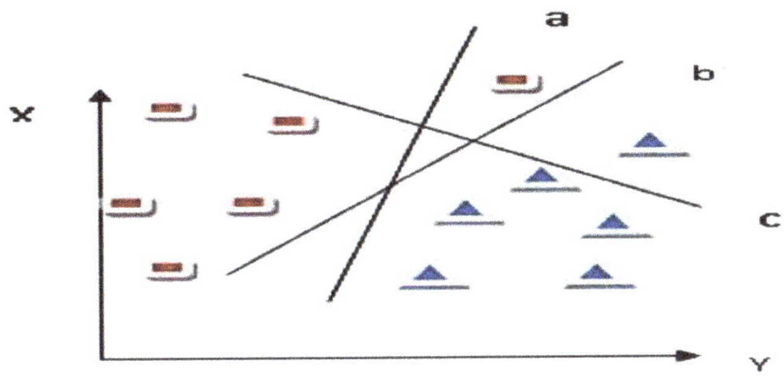

As I have already mentioned, one star at other end is like an outlier for star class. SVM has a feature to ignore outliers and find the hyperplane that has maximum margin. Hence, we can say, SVM is robust to outliers (Figure 7).

Figure 6. Segregation of the two classes using a straight line

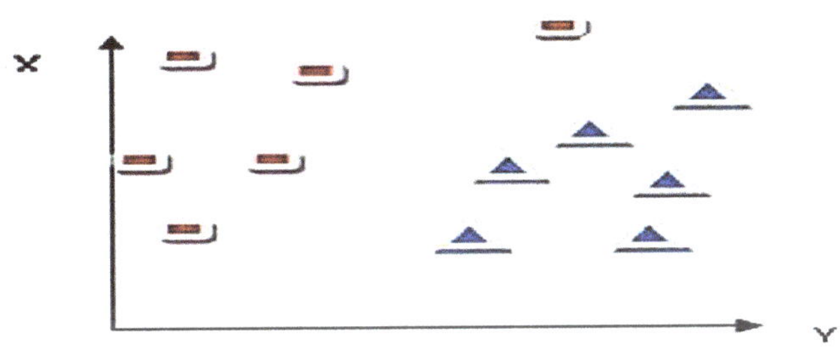

Figure 7. Hyper-plane with maximum margin

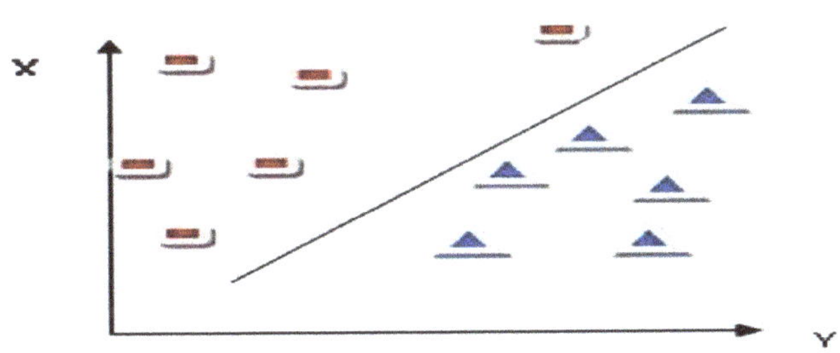

K-Nearest Neighbor

K-nearest neighbor is a non-parameterized technique (Figure 8). For statistical estimation and pattern recognition. The outcome of K- nearest neighbor is based on its evidence and the nearest neighbor is selected based on the distance. It can use in many applications like agriculture, health, text mining and financial (Razmjooy et al., 2012).

To understand K-Nearest Neighbour classifier working, an example is given in Table 1 is to calculate the values. Here, a distance of different student from Rohan (student) is calculated by K-Nearest Neighbour classifier method.

Sanjay D. Sawaitul et al. reported on the observations on the information about weather. These recorded parameters like wind speed and direction, zonal temperature,

Figure 8. K- nearest neighbor is a non-parameterized technique

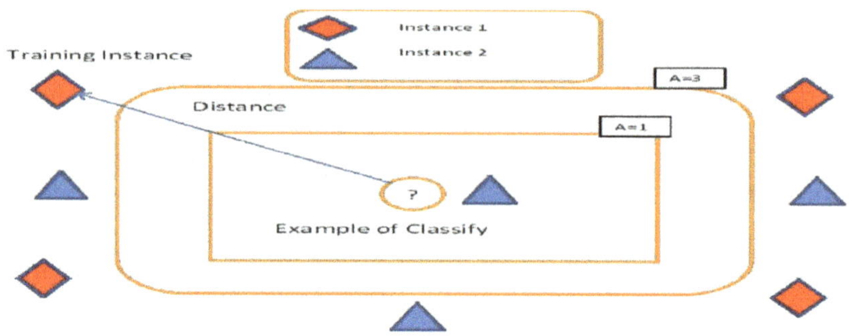

Table 1. To calculate K-Nearest Neighbour classifier

K-Nearest Neighbor Classifier, A = 3					
Student	**Age**	**Income(K)**	**No. of Cards**	**Response**	**Distance from Rohan**
Montee	55	40	4	No	$\sqrt{(40-38)^2-(40-55)+(4-3)}=x$
Jaismine	28	55	3	Yes	$\sqrt{(28-38)^2-(55-55)+(3-2)}=y$
Dinesh	62	199	2	Yes	$\sqrt{(62-38)^2-(199-55)+(2-1)}=z$
Anisha	35	165	1	Yes	$\sqrt{(35-38)^2-(165-55)+(1-5)}=a$
Tom	25	35	5	No	$\sqrt{(25-38)^2-(35-55)+(5-1)}=b$
Rohan	38	55	1	Yes	-

humidity in the atmosphere, rainfall measures and other climatic conditions are used to forecast weather, can be predicted using artificial neural networks, backpropagation techniques (Sawaitul et al., 2012) and change can be measured acutely and calamities can be prevented for future prospects. Somvanshi, V.K. et al., explained the modeling and prediction of rainfall using ANN techniques and Box- Jenkins methodology.

Artificial neural networks are also applicable in hydrology (Somvanshi et al., 2006). N. Q. Hung et al. used an ANN model for rainfall forecasting in Bangkok, Thailand. This model was predicted the rainfall and management of flood-like situation before 6 hours. This was used in Bangkok for rainfall prediction and managing the passing of flood water. This model predicts rainfall only but can't predict the temperature, humidity (Hung et al., 2009).

Sharma A et al. used the concept of soft computing by ANN Technique (Baboo et al., 2010) for weather forecasting. Ch. Jyosthna Devi, B et al., used artificial neural networks approach to predict the weather by use of back propagation (Sharma and Datta,2007). Pielke R.A. worked on a comprehensive meteorological modeling system RAMS using AAN method (Devi, et al., 2012).

APPLICATIONS

The weather, climatic conditions, its forecasts and its impact on agricultural crop field and the study soil of characteristics like humidity, soil fertility etc. better crop production can be marked using this technique. For e.g., the K-Means algorithm, the K Nearest Neighbor (KNN) is applied for simulating daily precipitations and different possible changes of the weather and environmental changes are analyzed.

The K-Means approach is used for classifying and characterizes soils in combination with GPS-based technologies. This decides the flora for a specific soil and can be classified for the betterment of crop. Thus, predictions can be done for their growth and high percentage yield and can be implemented. The agro-climatic conditions for crop yield can easily find out by use of data mining technique.

Applications in the Agriculture Field

Selection of Crop and Prediction of Crop Yield

The increase the percentage yield of crops, the selective of the crop is very important for a particular landscape area. This becomes are important criteria for the proper growth of crops and increase the quantitative estimate in the year. This depends upon the soil type, soil composition, climatic conditions, geographical areas, market economy etc. All the classification techniques are employed with respect to crop selection and to understand the ways for better crop yield in a particular area. The plant nutrient management system was proposed based on machine learning methods to meet all requirements like soil profile, fertile landscape etc for increasing percentage yield for a particular type of crops. CSM term is used for defining the crop selection method.

Weather Forecast

India has humid subtropical climate and agriculture mostly depend upon rainfall for irrigation. Therefore, here weather forecasting becomes a very important factor for proper crop selection, harvesting and crop production. Several works are going on to make a more simple approach digitally to understand the dynamic changes in weather over time. The use of ANN is important here. SVM algorithm approach is also used for such forecasting.

Smart Irrigation System

The water consumption is the highest in India by farmers for crop growth. Smart agriculture methods in the farming field are becoming a major factor for development in this sector. The ground level water is lowering down due to continuous climatic changes.

Therefore, the use of river water for crop productivity, harvesting is becoming an issue and can create disputes in the country for the preservation of water. To combat the scarcity of water, many companies have come up with sensor-based technology for smart farming which uses sensors to monitor the water level, nutrient content, weather forecast reports, and soil temperature.

Such devices are designed on the principles of machine learning. All content of the soil, nutritive values of soil are recorded using these sensors and indirectly use in soil using smart irrigation systems. This finally reduces labor, field cost etc.

Crop Yield Prediction

Data mining technique is used in the agricultural field and is of great demand. It analyzes large data sets for example like crop yield production and classifies data sets. It extracts information's from all data sets and transform into a useful structure. Figure 9 explained that the work flowchart of proposed crop yield prediction.

Figure 9. Flowchart for crop yield production

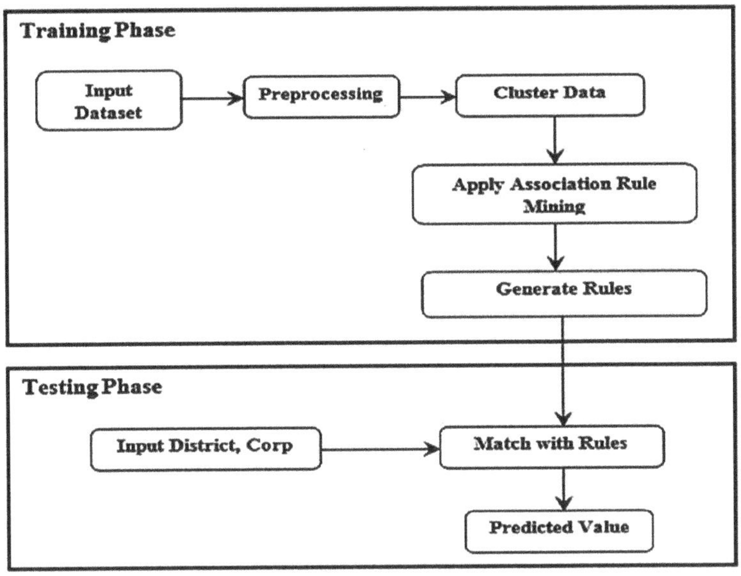

Methodology for Data Mining

There are two phases of methodology:

1. **Training Phase:** Data was collected and preprocessed by data clustering via k-means clustering algorithm followed by the association rule mining process. The several rules are then generated.
2. **Test Phase:** It predicts the values based on rules generated in the training phase.

Some data values are removed with are not suitable for crop yield and its production.

DATA CLUSTERING

The data clustering of data is performed by the k-means clustering algorithm. The only condition before clustering is that data should be first preprocessed. The k value signifies the number of clusters in which data are partitioned. Clusters are represented by their centers. The clusters are represented by their centers. The basic idea is that each sample should be closer to the center of its own cluster. If this is not verified, then the partition is modified, until each sample is closer to the center of the cluster it belongs to. The distance function between samples plays an important role since a sample can migrate from a cluster to another one based on the values provided by the distance function.

DATA CONVERSION

The clustered data was converted into 0's and 1's. The data conversion process is mainly used for association rule mining step. The data variables Area, Tanks, Bore Wells, Open Wells, Production and Yield are converted into vice versa.

Future Prospects

Application of information technology in agriculture field has gained attention recently for the production of good crops and agro social-economic benefits for farmers. This is the most significant application used in developing countries which can change the situation of decision making, the percentage increase in growth rate crop with improved quality and yield. This help in different crop compositions which may help in near future to grow a better-quality crop and different agricultural problems.

CHALLENGES AND ACCOMPLISHMENTS

Since we have already discussed the data processing using data mining techniques like- clustering, classification, machine learning, Support Vector Machines, Regression, Association Rule etc., for the betterment of agricultural development and production. But the major challenge which should be considered here is the data collection and modeling.

Grabbing data from multiple sources and managing them together for the creation of large datasets for data mining is our supreme requirement in analysis and prediction. We can't rely on public datasets available worldwide due to bio-geo-socio-economic uncertainties among various regions. Thus, for overcoming this issue, pre-planning is required before drawing any conclusion.

Agricultural sensors are having equally importance in the primary stage for aggregation of the crop, weather, and environmental behavior data. For instance, predicting the crop management policy for a particular region is determined only after getting some initial information about that place by short- or long-term physical inspections through sensors.

Those inspected data can be better utilized in modeling datasets which further used in data mining to get the predicted results through smart statistical algorithms. How-so-ever surveying results can be proven as a boon for data mining purpose and after digitalization, same could be proven great for initial predictions with moderate accuracy.

Also, the major accomplishments are the enhancement in this accuracy using proper rule-sets and data mining techniques. Additionally, the whole system is economical for local government as less computation power is needed for predicting purpose using the specialized algorithms and thus helps local farmers to grow and improvise their livelihoods. Using these techniques crop policies can be made publicly available for the farmers in their native languages for better understanding.

Another major accomplishment is the similar prototype application could be used for all type of datasets. Regardless of the data type supplied by the models, all should go through similar sort of algorithms.

This will help government or policymakers in designing a single software controlled centrally for all states and regions accepting all data collections i.e., soil, weather, humidity, income etc., through various methods, process them parallelly with similar but customised data mining techniques and relate the result at a single place giving productivity to the information.

CONCLUSION

The attempt has been made to provide an overview of data mining technique in agriculture. The use of Crop yield prediction model in agricultural mining estimates the agricultural production in different weather, soil, environmental conditions, and crop management. These techniques are used in the agricultural field to increase the income of the farmer, reduce the transportability cost as well as to predict the climate change by different of data's collected. This not only forecasts the crop growth but also predicts the better conditions in which the crop production can be increased. Thus it is concluded that the agricultural data mining technique becomes highly active research area in future progress in the field of agriculture.

REFERENCES

Baboo, S. S., & Shereef, I. K. (2010). An efficient weather forecasting system using artificial neural network. *International Journal of Environmental Sciences and Development*, *1*(4), 321–326. doi:10.7763/IJESD.2010.V1.63

Devi, C. J., Reddy, B. S. P., Kumar, K. V., Reddy, B. M., & Nayak, N. R. (2012). ANN approach for weather prediction using back propagation. *International Journal of Engineering Trends and Technology*, *3*(1), 19–23.

Hung, N. Q., Babel, M. S., Weesakul, S., & Tripathi, N. K. (2009). An artificial neural network model for rainfall forecasting in Bangkok, Thailand. *Hydrology and Earth System Sciences*, *13*(8), 1413–1425. doi:10.5194/hess-13-1413-2009

Moallem, P., Razmjooy, N., & Ashourian, M. (2013). Computer vision-based potato defect detection using neural networks and support vector machine. *International Journal of Robotics and Automation*, *28*(2), 137–145. doi:10.2316/Journal.206.2013.2.206-3746

Mousavi, B. S., Soleymani, F., & Razmjooy, N. (2013). Color image segmentation using neuro-fuzzy system in a novel optimized color space. *Neural Computing & Applications*, *23*(5), 1513–1520. doi:10.100700521-012-1102-3

Ramesh, V., & Ramar, K. (2011). Classification of agricultural land soils: A data mining approach. *Agricultural Journal*, *6*(3), 82–86. doi:10.3923/aj.2011.82.86

Razmjooy, N., Mousavi, B. S., & Soleymani, F. (2012). A real-time mathematical computer method for potato inspection using machine vision. *Computers & Mathematics with Applications (Oxford, England)*, *63*(1), 268–279. doi:10.1016/j.camwa.2011.11.019

Razmjooy, N., Mousavi, B. S., & Soleymani, F. (2013). A hybrid neural network Imperialist Competitive Algorithm for skin color segmentation. *Mathematical and Computer Modelling*, *57*(3-4), 848–856. doi:10.1016/j.mcm.2012.09.013

Sawaitul, S. D., Wagh, K. P., & Chatur, P. N. (2012). Classification and prediction of future weather by using back propagation algorithm-an approach. *International Journal of Emerging Technology and Advanced Engineering*, *2*(1), 110–113.

Sharma, A., & Datta, U. (2007, July). A Weather Forecasting System using concept of Soft Computing: A new approach. In AIP Conference Proceedings: Vol. 923. *No. 1* (pp. 275–287). AIP. doi:10.1063/1.2767047

Somvanshi, V. K., Pandey, O. P., Agrawal, P. K., Kalanker, N. V., Prakash, M. R., & Chand, R. (2006). Modeling and prediction of rainfall using artificial neural network and ARIMA techniques. *J. Ind. Geophys. Union*, *10*(2), 141–151.

Chapter 7
Examination of Plant/Weed Image Dataset Using a Hybrid Image Processing Tool

V. Rajinikanth
https://orcid.org/0000-0003-3897-4460
St. Joseph's College of Engineering, India

N. Sri Madhava Raja
St. Joseph's College of Engineering, India

S. Arunmozhi
Manakula Vinayagar Institute of Technology, India

B. Parvatha Varthini
St. Joseph's College of Engineering, India

K. Palani Thanaraj
St. Joseph's College of Engineering, India

ABSTRACT

Image evaluation procedures are widely employed in various domains to extract the useful information to make the necessary decision. This chapter implements a soft-computing tool to examine the benchmark plant/weed (BPW) pictures of computer vision problems in plant phenotyping (CVPPP2014) challenge database. The proposed work implements a hybrid image evaluation procedure based on social-group optimization algorithm (SGOA) and Shannon's entropy (SE)-based multi-thresholding and the Chan-Vese segmentation (CVS)-based extraction procedure. After extracting the crop/weed regions of BPW pictures, the superiority of the proposed tool is then assessed by implementing a comparative study between the extracted plant/weed region and its equivalent ground-truth. The results of this study substantiate that the proposed system is proficient in examining the BPW pictures and in future. This procedure can be considered to inspect the crop/weed pictures obtained with field supervising drones.

DOI: 10.4018/978-1-5225-8027-0.ch007

INTRODUCTION

In the modern era, most of the country's Gross Domestic Product (GDP) rate depends mainly on the agriculture sector. In order to get the better yield in farming, a substantial amount of modern cultivation procedures are already implemented in agricultural industries of economically grown countries (Bakhshipour et al., 2017). Modern techniques can be used to monitor the entire plant stages from the seed germination process to the harvesting process (Aitkenhead et al., 2003; Razmjooy et al., 2011).

Smart and Supervised Cultivation (SSC) procedures are widely implemented in most of the countries when there is a demand for the labors and the demand for agricultural products. In SSC, the entire agriculture process, such as seeding, watering, plant monitoring, and harvesting stage monitoring will be carried out with the help of dedicated monitoring devices. When an appropriate network with all the possible sensor and actuators are implemented in this system, the cultivator can monitor the entire process using a dedicated monitoring system such as the personal computer or a mobile phone and he can also initiate or modify the scheduled task based on the climatic condition, crop's stage, etc (Haug et al., 2014; Herrera et al., 2014).

The main objective of this chapter is to develop an image processing procedure to extract the plant/weed section from the RGB scale pictures recorded using a dedicated digital camera. In most of the plant monitoring cases, a remote digital camera based on a programmed drone or an autonomous robot can be employed to collect the essential plant images from the field based on a chosen resolution. The captured image is simultaneously sent to the monitoring station, which is responsible to perform an image analysis to identify the growth rate of the plant/weed. After the analysis, the outcome is then sent to the operator, who can take the decision regarding the next action to be implemented on the field to protect the plant from the weed and other intimidation (Herrera et al., 2014).

Due to its significance, a considerable number of SSC procedures are proposed and realized to improve the outcome of the agriculture industry. The image-based analysis of plant/weed section is the heart of the SSC process and the ultimate outcome relies mainly on the decision made during the image examination. Hence, in recent years, a number of procedures are proposed by the researchers to examine the plant images recorded with various techniques (Razmjooy et al., 2012; Moallem et al., 2013; Moallem et al., 2014).

The work of this chapter proposes a hybrid image processing technique to extract and examine the plant/weed section from a chosen RGB scale image. The work adopts the Benchmark Crop-Weed (BCW) database supplied by Haug and Ostermann (CVPPP2014) for the investigation (2014). The detailed methodology implemented during the image collection and the associated earlier research works on the considered images can be found in (Haug, S. & Ostermann, 2016). The work of the proposed chapter implements a technique by integrating the pre-processing and post-processing technique to examine the plant/weed region of the CVPPP2014 database. Initially, a multi-level thresholding with Social-Group-Optimization-Algorithm (SGOA) and Shannon's Entropy (SE) is employed to separate the plant/ weed from the background and later the segmentation based on the Chan-Vese is implemented to extract the plant/weed section. Finally, the advantage of the proposed image processing technique is confirmed with relative examination between the extracted region and the ground-truth available in the database.

In this chapter, a detailed discussion on the multi-level thresholding with various chosen thresholds (Th= 2-5) is initially implemented to chose the appropriate threshold level to separate the plant/weed section from the picture's background. A detailed analysis based on the well-known picture quality measures are implemented to chose the threshold level for the proposed scheme. Since, the increase threshold level may increase the complexity and the CPU run time of the pre-processing stage. If, an appropriate threshold level is fixed, then it is possible to extract and evaluate the plant/weed regions with minimal computation cost. The initial experimental work is demonstrated on eight RGB scale images of CVPPP2014 with a picture dimension of 1296x966 pixels. This dataset consist around 60 images with the associated ground-truth (GT). The experimental result of this study confirms that, proposed work offers an average image similarity of >90% when tested with the CVPPP2014 database.

BACKGROUND

In Smart and Supervised Cultivation (SSC), the entire process is monitored and controlled using dedicated semi/fully automated systems, in which the image processing task plays an essential role in plant/weed analysis. In SSC literature, a considerable plant/weed examinations based on the image assisting techniques are already proposed and implemented by the researchers. The work of Burgos et al. (2011) employs Otsu's thresholding practice on real-time assessment of crop-weed pictures of maize fields. Tellaeche et al. (2011) implemented Support-Vector-Machine assisted automated technique to recognize the weed part with the aid of RGB image segmentation. Montalvo et al. (2013) discussed Otsu's technique for the weed/crop identification in maize fields. Aitkenhead et al. (2003) employed a Neural-Network technique to assess the images collected from the cultivation field. The recent work of Herrera et al. (2014) converse the shape descriptor and fuzzy based weed categorization practice and this technique offered considerable accuracy in the classification task.

The proposed chapter adopts the Benchmark Crop-Weed (BCW) database supplied by Haug and Ostermann (2014) for the assessment. The chosen images of the CVPPP2014 dataset is then examined using the hybrid image processing approach recently discussed for the medical image examination task (Dey et al., 2018; Rajinikanth et al., 2017). The image processing scheme consists of an initial pre-processing stage to enhance the plat/weed section and the post-processing scheme extracts the required section from the pre-processed RGB scale picture. Finally, a relative assessment with the GT is executed and the superiority is then confirmed by computing the image similarity parameters.

MAIN FOCUS OF THE CHAPTER

The major focus of the proposed chapter is to implement a suitable image processing technique to extract the plant/weed section from the RGB test pictures CVPPP2014 dataset with better accuracy. The recent research of Rajinikanth et al. (2017, 2018) reported that a hybrid image examination technique will help to accomplish improved result compared to the traditional approaches. Their work also achieved better with the existing medical image datasets. Hence, in the proposed study, a technique based on the combination of Shannon's-Entropy (SE) multi-thresholding and the Chan-Vese region segmentation is executed to inspect the RGB scale plant/weed pictures.

Overview of Methodology

The overview of the proposed technique is clearly depicted in Figure 1.

Figure 1 depicts the configuration to be considered to implement the SSC system. Normally, the required field images are collected from the field using a pre-programmed autonomous imaging system. After capturing the image from the required region, it is then transferred to the monitoring hub. The main procedures, such as image processing, classification, and the decision making is done in the monitoring hub and the final decision is implemented in the agriculture field in order to maintain a closed loop monitoring and control to achieve a better yield. The work considered in this chapter is limited only with the image processing, in which the raw image obtained from the field is processed and the essential plant/weed section is extracted and examined. After extracting the required section from the field picture, a texture based evaluation and classification can be employed to know the healthiness and the growth rate of the plant compared to the weed. The classifier system can be trained and tested with the outcome obtained from the image processing section. Based on the outcome of the classifier section, the operator or

Figure 1. Overview of proposed plant/weed image assessment system

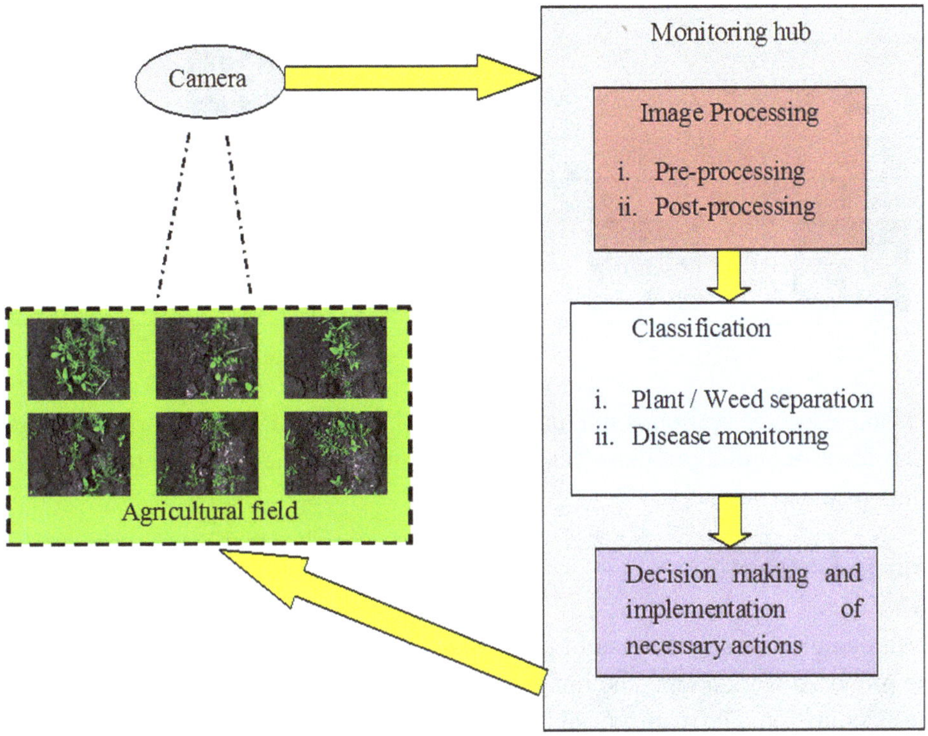

the monitoring computer may recommend some operations to be implemented in the field to improve the plant condition. From Figure 1, it can be noted that the reliability of SSC depends mainly on the image processing section. If a poor image examination section is implemented, it will affect the subsequent sections, such as the classification and decision making. Hence, in the proposed work, a hybrid image examination technique based on the combination of the pre- and post-processing is implemented and its outcome is validated against the GT offered by an image expert.

Figure 2 shows the pictorial representation of the image processing scheme implemented in the proposed chapter. Initially, the test image to be examined is collected from the CVPPP2014 and is then pre-processed based on the HSOA+SE multi-thresholding. During the thresholding, the test-image is enhanced based on the chosen threshold and in this work, in order to identify the most possible threshold level, the appropriate threshold value (Th= 2-5) is selected as discussed in the previous image processing works (Satapathy et al., 2018; Monisha et al., 2019; Razmjooy et al., 2011). After the enhancement, the plant/weed section is then mined using the Chan-Vese segmentation procedure and finally, a relative assessment between the

Figure 2. Outline of the proposed image processing scheme

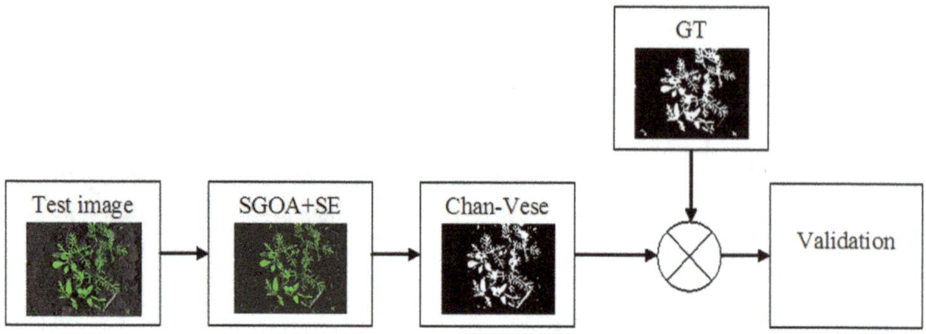

GT and extracted section is implemented and the image similarity parameters are computed. According to these values, the advantage of the proposed technique is confirmed and validated.

Image Dataset

Even though a notable number of plant/weed images are available in the literature, the proposed work adopts the images offered by Haug and Ostermann (2014) for the examination. The main advantage of the test images existing in the CVPPP2014 dataset is, it consists of high-resolution RGB images of the size of 1296x966 pixels. Every image is carefully recorded in a controlled environment and every image of this dataset is associated with the expert's GT and annotations. Further, every image has its own binary mask, which may help the researchers to identify the shape of the section to be extracted during the experimental investigation.

Social Group Optimization Algorithm

Generally, evolutionary optimization algorithms is to provide a solution within a reasonable time frame that is appropriate to solve the problem, the exploratory algorithm may not be the best solution to solve the problem, but can be the closest to the solution (Razmjooy et al., 2014; 2016;2018).

SGOA was proposed by Satapathy and Naik (2016) to find the better solution for the numerical optimization-problem. This involves the mimic of the group action and information allocation existing among humans. Traditional SGOA includes two phases, such as the civilizing phase to synchronize the locality of people based on the cost function and gaining phase permits the people to realize the optimal solution for the selected task.

The mimic of SGO is presented below:

$$Globalbest_j = max \{ f(X_u) \; for \; u = 1, 2, \ldots W \} \tag{1}$$

where X_u=initial information of people in the cluster, $u = 1, 2, \ldots, W$ denotes the entire people in a cluster, and f_v=fitness cost.

The gaining phase is considered to modernize the location of every human as in Equation (2):

$$X_{newu,v} = \Im * X_{oldu,v} + R * (Globalbest_v - X_{oldu,v}) \tag{2}$$

where X_{new}=reorganized location, X_{old}=premature location, *Globalbest*=best position, R=random numeral [0,1] and $\Im$ =self-introspection constraint [0,1]. The $\Im$ is chosen as 0.2 based on Satapathy and Naik (2016).

In acquiring stage, people will decide the collective resolution according to the information adjusting practice as shown in Eqn (3).

$$Xnew_{u,v} = Xold_{u,v} + R_a * (X_{u,v} - X_{r,v}) + R_b * (Gbest_v - X_{u,v}) \tag{3}$$

where R_a & R_b=random values [0,1] and $X_{r,v}$ =randomly selected position of an individual of group. Other details regarding the SGOA can be found in the work of Naik et al. (2016). The recent work of Monisha et al. (2018) and Rajinikanth and Satapathy (2018) also describe the outline of the SGOA. The research article by Dey et al. (2018) describes the working of the SGOA with an appropriate graphical representation.

Shannon's Entropy

A detailed description of the SE can be found in the article by Kannappan (1972). The recent work implemented on the medical and the general image processing schemes are also confirmed the superiority of the SE compared to other entropies (Paul et al., 2014; Rajinikanth et al., 2017; 2019; Raja et al. 2019).

In the SE, let us chose a test-picture of size A*B. The pixel association in test-picture (h,v) is defined as G(h,v), for $h \in \{1, 2, \ldots, A\}$ & $v \in \{1, 2, \ldots, B\}$. Let L is number of gray-levels of the test-picture and the set of every gray values {0,1,2,…, L-1} can be denoted as O, in such a way that:

$$G(h,v) \in O \;\; \forall (h,v) \in picture \tag{4}$$

Then, the normalized histogram will be;

$$Z = \{Th_1, Th_2, \ldots, Th_{L-1}\}. \tag{5}$$

For a bi-level thresholding case, Equation (5) becomes;

$$Z(Th) = x_0(Th_1) + x_1(Th_2) \tag{6}$$

$$F(Th) = \max_{Th} \{Z(Th)\} \tag{7}$$

where, Th = $\{Th_1, Th_2, \ldots, Th_L\}$ is the threshold value, Z $=\{x_0, x_1, \ldots, x_{L-1}\}$ is the normalized histogram, and $F(Th)$ is the optimal threshold. For an RGB image case, the above technique is separately implemented for the R, G, and B threshold cases.

Chan-Vese Segmentation

The Chan-Vese segmentation technique is a proven segmentation procedure initially proposed by Chan and Vese (1990). It is a modified form of the active-contour technique without any edge constraints. The Chan-Vese technique considered in the proposed work is initialized with a bounding box, which surrounds the plant/weed section. The X and Y coordinates are initially set by the operator, hence it comes under the category of the semi-automated segmentation technique. The recent works implemented in the medical image analysis (Rajinikanth et al., 2017; Manic et al., 2019) confirms the superiority of this segmentation procedure compared to other existing techniques. A detailed description and its mathematical expression can be found in the articles (Chann et al., 2000; Chan and Vese, 2001; Getreuer, 2012).

Assessment

The objective of this chapter is to develop a hybrid image processing practice to extract the plant/weed section from the RGB scale test images for further examinations, such as classification and decision making.

Initially, the performance of the pre-processing scheme is assessed based on a comparative study between the test-picture and thresholded picture. During this analysis, the image quality measures, like Normalised-Absolute-Error (NAE), Root-Mean-Square-Error (RMSE), Peak Signal to Noise Ratio (PSNR) and Normalised-Cross-Correlation (NCC) and Structural-Similarity-Index-Matrix (SSIM) are computed and tabulated. These values are then analyzed to fix an appropriate threshold level (Th) for the pre-processing technique. The expressions for this parameter can be found in Satapathy et al. (2018).

Further, an assessment among segmented plant/weed and GT is also implemented to find the superiority of the proposed post-processing scheme. Essential picture similarity values, such as precision, recall, sensitivity, F-measure, specificity, Balanced-Classification-Rate (BCR), Balanced-Error-Rate (BER) and accuracy are noted. The experimental investigation also confirms that recall and the sensitivity are similar. A detailed description of these parameters can be found in Grgic et al. (2004) and Lu et al. (2004). Further, recent articles, such as Manic et al. (2018) and Rajinikanth et al. (2018; 2019) also present the mathematical description of these parameters.

SOLUTIONS AND RECOMMENDATIONS

The proposed work of this chapter aims to extract the plant/weed section of RGB scale pictures obtained from the CVPPP2014 database. In this work, a hybrid image evaluation procedure is implemented based on the SGOA+SE and Chan-Vese segmentation and its superiority are confirmed using the image quality and similarity parameters.

Initially, the algorithm parameters are assigned as follows, number of people in SGOA=30, the dimension of the search =Th, number of iteration in pre-processing = 1500, no. of iterations in post-processing = 1500, and stopping criteria = maximized SE.

Figure 3 depicts the sample test images adopted from the CVPPP2014. In this Figure 3(a) represents the pseudo name, Figure 3(b) denotes the RGB scaled sample test-pictures adopted for the demonstration and Figures 3(c) and (d) depicts the binary mask and the GT respectively. In this work, initially the RGB scaled GT is converted into a grayscale picture and then it is converted as the binary image (black=0 and white=1). The converted binary GT is adopted to find the picture similarity parameters.

Initially, the test image with pseudo-name 001 is considered for the examination and its pre-processed result with Th = 2,3,4 & 5 is depicted in Figure 4. The visual examination of this figure confirms that, the outcome when Th=2 shows the perfect separation of the plant/weed with respect its background. The previous research work of Satapathy et al. (2018) points that, the computer's CPU runtime mainly depends on the 'Th' and the pre-processing time will rapidly increase with an increase in 'Th'. Hence, it is necessary to implement a pre-processing technique, which will offer the required outcome with lesser CPU time. In Figure 4, the images Figures

Figure 3. Outline of the proposed image processing scheme

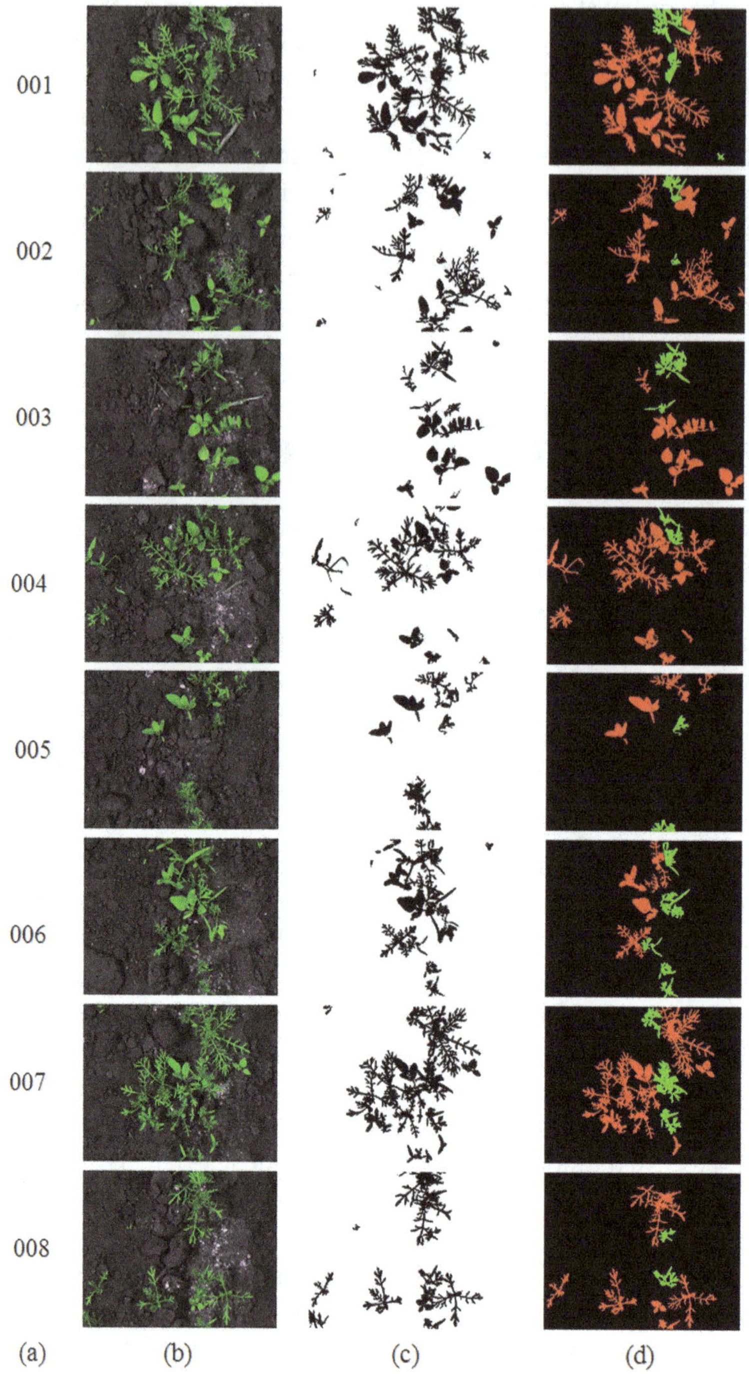

Figure 4. Result obtained with SGOA+SE thresholding

(a) (b) (c) (d)

4(a) to 4(d) shows the result when the multi-level thresholding is implemented with SGOA+SE with a chosen value of Th = 2,3,4, and 5.

Table 1 presents the computed image quality measures for various threshold values. The visual examination, as well as this table value, confirms that Th=2 will help to achieve a satisfactory result compared to other values. Hence, in the proposed scheme, Th=2 is fixed for other images available in the chosen database.

After the possible enhancement of the test-picture with the SGOA+SE thresholding, the Chan-Vese segmentation is then implemented with a carefully assigned bounding box. The execution of the post-processing and its result is clearly depicted in Figure 5.

In this procedure, initially, the Chan-Vese is performed with a bounding box, which will converge towards the image section when the iteration value increases. When all the similar pixel group is identified/covered the curve of the Chan-Vese, the execution stops and displays the binary form of the extracted image region. In this image, the execution stops when the iteration value is 521. When the Chan-Vese search stops, it will present the extracted plant/weed section for the further evaluation. A similar procedure is implemented for all the images of Figure 4 and its extracted binary image is depicted in Figure 6.

To assess the superiority of the proposed scheme, a comparative study among the mined section and the GT is executed as in Figure 7 and the corresponding outcomes are recorded for further examination.

Table 1. Image quality measures obtained for test–picture 001

Thresh	NAE	RMSE	PSNR (dB)	NCC	SSIM
2	0.1554	14.5327	24.8839	0.9717	0.6144
3	0.0953	8.9252	29.1184	0.9853	0.8065
4	0.0892	8.3422	29.7051	0.9880	0.8193
5	0.0879	7.9984	30.0708	0.9907	0.8237

Figure 5. Execution of Chan-Vese segmentation to extract the plant/weed section

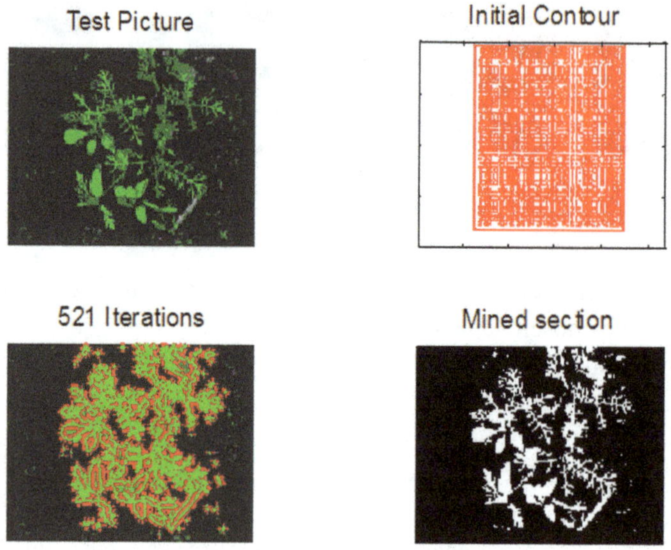

Figure 6. Mined plant/weed region with Chan-Vese technique

(a) (b) (c) (d)

Figure 7. Comparison of extracted plant/weed part with GT

The GT and mined section comparison are executed with Figure 6 images for all the threshold levels and the results are tabulated for further assessment. From Table 2, it can be noted that the similarity parameters attained for all the image cases with Th=2 is better compared to other cases. Hence, in the proposed chapter,

Table 2. Image similarity parameters obtained for test–picture 001

Thresh	Precision	Recall (Sensitivity)	F-measure	Specificity	BCR	BER	Accuracy
2	0.9565	0.9781	0.9672	0.7938	0.8860	0.1140	0.8812
3	0.9503	0.9731	0.9616	0.7642	0.8687	0.1313	0.8623
4	0.9327	0.9731	0.9525	0.6749	0.8240	0.1760	0.8104
5	0.9315	0.9712	0.9510	0.6690	0.8201	0.1798	0.8061

for every test-picture, the threshold value is assigned as 2 (ie. Th=2) dung the pre-processing task and it gives a better result with lesser CPU runtime compared to Th=3, Th=4, and Th=5.

This technique is implemented in other test image cases and the corresponding outcomes are presented in Figure 8. In this, Figure 8(a) and Figure 8(b) represents the pseudo-name and sample test-pictures respectively. Further, Figures 8(c) and (d) depicts the GT and mined sections. From this, it can be noted that the extracted plant/weed section is approximately similar in most of the cases. In order to confirm this, the similarity measures for the pictures of the CVPPP2014 dataset is computed and assessed.

Table 3 presents the computed values of the test-pictures. Where the Average1 is the mean value computed for the images with pseudo-name 001- 008 and the Average2 is the mean value obtained for the CVPPP2014 database (60 images). This table confirms that, except the BER, other performance measures are closer

Table 3. Image performance values achieved for the chosen test images

Image	Precision	Recall (Sensitivity)	F-measure	Specificity	BCR	BER	Accuracy
001	0.9565	0.9781	0.9672	0.7938	0.8860	0.1140	0.8812
002	1.0000	0.9993	0.9996	1.0000	0.9996	0.0360	0.9996
003	0.9985	1.0000	0.9992	0.9835	0.9917	0.8254	0.9917
004	1.0000	0.9969	0.9984	1.0000	0.9984	0.1548	0.9984
005	1.0000	0.9972	0.9986	1.0000	0.9986	0.1422	0.9986
006	0.9631	0.9934	0.9781	0.6612	0.8273	0.1727	0.8105
007	0.9622	0.9735	0.9678	0.8034	0.8885	0.1115	0.8844
008	0.9884	1.0000	0.9942	0.8903	0.9452	0.5482	0.9436
Average1	0.9836	0.9923	0.9879	0.8915	0.9419	0.2631	0.9385
Average2	0.9818	0.9917	0.9864	0.9022	0.9550	0.1819	0.9573

Figure 8. Extracted plant/weed pictures for the sample test-pictures

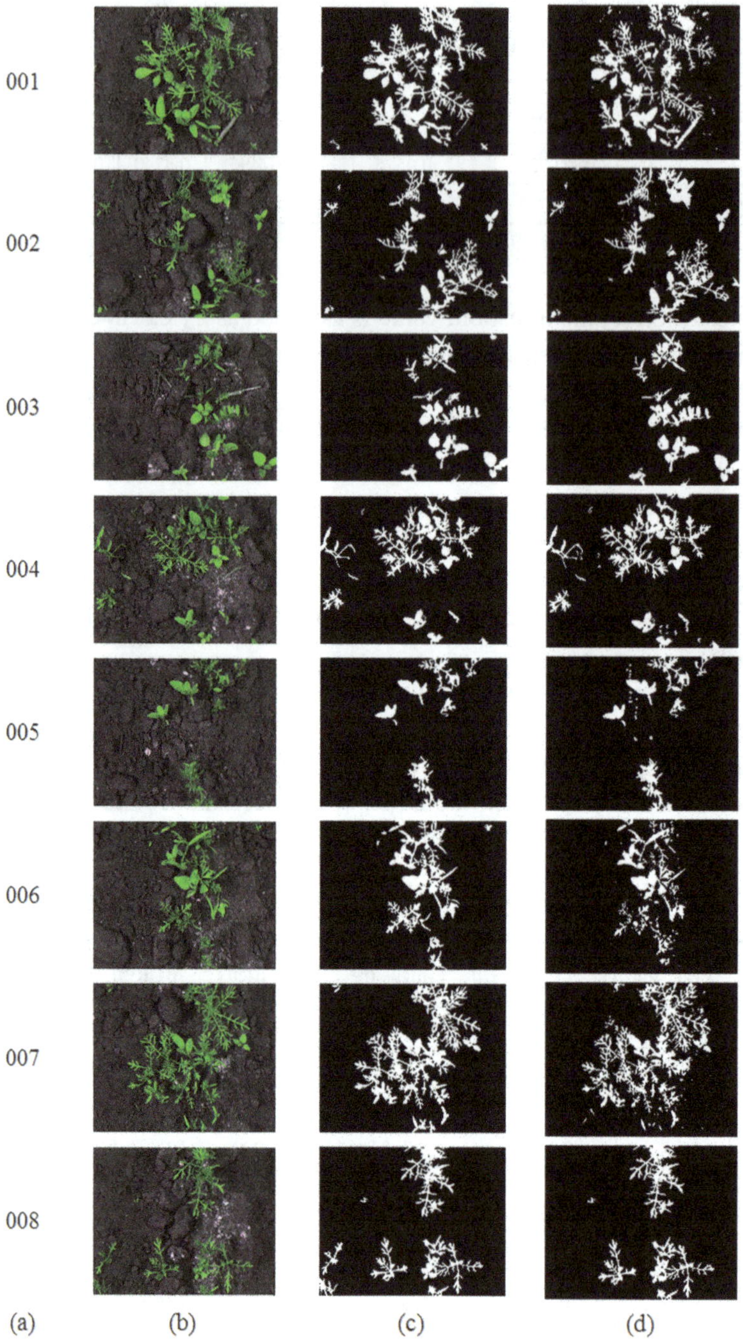

to unity (100%), which authenticates the superiority of the proposed technique in the agriculture image analysis.

Figure 9 presents the graphical representation of Average2 value and this image also confirms that the existing performance values are >90% in all the considered cased. This result confirms that the proposed technique has practical significance and in future, it can be considered to examine the real-time plant/weed pictures collected from a real cultivation field.

FUTURE RESEARCH DIRECTIONS

The automated cultivation process is always needed to improve the production rate with more profit. But, the initial cost required is moderate compared to the labor cost in agriculture. If an automated monitoring and control system is employed, it is possible to monitor the agriculture fields from a remote location. In the practical SSC systems, a substantial amount of transitional phases are available from the field level to the monitoring station. In smart agriculture system, different semi/automated assessment actions are executed to discover a range of cultivation field requirement, ranging from the germination to harvest. If the weed region and plant disease harshness are recognized, then it is possible for the operator to plan for an added treatment practice. The chief aim of the work discussed in this chapter is to develop an imaging procedure to examine the plant/weed section. In this work, an approach based on the SGOA assisted SE and Chan-Vese is proposed to examine the RGB scaled pictures. This work focuses on extracting and analyzing the chosen section of the CVPPP2014 database.

Figure 9. Evaluation of average similarity measures obtained for CVPPP2014

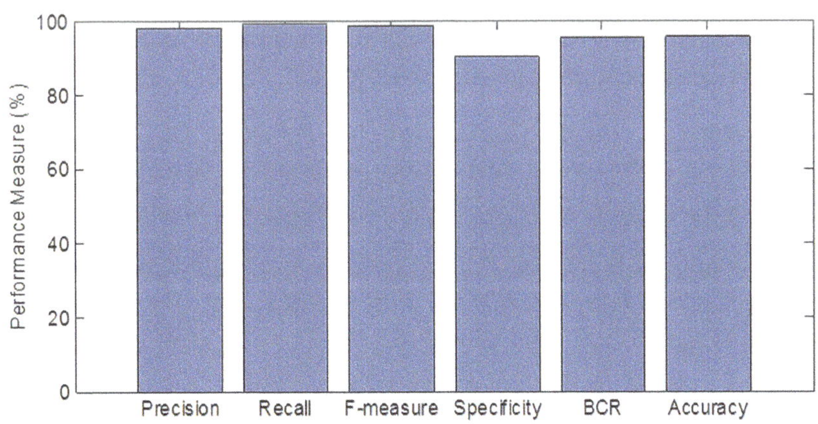

In future, the proposed work can be directed as follows: i) Extraction of the texture features for the plant/weed region based on the Gray Level Co-occurrence Matrix parameters. Execution of recent heuristic procedures to enhance the pre-processing section, iv) Extracting the features of the plant/weed region to implement the classifier units for the automated detection and diagnose in plant abnormalities.

CONCLUSION

This work executes an assessment scheme by integrating Shannon's function with Chan-Vese extraction practice to enhance the performance of plant/weed abnormality inspection. This work considers a CVPPP2014 database. This approach is tested using the benchmark datasets and the essential sections, such as plant/weed are mined are examined. The considered datasets are initially enhanced using multi-level thresholding with SGOA + SE. Later, the essential elements are then extracted using the Chan-Vese. The outcome of the Chan-Vese approach is validated against the GT picture. The results confirm that the proposed procedure help to achieve better similarity values. In future, a classifier system can be employed to classify the images according to the healthiness of the plant section.

REFERENCES

Aitkenhead, M. J., Dalgetty, I. A., Mullins, C. E., McDonald, A. J. S., & Strachan, N. J. C. (2003). Weed and crop discrimination using image analysis and artificial intelligence methods. *Computers and Electronics in Agriculture*, *39*(3), 157–171. doi:10.1016/S0168-1699(03)00076-0

Bakhshipour, A., Jafari, A., Nassiri, S. M., & Zare, D. (2017). Weed segmentation using texture features extracted from wavelet sub-images. *Biosystems Engineering*, *157*, 1–12. doi:10.1016/j.biosystemseng.2017.02.002

Burgos-Artizzu, X. P., Ribeiro, A., Guijarro, M., & Pajares, G. (2011). Real-time image processing for crop/weed discrimination in maize fields. *Computers and Electronics in Agriculture*, *75*(2), 337–346. doi:10.1016/j.compag.2010.12.011

Chan, T. F., Sandberg, B., & Vese, L. A. (2000). Active contours without edges for vector-valued images. *Journal of Visual Communication and Image Representation*, *11*(2), 130–141. doi:10.1006/jvci.1999.0442

Chan, T. F., & Vese, L. A. (1990). An active contour model without edges. *Lecture Notes in Computer Science*, *1682*, 141–151. doi:10.1007/3-540-48236-9_13

Chan, T. F., & Vese, L. A. (2001). Active contours without edges. *IEEE Transactions on Image Processing*, *10*(1), 266–277. doi:10.1109/83.902291 PMID:18249617

Dey, N., Rajinikanth, V., Amira, S., Ashour, A. S., & Tavares, J. M. R. S. (2018). Social Group Optimization Supported Segmentation and Evaluation of Skin Melanoma Images. *Symmetry*, *10*(2), 51. doi:10.3390ym10020051

Getreuer, P. (2012). Chan–Vese Segmentation. *Image Processing On Line*, *2*, 214–224. doi:10.5201/ipol.2012.g-cv

Grgic, S., Grgic, M., & Mrak, M. (2004). Reliability of objective picture quality measures. *Journal of Electrical Engineering*, *55*(1-2), 3–10.

Haug, S., Michaels, A., Biber, P., & Ostermann, J. (2014). Plant classification system for crop/weed discrimination without segmentation. *IEEE Winter Conference on Applications of Computer Vision (WACV)*, 1142-1149. 10.1109/WACV.2014.6835733

Haug, S., & Ostermann, J. (2014). A Crop/Weed field image dataset for the evaluation of computer vision based precision agriculture tasks. *Lecture Notes in Computer Science*, *8928*, 105–116. doi:10.1007/978-3-319-16220-1_8

Haug, S., & Ostermann, J. (2016). Plant classification for field robots: A machine vision approach. *Computer Vision and Pattern Recognition in Environmental Informatics*, *248*, 248–272. doi:10.4018/978-1-4666-9435-4.ch012

Herrera, P. J., Dorado, J., & Ribeiro, A. (2014). A novel approach for weed type classification based on shape descriptors and a fuzzy decision-making method. *Sensors (Basel)*, *14*(8), 15304–15324. doi:10.3390140815304 PMID:25195854

Kannappan, P. L. (1972). On Shannon's entropy, directed divergence and inaccuracy. *Probability Theory and Related Fields*, *22*(2), 95–100. doi:10.1007/BF00532728

Kowsalya, N., Kalyani, A., Chalcedony, C. J., Sivakumar, R., Janani, M., & Rajinikanth, V. (2018). An approach to extract optic-disc from retinal image using k-means clustering. *Fourth International Conference on Biosignals, Images and Instrumentation (ICBSII)*, 206-212. 10.1109/ICBSII.2018.8524655

Lu, H., Kot, A. C., & Shi, Y. Q. (2004). Distance-reciprocal distortion measure for binary document images. *IEEE Signal Processing Letters*, *11*(2), 228–231. doi:10.1109/LSP.2003.821748

Manic, K. S., Hasoon, F. N., Shibli, N. A., Satapathy, S. C., & Rajinikanth, V. (2019). An approach to examine brain tumor based on Kapur's entropy and Chan–Vese algorithm. *Advances in Intelligent Systems and Computing*, *797*, 901–909. doi:10.1007/978-981-13-1165-9_81

Manic, K.S., Naimi, I.S.A., Hasoon, F.N. & Rajinikanth, V. (2018). Jaya algorithm-assisted evaluation of tooth elements using digital bitewing radiography images. *Computational Techniques for Dental Image Analysis*, 107-128. doi:10.4018/978-1-5225-6243-6.ch005

Monisha, R., Mrinalini, R., Britto, M. N., Ramakrishnan, R., & Rajinikanth, V. (2019). Social group optimization and Shannon's function-based RGB image multi-level thresholding. *Smart Innovation. Systems and Technologies*, *105*, 123–132. doi:10.1007/978-981-13-1927-3_13

Montalvo, M., Guerrero, J. M., Romeo, J., Emmi, L., Guijarro, M., & Pajares, G. (2013). Automatic expert system for weeds/crops identification in images from maize fields. *Expert Systems with Applications*, *40*(1), 75–82. doi:10.1016/j.eswa.2012.07.034

Naik, A., Satapathy, S. C., Ashour, A. S., & Dey, N. (2016). Social group optimization for global optimization of multimodal functions and data clustering problems. *Neural Computing & Applications*. doi:10.100700521-016-2686-9

Paul, S., & Bandyopadhyay, B. (2014). A novel approach for image compression based on multi-level image thresholding using Shannon entropy and differential evolution. *Students' Technology Symposium (TechSym)*, 56-61. doi: 10.1109/TechSym.2014.6807914

Raja, N.S.M., Arunmozhi, S., Lin, H., Dey, N. & Rajinikanth, V. (2019). A study on segmentation of leukocyte image with Shannon's entropy. *Histopathological Image Analysis in Medical Decision Making*, 1-27. doi:10.4018/978-1-5225-6316-7.ch001

Rajinikanth, V., Raja, N. S. M., Satapathy, S. C., & Fernandes, S. L. (2017). Otsu's multi-thresholding and active contour snake model to segment dermoscopy images. *Journal of Medical Imaging and Health Informatics*, *7*(8), 1837–1840. doi:10.1166/jmihi.2017.2265

Rajinikanth, V., & Satapathy, S. C. (2018). Segmentation of ischemic stroke lesion in brain MRI based on social group optimization and fuzzy-Tsallis entropy. *Arabian Journal for Science and Engineering*, *43*(8), 4365–4378. doi:10.100713369-017-3053-6

Rajinikanth, V., Satapathy, S. C., Dey, N., Fernandes, S. L., & Manic, K. S. (2019). Skin melanoma assessment using Kapur's entropy and level set—A study with bat algorithm. *Smart Innovation. Systems and Technologies*, *104*, 193–202. doi:10.1007/978-981-13-1921-1_19

Rajinikanth, V., Satapathy, S. C., Dey, N., & Vijayarajan, R. (2018). DWT-PCA image fusion technique to improve segmentation accuracy in brain tumor analysis. *Lecture Notes in Electrical Engineering*, *471*, 453–462. doi:10.1007/978-981-10-7329-8_46

Rajinikanth, V., Satapathy, S. C., Fernandes, S. L., & Nachiappan, S. (2017). Entropy based segmentation of tumor from brain MR images–A study with teaching learning based optimization. *Pattern Recognition Letters*, *94*, 87–94. doi:10.1016/j.patrec.2017.05.028

Rajinikanth, V., Thanaraj, K. P., Satapathy, S. C., Fernandes, S. L., & Dey, N. (2019). Shannon's entropy and watershed algorithm based technique to inspect ischemic stroke wound. *Smart Innovation. Systems and Technologies*, *105*, 23–31. doi:10.1007/978-981-13-1927-3_3

Satapathy, S., & Naik, A. (2016). Social group optimization (SGO): A new population evolutionary optimization technique. *Complex & Intelligent Systems*, *2*(3), 173–203. doi:10.100740747-016-0022-8

Satapathy, S. C., Raja, N. S. M., Rajinikanth, V., Ashour, A. S., & Dey, N. (2018). Multi-level image thresholding using Otsu and chaotic bat algorithm. *Neural Computing & Applications*, *29*(12), 1285–1307. doi:10.100700521-016-2645-5

Tellaeche, A., Pajares, G., Burgos-Artizzu, X. P., & Ribeiro, A. (2011). A computer vision approach for weeds identification through Support Vector Machines. *Applied Soft Computing*, *11*(1), 908–915. doi:10.1016/j.asoc.2010.01.011

ADDITIONAL READING

Ahmed, S., Dey, N., Ashour, A. S., Sifaki-Pistolla, D., Bălas-Timar, D., Balas, V. E., & Tavares, J. M. R. S. (2017). Effect of fuzzy partitioning in Crohn's disease classification: A neuro-fuzzy-based approach. *Medical & Biological Engineering & Computing*, *55*(1), 101–115. doi:10.100711517-016-1508-7 PMID:27106754

Anitha, P., Bindhiya, S., Abinaya, A., Satapathy, S. C., Dey, N., & Rajinikanth, V. (2017). RGB image multi-thresholding based on Kapur's entropy—A study with heuristic algorithms. *In. Second International Conference on Electrical, Computer and Communication Technologies (ICECCT)*. Doi: 10.1109/ICECCT.2017.8117823

Ashour, A. S., Samanta, S., Dey, N., Kausar, N., Abdessalemkaraa, W. B., & Hassanien, A. E. (2015). Computed tomography image enhancement using cuckoo search: A log transform based approach. *Journal of Signal and Information Processing*, *6*(3), 244–257. doi:10.4236/jsip.2015.63023

Bakiya, A., Kamalanand, K., & Rajinikanth, V. (2018). Assessment of electromyograms using genetic algorithm and artificial neural networks. *Evolutionary Intelligence*, 1–11. doi:10.100712065-018-0174-0

Bakiya, A., Kamalanand, K., Rajinikanth, V., Nayak, R. S., & Kadry, S. (2018). Deep neural network assisted diagnosis of time-frequency transformed electromyograms. *Multimedia Tools and Applications*, 1–17. doi:10.100711042-018-6561-9

Balan, N. S., Kumar, A. S., Raja, N. S. M., & Rajinikanth, V. (2016). Optimal multilevel image thresholding to improve the visibility of plasmodium sp. in blood smear images. *Advances in Intelligent Systems and Computing*, *397*, 563–571. doi:10.1007/978-81-322-2671-0_54

Borra, S., Thanki, R., Dey, N., & Borisagar, K. (2018). *Secure transmission and integrity verification of color radiological images using fast discrete curvelet transform and compressive sensing*. Smart Health. doi:10.1016/j.smhl.2018.02.001

Chakraborty, S., Chatterjee, S., Dey, N., Ashour, A. S., Ashour, A. S., Shi, F., & Mali, K. (2017). Modified cuckoo search algorithm in microscopic image segmentation of hippocampus. *Microscopy Research and Technique*, *80*(10), 1051–1072. doi:10.1002/jemt.22900 PMID:28557041

Chakraborty, S., Dey, N., Samanta, S., Ashour, A. S., Barna, C., & Balas, M. M. (2017). Optimization of non-rigid demons registration using cuckoo search algorithm. *Cognitive Computation*, *9*(6), 817–826. doi:10.100712559-017-9508-y

Chatterjee, S., Dey, N., Ashour, A. S., & Drugarin, C. V. A. (2017). Electrical energy output prediction using cuckoo search based artificial neural network. *Lecture Notes in Networks and Systems, 18*, 277–285. doi:10.1007/978-981-10-6916-1_26

Dey, N., Ashour, A. S., Shi, F., Fong, S. J., & Sherratt, R. S. (2017). Developing residential wireless sensor networks for ECG healthcare monitoring. *IEEE Transactions on Consumer Electronics, 63*(4), 442–449. doi:10.1109/TCE.2017.015063

Dey, N., Ashour, A. S., Shi, F., Fong, S. J., & Tavares, J. M. R. S. (2018). Medical cyber-physical systems: A survey. *Journal of Medical Systems, 42*(4), 74. doi:10.100710916-018-0921-x PMID:29525900

Dey, N., Ashour, A. S., Shi, F., Fong, S. J., & Tavares, J. M. R. S. (2018). Medical cyber-physical systems: A survey. *Journal of Medical Systems, 42*(4), 74. doi:10.100710916-018-0921-x PMID:29525900

Goswami, S., Roy, P., Dey, N., & Chakraborty, S. (2016). Wireless body area networks combined with mobile cloud computing in healthcare: A survey. Class. *Clust. Biomed. Sign. Proc., 7*, 388.

Hore, S., Chakroborty, S., Ashour, A. S., Dey, N., Ashour, A. S., Sifaki-Pistolla, D., ... Chaudhuri, S. R. (2015). Finding contours of hippocampus brain cell using microscopic image analysis. *Journal of Advanced Microscopy Research, 10*(2), 93–103. doi:10.1166/jamr.2015.1245

Lakshmi, V.S., Tebby, S.G., Shriranjani, D. & Rajinikanth, V. (2016). Chaotic cuckoo search and Kapur/Tsallis approach in segmentation of T.cruzi from blood smear images. *Int. J. Comp. Sci. Infor. Sec. (IJCSIS)*, 14 (CIC 2016), 51-56.

Li, Z., Dey, N., Ashour, A. S., & Tang, Q. (2017). Discrete cuckoo search algorithms for two-sided robotic assembly line balancing problem. *Neural Computing & Applications*, 1–12. doi:10.100700521-017-2855-5

Lin, H., & Rajinikanth, V. (2018). Normality evaluation of EEG signals based on amplitude level and entropy values. *International Journal of Signal Processing Systems, 6*(3), 22–26. doi:10.18178/ijsps.6.3.22-26

Liu, T., Xu, H., Jin, W., Liu, Z., Zhao, Y. & Tian, W. (2014). Medical image segmentation based on a hybrid region-based active contour mode. *Computational and Mathematical Methods in Medicin,* 2014, Article ID 890725, 10 pages. doi:. doi:10.1155/2014/890725

Mani, M. S., Manisha, S., Thanaraj, K. P., & Rajinikanth, V. (2018). Automated segmentation of Giemsa stained microscopic images based on entropy value. *In. International Conference on Intelligent Computing, Instrumentation and Control Technologies (ICICICT)*, IEEE, 1124 - 1128. Doi: 10.1109/ICICICT1.2017.8342727

Manickavasagam, K., Sutha, S., & Kamalanand, K. (2014). Development of systems for classification of different plasmodium species in thin blood smear microscopic images. *Journal of Advanced Microscopy Research*, *9*(2), 86–92. doi:10.1166/jamr.2014.1194

Moallem, P., Razmjooy, N., & Ashourian, M. (2013). Computer vision-based potato defect detection using neural networks and support vector machine. *International Journal of Robotics and Automation*, *28*(2), 137–145. doi:10.2316/Journal.206.2013.2.206-3746

Moallem, P., Razmjooy, N., & Mousavi, B. S. (2014). Robust potato color image segmentation using adaptive fuzzy inference system. *Iranian Journal of Fuzzy Systems*, *11*(6), 47–65.

Palani, T., & Parvathavarthini, B. (2017). Multichannel interictal spike activity detection using time–frequency entropy measure. *Australasian Physical & Engineering Sciences in Medicine*, *40*(2), 413–425. doi:10.100713246-017-0550-6 PMID:28409335

Palani, T. K., Parvathavarthini, B., & Chitra, K. (2016). Segmentation of brain regions by integrating meta heuristic multilevel threshold with Markov random field. *Current Medical Imaging Reviews*, *12*(1), 4–12. doi:10.2174/1573394711666150827203434

Raj, S. P. S., Raja, N. S. M., Madhumitha, M. R., & Rajinikanth, V. (2018). Examination of Digital Mammogram Using Otsu's Function and Watershed Segmentation. *In. Fourth International Conference on Biosignals, Images and Instrumentation (ICBSII)*, IEEE, 206-212. Doi: 10.1109/ICBSII.2018.8524794

Raja, N. S. M., Fernandes, S. L., Dey, N., Satapathy, S. C., & Rajinikanth, V. (2018). Contrast enhanced medical MRI evaluation using Tsallis entropy and region growing segmentation. *Journal of Ambient Intelligence and Humanized Computing*, 1–12. doi:10.100712652-018-0854-8

Raja, N. S. M., Rajinikanth, V., Fernandes, S. L., & Satapathy, S. C. (2017). Segmentation of breast thermal images using Kapur's entropy and hidden Markov random field. *Journal of Medical Imaging and Health Informatics*, *7*(8), 1825–1829. doi:10.1166/jmihi.2017.2267

Raja, N.S.M., Rajinikanth, V. & Latha, K. (2014). Otsu based optimal multilevel image thresholding using firefly algorithm. *Modelling and Simulation in Engineering*, Vol. 2014, Article ID 794574, 17 pages. Doi: . doi:10.1155/2014/794574

Rajinikanth, V., & Couceiro, M. S. (2014). Multilevel segmentation of color image using Lévy driven BFO algorithm. *In. Proceedings of the 2014 International Conference on Interdisciplinary Advances in Applied Computing*, ACM. 19 pages. Doi. 10.1145/2660859.2660925

Rajinikanth, V., & Couceiro, M. S. (2015). RGB histogram based color image segmentation using firefly algorithm. *Procedia Computer Science*, *46*, 1449–1457. doi:10.1016/j.procs.2015.02.064

Rajinikanth, V., Dey, N., Satapathy, S. C., & Ashour, A. S. (2018). An approach to examine magnetic resonance angiography based on Tsallis entropy and deformable snake model. *Future Generation Computer Systems*, *85*, 160–172. doi:10.1016/j.future.2018.03.025

Rajinikanth, V., Fernandes, S. L., Bhushan, B., & Sunder, N. R. (2018). Segmentation and analysis of brain tumor using Tsallis entropy and regularised level set. *Lecture Notes in Electrical Engineering*, *434*, 313–321. doi:10.1007/978-981-10-4280-5_33

Rajinikanth, V., Raja, N. S. M., & Kamalanand, K. (2017). Firefly algorithm assisted segmentation of tumor from brain MRI using Tsallis function and Markov random field. *Journal of Control Engineering and Applied Informatics*, *19*(3), 97–106.

Rajinikanth, V., Raja, N. S. M., & Latha, K. (2014). Optimal multilevel image thresholding: An analysis with PSO and BFO algorithms. *Australian Journal of Basic and Applied Sciences*, *8*, 443–454.

Rajinikanth, V., Raja, N. S. M., & Satapathy, S. C. (2016). Robust color image multi-thresholding using between-class variance and cuckoo search algorithm. *Advances in Intelligent Systems and Computing*, *433*, 379–386. doi:10.1007/978-81-322-2755-7_40

Rajinikanth, V., Raja, N. S. M., Satapathy, S. C., Dey, N., & Devadhas, G. G. (2018). Thermogram assisted detection and analysis of ductal carcinoma in situ (DCIS). *In. International Conference on Intelligent Computing, Instrumentation and Control Technologies (ICICICT)*, IEEE, 1641-1646. Doi: 10.1109/ICICICT1.2017.8342817

Rani, J., Kumar, R., Talukdar, F. A., & Dey, N. (2017). The Brain tumor segmentation using Fuzzy C-Means technique: A study, Recent Advances in Applied Thermal Imaging for Industrial Applications, IGI global, 40-61.

Razmjooy, M. (2012). A multi layer perceptron neural network trained by invasive weed optimization for potato color image segmentation. *Trends in Applied Sciences Research*, *7*(6), 445–455. doi:10.3923/tasr.2012.445.455

Razmjooy, N., Khalilpour, M., & Ramezani, M. (2016). A new meta-heuristic optimization algorithm inspired by FIFA world cup competitions: Theory and its application in PID designing for AVR system. *Journal of Control. Automation and Electrical Systems*, *27*(4), 419–440. doi:10.100740313-016-0242-6

Razmjooy, N., Mousavi, B. S., Sargolzaei, P., & Soleymani, F. (2011). Image thresholding based on evolutionary algorithms. *International Journal of Physical Sciences*, *6*(31), 7203–7211.

Razmjooy, N., Mousavi, B. S., & Soleymani, F. (2012). A real-time mathematical computer method for potato inspection using machine vision. *Computers & Mathematics with Applications (Oxford, England)*, *63*(1), 268–279. doi:10.1016/j.camwa.2011.11.019

Razmjooy, N., & Ramezani, M. (2014). An improved quantum evolutionary algorithm based on invasive weed optimization. *Indian Journal of Social Research*, *4*(2), 413–422.

Razmjooy, N., Sheykhahmad, F. R., & Ghadimi, N. (2018). A hybrid neural network–world cup optimization algorithm for melanoma detection. *Open Medicine: a Peer-Reviewed, Independent, Open-Access Journal*, *13*(1), 9–16. doi:10.1515/med-2018-0002 PMID:29577090

Revanth, K., Shree, T. D. V., Raja, N. S. M., & Rajinikanth, V. (2018). Computational investigation of stroke lesion segmentation from Flair/DW modality MRI. *In. Fourth International Conference on Biosignals, Images and Instrumentation (ICBSII)*, IEEE, 206-212. Doi: 10.1109/ICBSII.2018.8524617

Roopini, I. T., Vasanthi, M., Rajinikanth, V., Rekha, M., & Sangeetha, M. (2018). Segmentation of tumor from brain MRI using fuzzy entropy and distance regularised level set. *Lecture Notes in Electrical Engineering*, *490*, 297–304. doi:10.1007/978-981-10-8354-9_27

Satapathy, S.C. & Rajinikanth, V. (2018). Jaya algorithm guided procedure to segment tumor from brain MRI. *Journal of Optimization*, Vol.2018, Article ID 3738049, 12 pages. Doi: . doi:10.1155/2018/3738049

Shree, V. T. D., Revanth, K., Raja, N. S. M., & Rajinikanth, V. (2018). A hybrid image processing approach to examine abnormality in retinal optic disc. *Procedia Computer Science*, *125*, 157–164. doi:10.1016/j.procs.2017.12.022

Shriranjani, D., Tebby, S. G., Satapathy, S. C., Dey, N., & Rajinikanth, V. (2018). Kapur's entropy and active contour-based segmentation and analysis of retinal optic disc. *Lecture Notes in Electrical Engineering*, *490*, 287–295. doi:10.1007/978-981-10-8354-9_26

Sudhan, G. H. H., Aravind, R. G., Gowri, K., & Rajinikanth, V. (2017). Optic disc segmentation based on Otsu's thresholding and level set. *In. International Conference on Computer Communication and Informatics (ICCCI).* 1-5. Doi: 10.1109/ICCCI.2017.8117688

Tuan, T. M., & … . (2018). Dental diagnosis from X-Ray images: An expert system based on fuzzy computing. *Biomedical Signal Processing and Control*, *39*, 64–73. doi:10.1016/j.bspc.2017.07.005

Vaishnavi, G. K., Jeevananthan, K., Begum, S. R., & Kamalanand, K. (2014). Geometrical analysis of schistosome egg images using distance regularized level set method for automated species identification. *J. Bioinformatics Intell. Cont*, *3*(2), 147–152. doi:10.1166/jbic.2014.1080

Vishnupriya, R., Raja, N. S. M., & Rajinikanth, V. (2017). An efficient clustering technique and analysis of infrared thermograms, *In. Third International Conference on Biosignals, Images and Instrumentation (ICBSII)*, IEEE, 1-5. 10.1109/ICBSII.2017.8082275

Wang, Chen, Y., Yang, N., Zheng, L., Dey, N., Ashour, A. S., ... Shi, F. (2019). Classification of mice hepatic granuloma microscopic images based on a deep convolutional neural network. *Applied Soft Computing*, *74*, 40–50. doi:10.1016/j.asoc.2018.10.006

Chapter 8
Application of Software in Soil and Groundwater Recharge Estimation in Ilorin, Nigeria

Afolabi M. Asani
Ladoke Akintola University, Nigeria

Salihu Lukman
University of Hafr Al-Batin, Saudi Arabia

Isaiah Adesola Oke
https://orcid.org/0000-0002-7082-7682
Obafemi Awolowo University, Nigeria

ABSTRACT

Rainfalls measured in a selected location in Ilorin, Nigeria and standard formula were used to fix the unknown parameters of the new numerical formula using Microsoft Excel Solver. The new numerical formula was used to estimate groundwater recharge from the rainfall. The accuracy of the new numerical formula was evaluated statistically and compared with the previous formulae in use using field groundwater recharge. Correlation between rainfall and estimated groundwater recharge was stablished. Annual cost benefit of groundwater recharge was computed. The study revealed that new formula provided the lowest relative error of 0.001%, the highest MSC of 17.747; the degree of accuracy of 99.999% correlation factor between rainfall and groundwater recharge using the new numerical formula was 0.1612 with correlation coefficient of 0.6079. The average annual cost benefit was1080.24 $ m-2 per year. It was concluded that modeling of groundwater recharge using the new numerical formula is a promising tool for estimating groundwater recharge with minimum error in water resources management.

DOI: 10.4018/978-1-5225-8027-0.ch008

INTRODUCTION

Groundwater (GW) is an essential natural resource for human life that can be analysed with the help of images provided by Geographical Information Systems (GISs). Excess withdrawal without renewing of this natural resource is tremendously increased along with the growing population (Jumadi, 2015). Management and maintenance efforts to assure the sustainability of this natural water resource are critical. One of the efforts to sustain the existence of GW is to conduct balance management between utilization and potential recharge through an accurate numerical formula which needs adequate information of the physical condition of the environment and socioeconomic condition of the surrounding population. Groundwater Recharge (GWR) includes the drive of water from the unsaturated zone into the saturated region of the soil below the water table. The water runs together with the associated flow pattern away from the water table within the saturated region (Umar *et al.,* 2018). GWR occurs at a point when water containing dissolved minerals flows beyond the Groundwater Level (GWL) and infiltrates into the saturated zone. Modifications in GW storage comprise various recharge and discharge progressions. Most important recharge sources are rainfall, recharge from rivers, recharge from ponds, recharge from irrigation fields, etc. The crucial discharge processes are evapotranspiration, pumping, and baseflow to rivers. Several factors can affect the existence and discrepancy of GWL in a region. The major and significant factors that influence GWRinclude topography, lithology, geological structures, depth of weathering, the extent of fractures, the primary and secondary porosity of the soil, slope and drainage patterns of the region, landform and land cover, environmental factors and climate (Umar *et al.,* 2018). It is well known and established that there is a wide range of direct and indirect methods to estimate GWRprocesses, with a degree of approximation depending on different spatio-temporal scales, have been proposed (Scanlon *et al.,* 2002, 2006, Qinghua *et al.,* 2016; Brian *et al.,* 2016). The methods include lysimeter measurements, soil moisture budgets, and effective infiltration coefficients, as well as water table rise, tracer, and remote sensing methods. At a regional scale, to estimate the endogenous and exogenous variables controlling GWRprocesses, multidisciplinary analyses of hydrological time series, hydrogeological and geomorphological data have been implemented in a GIS environment (Andreo *et al.,* 2008; Dripps and Bradbury, 2010). Moreover, conceptual and physically based models accounting for the spatial variability of parameters which control recharge have been proposed (Hartmann e*t al.,* 2012; 2013, Allocca e*t al.,* 2014).

These GWRmethods and models employed in the determination of the GWRare characterized into three groups: (i) physical models calculated from the base flow, (ii) chemical models for the measurement of water-soluble substances, and (iii) numerical models using mathematical and computational techniques. The numerical method approaches include the computer programs such as HELP, RORA, PULSE, PART, HY SEP and WellsWater-table fluctuations (Seyed *et al.*, 2013, Umar *et al.*, 2018). Chandra (1979) reported that the methods and techniques that are prevalently applied to estimate the natural are soil water balance, zero flux plane; one-dimensional soil-water flow model, inverse modeling technique, GWL fluctuation, hybrid water fluctuation, GW balance method; isotope and solute profile techniques. Kumar (2000) grouped the methods and the techniques of estimation of GWR into four groups as empirical methods; GW resource estimation; GW balance approach and soil moisture data-based methods (Seyed *et al.*, 2013; Islam *et al.*, 2014).

The empirical methods comprise the following equations modified by Chaturvedi, Kumar, and Seethapathi, UP and Rao formulae. Chaturvedi (1973) derived an empirical equation, which expresses recharge as a function of the annual precipitation as follows (Oke *et al.*, 2017):

$$R_r = 2.0(P-15)^{0.4}, \tag{1}$$

where R_r is the net recharge due to precipitation during the year (inches), and P is the annual precipitation (inches).

The Chaturvedi formula was restructured and modified by further work at the U.P. Irrigation Research Institute, Roorkee and the modified Chaturvedi (UP) form of the formula is as follows (Oke *et al.*, 2017):

$$R_r = 1.35(P-14)^{0.5}. \tag{2}$$

Kumar and Seethapathi (2002) established an empirical relationship between GWR and the corresponding values of rainfall in the monsoon season using the non-linear regression equation of (Oke *et al.*, 2017):

$$R_{rm} = 0.63(P_m - 15.28)^{0.76}, \tag{3}$$

where R_{rm} is the GWR from rainfall in monsoon season (inch), and P_m is the mean rainfall in monsoon season (inch).

Rao derived an empirical relationship to determine the GWR in limited climatological homogeneous areas as follows (Oke *et al.*, 2017, Umar *et al.*, 2018):

$$R_r = K(R-X), \tag{4}$$

where R_r is the recharge (mm), K is a constant; R is the precipitation (mm), and X is the number of point rainfall. The following boundary empirical equations were applied to different parts of Karnataka as follows:

$R_r = 0.20\ (P - 400)$, for regions with annual normal rainfall (P) between 400 and 600 mm;
$R_r = 0.25\ (P - 400)$, for zones with P between 600 and 1000 mm
$R_r = 0.35\ (P - 600)$, for regions with P above 2000 mm.

Islam *et al.* (2014) proposed an empirical relationship between recharge and the corresponding values of rainfall. The relationship is as follows (Oke *et al.,* 2017; Umar e*t al.,*2018):

$$R_r = 0.85\left(P - 27.51\right)^{0.56}. \tag{5}$$

Seyed et al., (2013) developed an empirical relationship between recharge and the corresponding values of rainfall and wind speed (W_s) at 1-meter as follows (Seyed *et al.,* 2014):

$$R_r = 5.377 + 0.071P \text{, and} \tag{6}$$

$$R_r = 5.170 + 0.072P + 0.014W_s. \tag{7}$$

More on GWR estimate can be found in literature such as (Kung *et al.*, 2013; Rana *et al.*, 2014); Adeleke *et al.,* 2015; Islam *et al.*, 2015; Sabri *et al.*, 2015; Oke *et al.* 2015; Oke *et al.* 2016; Oke *et al.* 2017; Umar *et al.,* 2018). Estimating recharge in Nigeria has been a difficult issue due to many factors, which comprise the paucity of data, non-availability of mathematical and numerical formulae applicable in the country. Literature highlighted the GWR importance in the national development and economics. These discoveries showed that there is the necessity to develop a numerical methodology to estimate GW in Nigeria and evaluated formula for GW estimate that will be applicable in Nigeria and elsewhere. This study key objective, therefore, is to propose an advanced numerical strategy to estimate GW that will be applicable in Nigeria, and evaluate its accuracy and applicability using the following metrics: relative error, total error, degree of accuracy, numerical reliability, Model of Selection (MSC) in addition to Akaike Information Criterion (AIC).

MATERIALS AND METHODS

A general non-linear regression model that represents a generalization of simple formulae, which relates rainfall to the GWR volume estimate was proposed as follows (Oke *et al.,* 2017, Umar *et al.,* 2018):

$$R_{gr} = A((P - B))^{c}, \tag{8}$$

where R_{gr} is the GW estimate (depth, mm); *A*, *B,* and *c* are the constants parameters for the model and *P* is the depth of rainfall (mm). The model was proposed and selected based on the simplicity and minimum number of fixed parameters to be determined. Rainfall data of Ilorin, Nigeria between 1916 and 2017 was obtained from literature (Akintola, 1986) and Nigerian Meteorological Agency (NiMet) meteorological station (Ilorin). The rainfall data were used to calculate the GWR from rainfall using standard empirical formulae. The parameters in the proposed new numerical model (*A*, *B* and *c*) were determined using Microsoft Excel Solver and GW estimated from rainfall. Microsoft Excel Solver was selected based on availability (at no additional cost) and accuracy in numerical solutions. The new numerical formula was used to estimate GWR from the rainfall from selected location in Nigeria (Ilorin). Annual cost benefit was computed using stanadard methods (TetraTech, 2016). Correlation factor and coefficient between rainfall and GWR were computed.

The accuracy of the developed numerical model was evaluated using the Analysis of Variance (ANOVA), relative error; a Model of Selection (MSC) and Akaike Information Criterion (AIC) and compared with the selected numerical model in use (Rao, Islam *et al,* 2014, Chaturvedi and modified Chaturvedi formulae). Although, Chaturvedi and the Kumar and Seethapathi formulae were developed using the imperial unit, the GWR estimated was converted to system international(SI) unit to ensure consistency in a unit of measurement. Procedures employed in the computations of model parameters using Microsoft Excel Solver are as follows (Oke *et al.,* 2017, Umar *et al.,* 2018):

- Microsoft Excel Solver was added in on the toolbar of Microsoft Excel;
- Target (limit) value of the iteration was set for the software based on square of difference as $\left[\sum_{t=1}^{n} R_t - \sum_{t=1}^{n} A(P_t - B)^{c}\right]^2 = 0$;

- Changing cells of the iterations were selected, number of iterations, degree of accuracy and maximum time for the iteration were set for the software to meet the target; and

The iteration started through Microsoft Excel Solver (Figure 1).

The model of selection criterion (MSC) is interpreted as the proportion of expected GWR variation that can be explained by the obtained GW recharge. A higher the value of the MSC indicates a higher accuracy, validity and the good fitness of the method. MSC was computed using equation (9) as follows (Oke, 2007, Umar *et al.*, 2018):

$$MSC = \ln \frac{\sum_{i=1}^{n} \left(Y_{obsi} - \overline{Y}_{obs}\right)^2}{\sum_{i=1}^{n} \left(Y_{obsi} - Y_{cali}\right)^2} - \frac{2p}{n} \tag{9}$$

where Y_{obsi} is the GWR estimated using the UP formula, $\overline{Y}_{obs}$ is the average GWR estimated using the UP formula, p is the total number of fixed parameters to be estimated in the equation; n is the total number of GWR estimated, and Y_{cali} is the GWR estimated using new model equation.

The AIC was derived from the Information Criterion of Akaike (1976). It allows a direct comparison among models with a different number of parameters. The AIC presents the information on a given set of parameter estimates by relating the coefficient of determination to the number of parameters. The AIC and relative error (R_{Err}) were determined using equation (8 and 9 respectively) as follows:

$$AIC = n\left(\ln \sum_{i=1}^{n} \left(Y_{obsi} - Y_{cali}\right)^2\right) + 2p \tag{10}$$

$$RErr(\%) = \frac{\sum_{i=1}^{n} 100\left(\frac{Y_{obsi} - Y_{cali}}{Y_{obsi}}\right)}{n}. \tag{11}$$

The annual cost benefit value of GWR was computed as follows (assuming that 365 Nigerian Naira(₦) is equivalent to 1 US$, TetraTech, 2016):

$$AV_{gr} = P_{gr}\left(\frac{i(1+r)^t}{(1+i)^t-1}\right)\sum_{i=1}^{n} R_i, \tag{12}$$

where AV_{gr} is the annual cost benefit value of the GWR (\$ m^{-2}), P_{gr} is the average per unit present value of GWR benefit (\$/m^2), t is the index number of the year (2018 as the base year, $t = X - 2018$), R_i is the monthly GW recharge, n is the number of month, r is the likely inflation rate per year (%) and i is the interest rate per year. The average per unit present value of GWR benefit is calculated as follows:

$$P_{gr} = B_d\left(\frac{(1+r)^t}{(1+i)^t}\right), \tag{13}$$

where P_{gr} is the average per unit present value of GWR benefit (\$/m^2), and B_d is the benefit value = estimated cost of treated bottled water per m^2.

Figure 1. Procedure for using Microsoft Excel Solver in the computation of the perimeters (Oke et al, 2017; Umar et al. 2018)

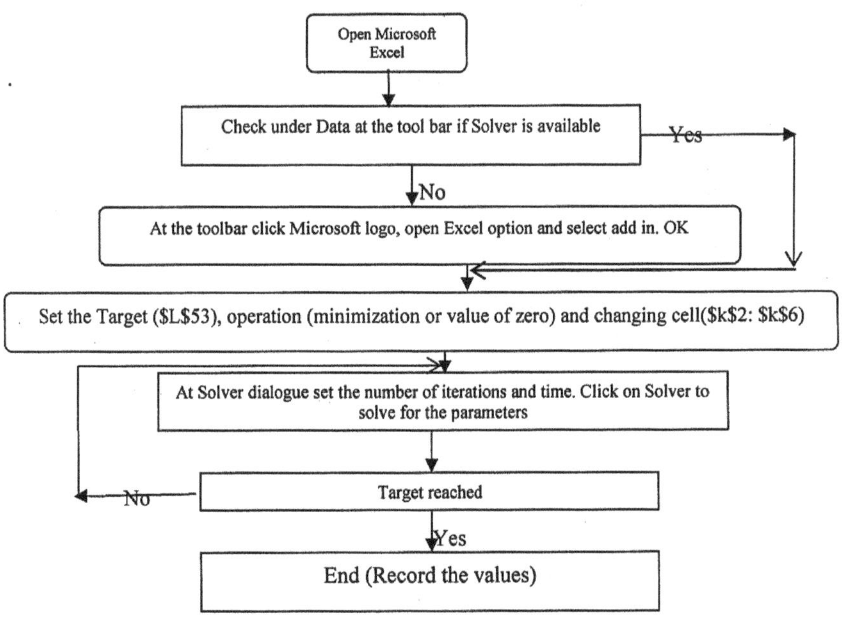

RESULTS AND DISCUSSION

Results from this study encompass the following categories: the rainfall and its statistical evaluation, the new numerical formula, the estimated GW recharges, the correlation factor and the coefficient between rainfall and GWR and, finally, the performance evaluation of the developed numerical formula.

Rainfall and Statistical Evaluation

Figure 2 presents monthly and annual rainfall of the location. From the figure the monthly rainfall varied with the month as well as the year, which indicated that heavy rainfall season in Ibadan starts from April and ends in the month of October every year. Although there was fluctuations in the amount of rainfall, but the range for the month remains valid. Tables 1 and 2 show summary of Analysis of Variance (ANOVA) for the monthly and annual rainfalls. The table revealed that there was a significant difference between the pattern of annual rainfall at 95% confidence level ($F_{101, 1111} = 1.427, p = 0.005$). The table also revealed that there was significant

Figure 2. Rainfall over a period of 102 years at Ilorin, Nigeria

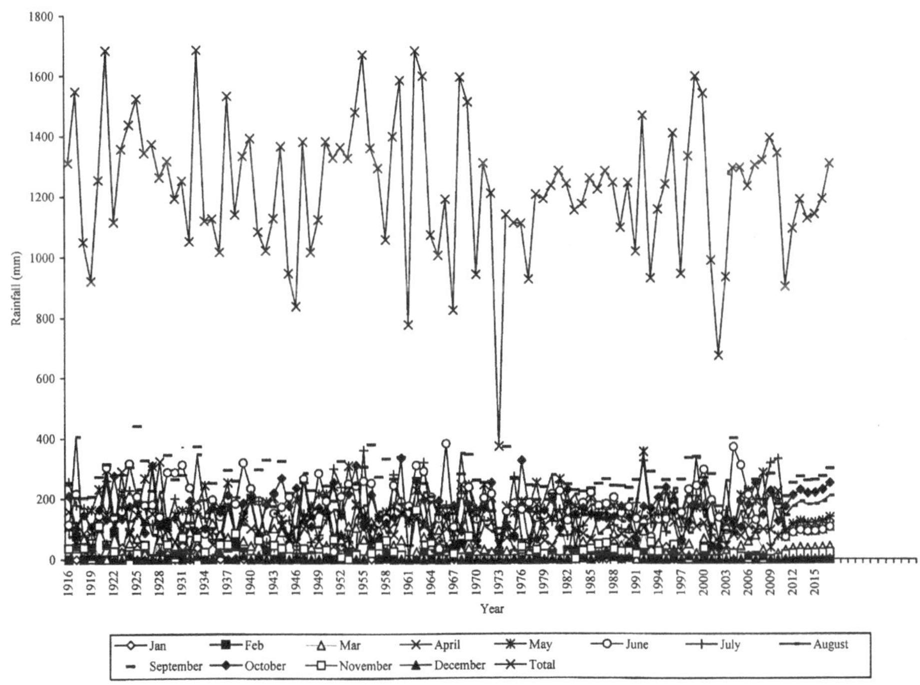

Table 1. ANOVA of the monthly Rainfall in Ilorin over a period of 102 years

Source of Variation	Sum of Squares	Degree of freedom	Mean Sum of Square	F- Value	P-value	F-Critical
Due to yearly rain	432921	101	4286.347	1.427	0.005	1.257
Due to Monthly rain	7103991	11	645817.389	214.989	0.000	1.797
Error	3337387	1111	3003.949			
Total	10874299	1223				

Table 2. ANOVA of the Total Rainfall in Ilorin over a period of 102 years

Source of Variation	Sum of Squares	Degree of freedom	Mean Sum of Square	F- Value	P-value	F-Critical
Within the Years	3397960	101	33643.170	0.044	1.000	1.391
Between the Years	76746358	100	767463.580			
Error	80144318	201				

difference between the pattern of the monthly rainfall for the selected period and location ($F_{11, 1111}$ = 214.989, p = 0.000). Table 2 revealed that there was no significant difference between total annual rainfall at 95% confidence level ($F_{101, 201}$ = 0.044, p = 1.0000).

The New Numerical Formula

The proposed numerical formula is as presented in equation (6). The value of the parameters are $A = 6.821$; $B = 356.152$ and $c = 0.500$. The value indicates that the new numerical formula for estimating GWR can be expressed as follows:

$$R_{gr} = 6.821\left(\left(P - 356.152\right)\right)^{0.500} \quad (13)$$

This result shows that the new numerical formula is similar to previous numerical formulae in literature such as Oke *et al.* (2016). Oke *et al.* (2016) documented that relationship between rainfall and GWR in Abeokuta and Ikeja, Nigeria can be expressed as follows:

$$R_{gr} = 0.15P - 102.8 \quad (14)$$

and

$$R_{gr} = 0.320P - 258 \tag{15}$$

From Oke *et al.* (2016) correlation coefficients (R^2) for Abeokuta [refer to equation (14)] and Ikeja [look at equation (15)] are 0.180 and 0.393 respectively. In Oke *et al.* (2017), the numerical formula for Abeokuta is presented in equation (16). The value of the parameters are $A = 1.673$; $B = 7.29$ and $c = 0.672$. The value indicated that the numerical equation for estimating GWR can be written as follows:

$$R_{gr} = 1.673((P - 7.219))^{0.672} \tag{16}$$

In Umar *et al.*, (2018), the value of the parameters are $A = 0.621$; $B = 1.019$ and $c = 0.814$. The value indicates that the new numerical equation for estimating GWR in Yola, Nigeria can be expressed as follows:

$$R_{gr} = 0.621((P - 1.019))^{0.814} \tag{17}$$

These results revealed that no numerical model is the same in values and in magnitude of the constants.

Estimated Groundwater Recharges

Figure 3 presents the statistical summary of GW recharges estimated using the new numerical formula and formulae that in use for all the location. Figure 3 revealed that GWR estimated using the relationship from (Islam *et al.*, 2014) had the lowest minimum and the mean in all the cases, while GW estimated using the Kumar and Seethapathi's formula had the highest maximum and the mean in all the cases. Figure 3 also revealed that the lowest standard deviation came from Chaturvedi's formula in all the case, while the highest standard deviation came from Rao formula. The result indicates that variation in the GWR estimated was higher in Rao's formula than any other formula. The highest median came from Rao's formula, and the lowest median came from Islam *et al.(2014)* formula. The result observed in term of median indicates that Rao's formula gave a higher value of GWR estimate than other formulae, while the work from (Islam *et al.*, 2014) gave the lowest GWR estimate. The results were different because Rao's formula an empirical relationship for GWR in limited climatological homogeneous areas and Kumar and Seethapathi's

Figure 3. Estimated groundwater recharge using various numerical formulae

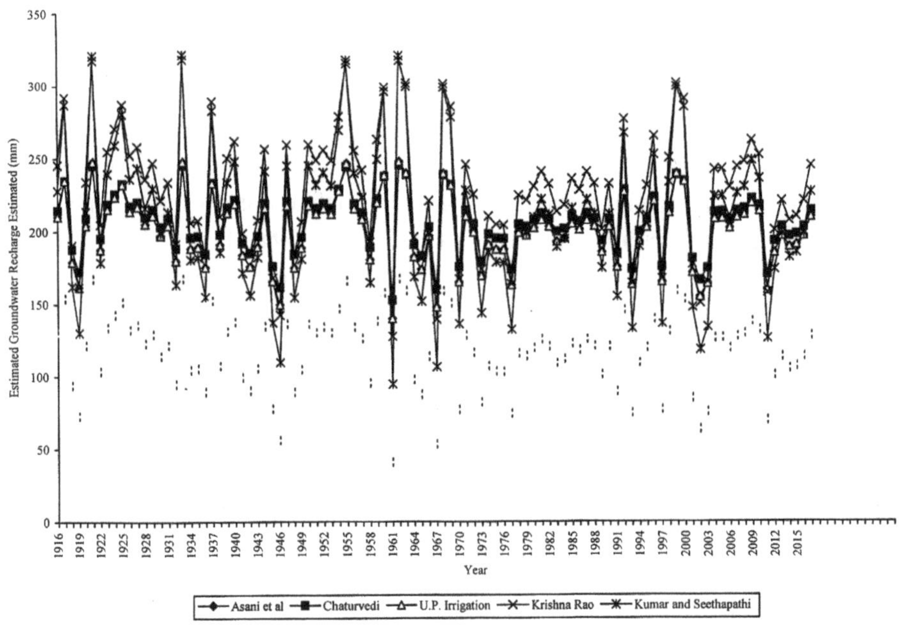

(2002) formula was for GWR from rainfall in the monsoon season (Oke *et al.*, 2017, Umar *et al.*, 2018). Three empirical formulae exhibit some similarities and closeness. The three empirical formulae are the new numerical formula, modified Chaturvedi's and Chaturvedi's formulae. The Figure shows that the estimated GW recharges vary with the location as well as the empirical model used. This result indicates that accuracy of these formulae is not the same. Analysis of variance of this estimated GW recharges reveal that there was a significant difference between the estimated GW recharges within the formulae and within the year (Table 3) in all

Table 3. ANOVA of the estimated groundwater recharges (Ilorin)

Source of Variation	Sum of Squares	Degree of freedom	Mean Sum of Square	F- Value	*P*-value	*F- Critical*
Within the Years	610677.64	101	6046.313	37.496	0.000	1.274
Between Methods	776645.76	5	155329.153	963.268	0.000	2.232
Error	81432.438	505	161.252			
Total	1468755.8	611				

the cases at 99% confidence level. In the analysis of variance (ANOVA) F – values were 30.237 and 634.33 with p- values of 4.4 x 10 $^{-179}$ and 6.4 x 10 $^{-228}$ for variation of GWR within the years and the formulae respectively.

Performance Evaluation of the Developed Numerical Model

Table 4 presented the performance evaluation of the model and the formulae. The Table revealed that the Islam *et al.* (2014) formula had the lowest MSC and accuracy, the highest AIC and errors (relative error, total and root square errors). From the Table, it was revealed that the highest MSC and Accuracy; lowest errors (total, relative and root square) and AIC were from the new model. Modified Chaturvedi and Chaturvedi were the next to the new model, which indicates that the new model is accurate and the next accurate formulae were modified Chaturvedi and Chaturvedi. A slight reduction in the accuracy of these two formulae may be attributed to the development of the formulae as an imperial unit formulae and conversion from imperial unit to system international unit. There are three empirical formulae that exhibit some similarities and closeness. The three empirical formulae are the new model, modified Chaturvedi and UP formulae. The Table shows the values of MSC as negative for three formulae (Kumar and Seethapathi formula, Rao formula and Islam *et al.* formula), which indicates that utilization of these three formulae in the GWR estimate should be at the lowest level. Figure 4 shows correlation factor and cofficient between rainfall and GW recharge. From the figure correlation factor was found to be 0.1612 with coeffcient of determinetion (R^2) of 0.6079 and correlation coefficent of 0.7789, which indicated that the correlation between these to parameters was fair (Loveday, 1980). Figure 5 presents annual cost benefit of GW recharge. From the figure, it can be seen that annual cost benefit of the process varies with the years. These results indicated that larger rainfall, higher surface area and regular

Table 4. Results of statistical evaluations of the models

-	New Numerical Model	Chaturvedi	Kumar and Seethapathi	Rao	Islam et al., (2014)
Relative Error (%)	0.006	3.08	13.65	11.32	42.58
Total Error	0.022	4301.47	118167.33	85583.88	714633.43
Root Square Error	2.12E-04	42.17	1158.50	839.06	7006.21
AIC	-385.282	859.40	1197.35	1164.44	1380.91
MSC	14.630	2.43	-0.89	-0.56	-2.69
Accuracy	99.994	96.923	86.353	88.679	57.418

Figure 4. Correlation between rainfall and estimated groundwater recharge

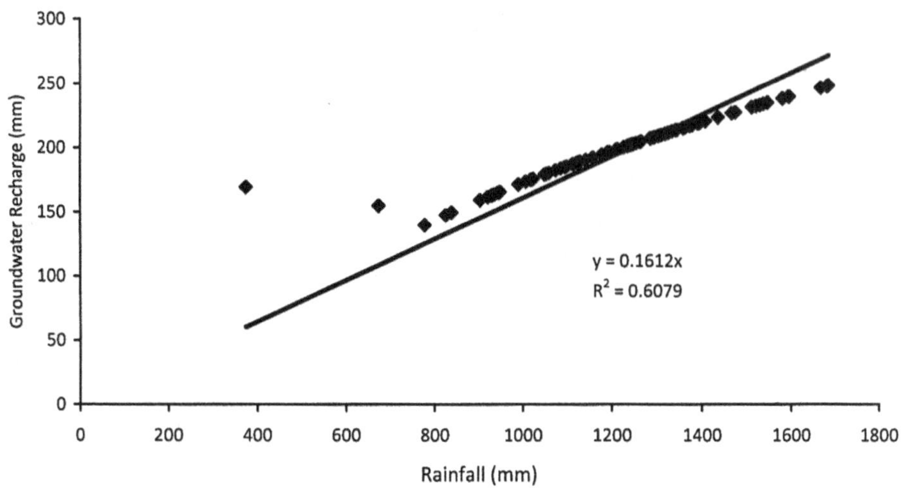

Figure 5. Annual cost benefit of the groundwater recharge

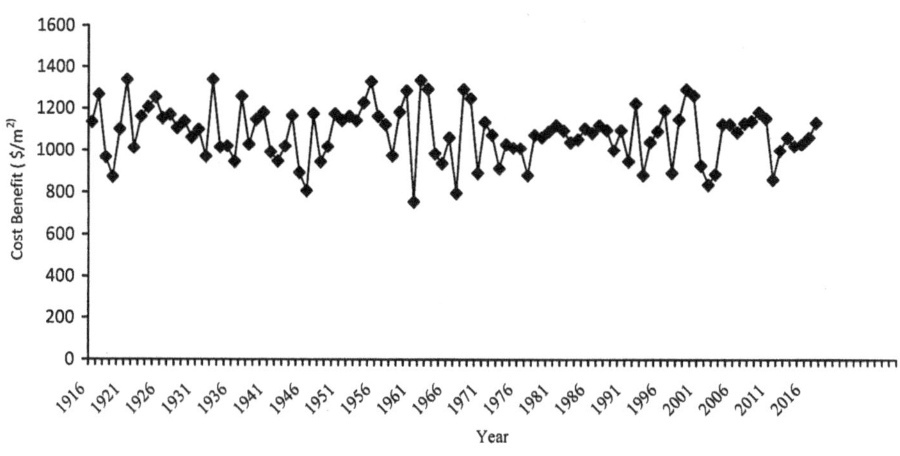

rainfall provides higher cost benefit of GW recharge. Correlation factor between rainfall and GWR using the new numerical formula was 0.1612 with correlation coefficient of 0.6079. The average annual cost benefit was 1080.24 \$ m^{-2} per year.

SOLUTIONS AND RECOMMENDATIONS

The quantification rates for both natural and anthropogenic-triggered GWR is an essential requirement for effective GW resource managing, which is principally crucial in dry and semiarid areas where resources are frequently vital to a cost-effective development. Still, the rate of aquifer renewal is one of the most problematic aspects to measure in the assessment of the GW resources.

Even though there are many deep-rooted approaches for the quantitative recharge estimation, scarce methods can be applied fruitfully in the field. Complications in reliably quantifying GWR stem from a variety of factors. These problems include the restricted capability to find and quantify the likely mechanisms of recharge and significant features influencing recharge for a specified place, the nonlinear recharge behavior with time, the highly inconstant surface distribution of GW recharge, the shortage of hydrogeological information, and the general intricacies of the hydrologic equilibrium.

Information Technology (IT) offers a rapid, worthy and analytical outcomes. Such tools range from simple spreadsheets to more complex programs, which again are contingent upon the requirements at the user end. Whatever the stream or problem domain, the most critical aspect is to conceptualization. A wide range of soft computing practices has been tested in Water Resources engineering to handle GW issues. The most extensively adopted forms of Soft Computing (SC) appear as front-end, database and back-end applications with further incorporation of advanced technologies like GIS.

Recently, SC techniques became widely used to estimate the uncertainty associated with model parameters, and variables. A Fuzzy Logic (FL) approach involves a many-valued logic (or probabilistic reasoning) which approximates rather than giving a fixed and precise outcome (Moallem wt al., 2014; Mousavi et al., 2014; Razmjooy et al., 2017). FL is a problem-solving methodology to implement systems ranging from embedded microcontrollers to large distributed and networked systems. FL was created as a better way of sorting and handling information but has proven to be an excellent selection for several control applications, since it imitates human decision-making and it and can help to assess GW recharge. It is very robust to data input uncertainty operator and frequently works with a little tuning when realized. The MATLAB Fuzzy Logic toolbox provides functions and to analyze, design, and simulate systems for many conventional SC methods, including fuzzy clustering and learning. It can model even complex systems behavior using plain logic rules and then realize them as a fuzzy inference structure to be used with GIS and Remote Sensing (RS) to extract useful evidence and find a fuzzy model for GWR estimate.

The steps involved in FL simulation can be developed as a soil water balance method to calculate GW recharge. Essential steps in this method are (i) the creation of membership functions for input and output variables, and (ii) the formulation of rules. The model has to be both calibrated and validated while considering precipitation and average temperature as inputs and recharge values as outputs. Input and output of soil water balance simulations were employed for the calibration of the FL model. The comparison has been made for both calibration and validation period between the results. The model can be calibrated using daily data from a rainfall, which is classified into several membership functions and mean temperatures into *N* membership functions. The output recharge quantity is classified into *K* membership functions. The subsequent points should be stressed:

1. The FL model can be calibrated from the data gathered during years, and the model tested for another year period. It is observed that the average error between the simulation and the FL model during the calibration period can be, for instance, 5% concerning the simulation model and -5% relating to the FL model.
2. During the validation and testing period of the model, an average error metric has to be determined concerning the simulation model and regarding the FL model.
3. The outcomes from the calibrated model can be displayed in sets of five years and balanced (and plotted) every two years separately.

Recently, several SC applications appeared in the field of hydraulic and water resources. Works are using Machine Learning (ML) (Alizamir *et al.*, 2018; Barzegar *et al.*, 2018; Roushangar *et al.*, 2017). Data-driven models appear in (Najafzadeh *et al.*, 2017). Artificial Neural Networks (ANNs) in addition to Support Vector Machines (SVMs) arise in (Haghiabi *et al.*, 2017; Adamala *et al.*, 2017; Londhe and Shah, 2017; Razmjooy et al., 2013). Particle Swarm Optimization (PSO) appear in (Niu *et al.*, 2018). Meta-Heuristic Algorithms gave interesting results in (Hosseini *et al.*, 2016). Bayesian networks were explored (Alizadeh *et al.*, 2017). Nevertheless, the natural phenomenon modeling with such line of attack should consider the kinds of problems in real-field applications (Barati, 2018; Barati, 2017) carefully.

Using two estimation scenarios for determining the spatial distribution of the GW Electro Conductivity (EC) or GWEC for short: (i) the triple input abstraction employs the longitude, latitude, and amount of months to model the input vector, and (ii) the quadruple input model with three inputs as well as the chlorine (Cl) ion. For the first topic, the GWL use as an input of the models can raise the performance of the established models for the estimation of the GWEC. The GWL pattern shows the GW flow direction, and it is well known that the GW quality declines during the GW motion from the aquifer entry to its outlet. So, the value of the GWEC grows in the direction of the GW movement. By bearing in mind that the GWL decreases from the aquifer entry to its outlet in an unconstrained aquifer, the GWL is a trustworthy input for the GWEC estimation.

The sampling points for the hydrochemical investigation of the GW correspond to *S* operational deep wells. The GWLs in the positions of the operation wells are estimated via *L* observations of the wells. The Inverse Distance Weighted (IDW) geostatistical scheme is used for the spatial interpolation of the GWL. The GW directions were also analyzed.

The relations between the GWEC given in µmho/cm and the latitude, the longitude, and the GWL (m) can follow the best linear fit along with the correlation coefficient value. As it is clear, the relationship between the GWEC and GWL is more significant than others (i.e., the GWEC can be correlated with the latitude or longitude). It should be noted that the relationships between the GWEC and the latitude and longitude are not linear, but it is considered linear for comparison purposes. The GWEC decreases with growing GWL. Moreover, the aquifer GW directions can be determined. In other words, its entrance and the aquifer outlet. The GWEC varies with the longitude. However, the relation between the EC and the latitude is more intricate. This evidence upholds that the value of the GWEC increases in the direction of the GW motion, which moves from higher values of GWL to lower values for an unconfined aquifer. Employing the latitude besides the longitude is necessary as an input in the application of data-driven policies for the spatial distribution estimation of the GWEC. GWL can also be another input to increase the modeling performance.

The relation between the GWEC and Cl was depicted by (Barati, 2018) where an excellent linear connection exists between the GWEC and the Cl for any case study because the GWEC and the Cl are directly linked to each other. Still, there is not any distinctive mathematical equation between the GWEC and Cl for all aquifers.

For the second circumstance, practically speaking, in the GW chemical analysis, the sample study to determine the GWEC becomes more straightforward than the analysis of a sample to find the GW anions and cations (comprising Cl). Therefore, with the Cl for a given sample, the value of the GWEC can also be determined with simple chemical analysis. Ignoring the Cl ion and using the GWL in the modeling process to estimate the GWEC may decrease the model accuracy, because the GWEC and Cl have a significant association, although the use of GWL, longitude, latitude, and the sum of months as the model input is the more practical approach.

FUTURE RESEARCH DIRECTIONS

As a consequence of these uncertainties, the project designs as well as management strategies have to be sufficiently malleable to avoid radical changes if the initial predictions are wrong or due to improper assumptions about recharge rates among other hydrogeological factors. The GWR estimation becomes an iterative process that permits progressive collection of the aquifer-response records and resource appraisal. Furthermore, more than one method is needed so that the results can be verified.

When appraising the GW recharge, one must start with a good different recharge model and other mechanisms related to the study area. The field data must be surveyed carefully to detect the likely types of flows. The recharge mechanism at a certain place depends on a variety of elements, which may differ from the influencing factors in a different place. Hence, if a method has efficaciously appraised the recharge in one area there is no guarantee that it can work in another region, even if the circumstances look similar. As soon as the recharge mechanisms have been identified, calculations can estimate the recharge. In addition to relying on a good concept, the choice of procedures should be conducted by the investigation goals, existing data and resources, and prospects of obtaining complementary data. Careful consideration to the following demands may simplify recharge estimation:

1. The acceptable aquifer recharge: A full aquifer will reject further water, which must then find another destination.
2. The amount of water that the unsaturated zone can transmit: High potential recharge rates (as, from streams or irrigation waterways) may not pass through low conductivity levels.
3. Other destinations for potential recharge, and their size.
4. The potential of recharge.
5. The real recharge: This phase reflects the balance as well as the destinations of all water from the source, established upon the results from first four questions.
6. The estimates analysis: As previously stated, more than one scheme should be used every possible time.

The main issue to decide on a recharge estimation procedure is the spatial and temporal scale of interest. If the foremost concern is to obtain decent recharge approximations over a limited extent (e.g., for waste dumping or local water source purposes), then the necessity for detailed knowledge is apparent. In this state of affairs, numerous site surveys, which also entail the identification of the preferential movement contributions, are required. Contrarywise, for ventures on a regional scale, or those demanding only introductory recharge estimates, GW-based techniques (for instance, those concerning the elucidation of fluctuations in GWLs) are pertinent and small-scale changeability in local recharge stops to be a problem.

More image processing and pattern recognition knowledge can be incorporated into GW estimation models. The more intelligent the algorithm the more demanding it is in terms of computational requirements. SC helps alleviating the burden from optimization procedures when used in place of hard computing strategies (Heamnth and Estrela, 2017).

The GWR information can come from in-situ or remote sensing sources (Alves *et al.*, 2011; Fernandes et al., 2014; de Jesus *et al.*, 2016; Estrela et al., 2019). Typically, GIS and RS demand different image modalities with different resolutions that have to be registered (Coelho *et al.*, 2012; de Jesus *et al.*, 2018; Estrela *et al.*, 2018). There are several PSO variants that can help to work with different image resolution according to the necessity (de Jesus *et al.*, 2018).

As knowledge is built, future system can fuse different sources of GW information and solve problems by means of visual semantic (Herrmann, et al., 2016).

CONCLUSION

Research works highlighted the GWR importance in the national growth and economics. GWR estimation in Nigeria has been a challenging subject because of many factors comprising the data scarcity, lack of mathematical models and numerical expressions pertinent to the country. Thus, the main objective of this investigation is to propose an advanced numerical strategy to estimate GW that can be appropriate to Nigeria, and evaluate its accuracy and applicability using the following metrics: relative error, total error, degree of accuracy, numerical reliability, Model of Selection (MSC) in addition to Akaike Information Criterion (AIC).

It can be concluded that although Islam *et al.* (2014) formulation, the proposed new numerical procedure, UP formula, Chaturvedi expression, Kumar and Seethapathi equation in addition to Rao's formula gave different GW recharges with a noteworthy difference in the recharges obtained. Attention should be taken when using the methodology from Islam *et al,* (2014), Rao Kumar and Seethapathi formulae based on the value of MSC, AIC and their accuracies. The new numerical formula is among the best for GWR estimate based on MSC, AIC and relative error coupled due to its simplicity.

REFERENCES

Adamala, S., Raghuwanshi, N. S., Mishra, A., & Singh, R. (2017). Generalized wavelet neural networks for evapotranspiration modeling in India. *ISH Journal of Hydraulic Engineering*, *29*(1), 1–13. doi:10.1080/09715010.2017.1327825

Adeleke, O. O., Makinde, V., Ayobami, O. E., Oluwaseun, F. D., Akintayo, O. O., & Taiwo, J. A. (2015). Estimation of Groundwater Recharges Using Empirical Formulae in Odeda Local Government Area, Ogun State, Nigeria. *Challenges*, *6*, 271–281. doi:10.3390/challe6020271

Akaike, H. (1976). An information criterion (AIC). *Math. Sci*, *14*, 1–5.

Alizadeh, M. J., Shahheydari, H., Kavianpour, M. R., Shamloo, H., & Barati, R. (2017). Prediction of longitudinal dispersion coefficient in natural rivers using a cluster-based Bayesian network. *Environmental Earth Sciences*, *76*(2), 86. doi:10.100712665-016-6379-6

Alizamir, M., Kisi, O., & Zounemat-Kermani, M. (2018). Modelling long-term groundwater fluctuations by extreme learning machine using hydro-climatic data. *Hydrological Sciences Journal*, *63*(1), 63–73. doi:10.1080/02626667.2017.1410891

Alves, F. D. S., Estrela, V. V., & Matos, L. J. (2011). *Hyperspectral Analysis of Remotely Sensed Images, Sustainable water management in the tropics and subtropics - and case studies in Brazil* (Vol. 2). University of Kassel.

Andreo, B., Vías, J., Durán, J. J., Jiménez, P., López-Geta, J. A., & Carrasco, F. (2008). Methodology for groundwater recharge assessment in carbonate aquifers: Application to pilot sites in southern Spain. *Hydrogeology Journal*, *16*(5), 911–925. doi:10.100710040-008-0274-5

Barati, R. (2017). Comment on "Evaluation of a physically based quasilinear and a conceptually based nonlinear Muskingum methods. *Journal of Hydrology (Amsterdam)*, *550*, 218–219. doi:10.1016/j.jhydrol.2017.05.006

Barati, R., Neyshabouri, S. S., & Ahmadi, G. (2014). Sphere drag revisited using shuffled complex evolution algorithm. River Flow, 345–353. doi:10.1201/b17133-51

Barzegar, R., Moghaddam, A. A., Deo, R., Fijani, E., & Tziritis, E. (2018). Mapping groundwater contamination risk of multiple aquifers using multi-model ensemble of machine learning algorithms. *The Science of the Total Environment*, *621*, 697–712. doi:10.1016/j.scitotenv.2017.11.185 PMID:29197289

Brian, F. T., Ali, B., & James, S. F. (2016). Precipitation Intensity Effects on Groundwater Recharge in the Southwestern United States. *Water (Basel)*, *8*(3), 90–105. doi:10.3390/w8030090

Chaturvedi, R. S. (1973). A note on the investigation of ground water resources in western districts of Uttar Pradesh. In *Annual Report* (pp. 86–122). U.P. Irrigation Research Institute.

Coelho, A. M., & Estrela, V. V. (2012). EM-based mixture models applied to video event detection. In P. Sanguansat (Ed.), *Principal component analysis—engineering applications. InTech Open*. doi:10.5772/38129

de Jesus, M. A., Estrela, V. V., & Saotome, O. (2016) Super-resolution in a nutshell. *Proceedings of the Brazilian technology symposium 2016* (BTSym 2016).

de Jesus, M. A., Estrela, V. V., Saotome, O., & Stutz, D. (2018). Super-Resolution via Particle Swarm Optimization Variants. In J. Hemanth & V. Balas (Eds.), *Biologically Rationalized Computing Techniques For Image Processing Applications. Lecture Notes in Computational Vision and Biomechanics* (Vol. 25). Cham: Springer. doi:10.1007/978-3-319-61316-1_14

Dripps, W. R., & Bradbury, K. R. (2010). The spatial and temporal variability of groundwater recharge in a forested basin in northern Wisconsin. *Hydrological Processes*, *24*, 383–392.

Estrela, V. V., Hemanth, J., Saotome, O., Grata, E. G. H., & Izario, D. R. F. (2019). Emergency Response Cyber-Physical System for Flood Prevention with Sustainable Electronics. In *Proceedings of the 3rd Brazilian Technology Symposium, BTSym 2017*. Springer. 10.1007/978-3-319-93112-8_33

Estrela, V. V., Saotome, O., Loschi, H. J., Hemanth, J., Farfan, W. S., Aroma, J., ... Grata, E. G. H. (2018). Emergency response cyber-physical framework for landslide avoidance with sustainable electronics. *Technologies*, *6*(2), 42. doi:10.3390/technologies6020042

Fernandes, S. R., Estrela, V. V., Magalhaes, H. A., & Saotome, O. (2014). On improving sub-pixel accuracy by means of B-Spline. *Proceedings of the 2014 IEEE international conference on imaging system and techniques (IST 2014)*. 10.1109/IST.2014.6958448

Haghiabi, A. H., Azamathulla, H. M., & Parsaie, A. (2017). Prediction of head loss on cascade weir using ANN and SVM. *ISH Journal of Hydraulic Engineering*, *23*(1), 102–110. doi:10.1080/09715010.2016.1241724

Hartmann, A., Lange, J., Weiler, M., Arbel, Y., & Greenbaum, N. (2012). A new approach to model the spatial and temporal variability of recharge to karst aquifers. *Hydrology and Earth System Sciences*, *16*(7), 2219–2231. doi:10.5194/hess-16-2219-2012

Hartmann, A., Weiler, M., Wagener, T., Lange, J., Kralik, M., Humer, F., ... Huggenberger, P. (2013). Process-based karst modelling to relate hydrodynamic and hydrochemical characteristics to system properties. *Hydrology and Earth System Sciences*, *17*(8), 3305–3321. doi:10.5194/hess-17-3305-2013

Hemanth, D. J., & Estrela, V. V. (2017). Deep Learning for Image Processing Applications (vol. 31). IOS Press.

Herrmann, A. E., & Estrela, V. V. (2016). Content-Based Image Retrieval (CBIR) in Remote Clinical Diagnosis and Healthcare. In M. Cruz-Cunha, I. Miranda, R. Martinho, & R. Rijo (Eds.), *Encyclopedia of E-Health and Telemedicine* (pp. 495–520). Hershey, PA: IGI Global. doi:10.4018/978-1-4666-9978-6.ch039

Hosseini, K., Nodoushan, E. J., Barati, R., & Shahheydari, H. (2016). Optimal design of labyrinth spillways using meta-heuristic algorithms. *KSCE Journal of Civil Engineering*, *20*(1), 468–477. doi:10.100712205-015-0462-5

Islam, M. R., Anupam, C., Kazi, M., Islam, M., & Ziaur, R. (2014). Development of rainfall recharge model for natural groundwater recharge estimation in Godagari Upazila of Rajshahi district, Bangladesh. *American Journal of Civil Engineering*, *2*(2), 48–52. doi:10.11648/j.ajce.20140202.16

Islam, S., Ram, K. S., & Roohul, A. K. (2015). Methods of Estimating Ground water Recharge. *International Journal of Engineering Associates*, *5*(2), 6–13.

Jumadi, Y. P. (2015). Development and Evaluation of Web GIS Application for Groundwater Management of Karanganyar Regency. *Forum Geografi*, *29*(2), 164–173.

Kumar, C. P. (2000). *Groundwater Assessment Methodology. 21*. National Institute of Hydrology.

Kumar, C. P., & Seethapathi, P. V. (2002). Assessment of natural ground water recharge in upper Ganga canal command area. *Journal of Applied Hydrology*, *15*(4), 13–20.

Kung, W.-J., Yeh, H.-F., Lin, H.-I., Chen, W.-P., & Lee, C.-H. (2013). An approach to evaluate groundwater recharge from streamflow and groundwater records. *Geosciences Journal, 17*(3), 353–362. doi:10.100712303-013-0027-x

Londhe, S. N., & Shah, S. (2017). A novel approach for knowledge extraction from artificial neural networks. *ISH Journal of Hydraulic Engineering*, 1–13. doi:10.1080/09715010.2017.1409667

Loveday, R. (1980). *Statistics, A Second Course in Statistics* (2nd ed.). New York: Cambridge University Press.

Moallem, P., Razmjooy, N., & Ashourian, M. (2013). Computer vision-based potato defect detection using neural networks and support vector machine. *International Journal of Robotics and Automation, 28*(2), 137–145. doi:10.2316/Journal.206.2013.2.206-3746

Moallem, P., Razmjooy, N., & Mousavi, B. S. (2014). Robust potato color image segmentation using adaptive fuzzy inference system. *Iranian Journal of Fuzzy Systems, 11*(6), 47–65.

Mousavi, B. S., Soleymani, F., & Razmjooy, N. (2013). Color image segmentation using neuro-fuzzy system in a novel optimized color space. *Neural Computing & Applications, 23*(5), 1513–1520. doi:10.100700521-012-1102-3

Mousavi, B. S., Soleymani, F., & Razmjooy, N. (2014). Semantic image classification by genetic algorithm using optimised fuzzy system based on Zernike moments. *Signal, Image and Video Processing, 8*(5), 831–842. doi:10.100711760-012-0311-7

Najafzadeh, M., Tafarojnoruz, A., & Lim, S. Y. (2017). Prediction of local scour depth downstream of sluice gates using data-driven models. *ISH Journal of Hydraulic Engineering, 23*(2), 195–202. doi:10.1080/09715010.2017.1286614

Niu, W. J., Feng, Z. K., Cheng, C. T., & Zhou, J. Z. (2018). Forecasting daily runoff by extreme learning machine based on quantum-behaved particle swarm optimization. *Journal of Hydrologic Engineering, 23*(3), 04018002. doi:10.1061/(ASCE)HE.1943-5584.0001625

Oke, Martins, Salako, & Idowu. (2016). Estimation of Groundwater Recharge in the Lower-Ogun River Basin using Two independent Methods. *Nigerian Journal of Hydrological Sciences*, 181 – 200

Oke, I. A. (2007). Reliability and Statistical Assessment of Methods For Pipe Network Analysis. *Environmental Engineering Science*, *24*(10), 1481–1489. doi:10.1089/ees.2007.0001

Oke, I. A., Lukman, S., & Ismail, A. (2017). Development And Performance Evaluation Of A New Numerical Model For Groundwater Recharge Estimation. *Nigeria Journal of Engineering*, *23*(2), 56–65.

Oke, M. O., Martins, O., Idowu, O. A., & Aiyelokun, O. (2015). Comparative Analysis of Groundwater Recharge Estimation Value Obtained Using Empirical Methods In Ogun and Oshun River Basins. *Ife Journal of Science*, *17*(1), 53–64.

Qinghua, W., Guiling, W., Wei, Z., Haodong, C., & Wei, Z. (2016). Estimation of Groundwater Recharge Using Tracers and Numerical Modeling in the North China Plain. *Water (Basel)*, *8*(8), 353–373. doi:10.3390/w8080353

Rana, C., & Ray, R. K. (2014). *Assessment of Ground Water Resources: A Review of International Practices A Review of International Practices*. Faridabad: Govt. of India Ministry of Water Resources Central Ground Water Board.

Razmjooy, N., Mousavi, B. S., & Soleymani, F. (2013). A hybrid neural network Imperialist Competitive Algorithm for skin color segmentation. *Mathematical and Computer Modelling*, *57*(3-4), 848–856. doi:10.1016/j.mcm.2012.09.013

Razmjooy, N., Ramezani, M., & Ghadimi, N. (2017). Imperialist competitive algorithm-based optimization of neuro-fuzzy system parameters for automatic red-eye removal. *International Journal of Fuzzy Systems*, *19*(4), 1144–1156. doi:10.100740815-017-0305-2

Roushangar, K., Alami, M. T., Majedi Asl, M., & Shiri, J. (2017). Modeling discharge coefficient of normal and inverted orientation labyrinth weirs using machine learning techniques. *ISH Journal of Hydraulic Engineering*, 1–10.

Scanlon, B. R., Healy, R. W., & Cook, P. G. (2002). Choosing appropriate techniques for quantifying groundwater recharge. *Hydrogeology Journal*, *10*(1), 18–39. doi:10.100710040-001-0176-2

Scanlon, B. R., Keese, K. E., Flint, A. L., Flint, L. E., Gaye, C. B., Edmunds, W. M., & Simmers, I. (2006). Global synthesis of groundwater recharge in semiarid and arid regions. *Hydrological Processes*, *20*(15), 3335–3370. doi:10.1002/hyp.6335

Seyed R.; Saghravani, I. Y., Sa'ar I, M., & Seyed, F. S. (2013). Estimating Groundwater Recharge Using Empirical Method: A Case Study in the Tropical Zone. *Sains Malaysiana, 42*(5), 553–560.

TetraTech. (2016). *Estimating Monetized Benefits of Groundwater Recharge from Stormwater Retention Practices.* Prepared for United States Environmental Protection Agency Nonpoint Source Control Branch (4503T).

Umaru, A. B., Lukman, S., Oke, I. A., & Umara, B. G. (2018). A New Numerical Method for Groundwater Recharge Estimation In Yola, Nigeria. *Arid Zone Journal of Engineering, Technology and Environment.*

Yeh, H. Y., Cheng-Haw, L., Kuo-Chin, H., & Po-Hsun, C. (2009). GIS for the assessment of the groundwater recharge potential zone. *Environ Geol., 58*(1), 185–195. doi:10.100700254-008-1504-9

KEY TERMS AND DEFINITIONS

Base Flow: The portion of streamflow delayed shallow subsurface flow.

Direct Recharge: Water added to the groundwater reservoir in excess of soil moisture and evapotranspiration deficits, by direct vertical percolation of water.

Groundwater Flow System: Regional groundwater flow systems are characterized by laterally extensive aquifers, which may be thicker than 300 meters, and recharge and discharge areas separated by distances of fifty or more kilometers.

Groundwater Level: Describe the depth or elevation above or below sea level at which the surface of groundwater stands.

Groundwater Recharge: A hydrologic process where water moves downward from surface water to groundwater.

Recharge Area: The place where water is able to seep into the ground and refill an aquifer because no confining layer is present.

Rejected Recharge: Water that infiltrates to the water table but then discharges because the aquifer is full and cannot accept it.

Chapter 9
Applications of Image Compression on Agricultural Image Data Analysis

K. Seetharaman
Annamalai University, India

ABSTRACT

With the advent of the cutting-edge technologies in information repository and communication, data storage is rapidly increased, and it is required to reduce the size of the data. Especially in the case of agricultural image data like types of plants, crops, seeds; kinds of diseases and their remedial pesticides; and the agricultural satellite; images require a huge volume of memory space to store. To avoid this problem, it is required to reduce the size of the data and redundancy of the data. To overcome this problem, this chapter proposes a compression method, based on an adaptive Gaussian Markov random field model for agricultural image data compression, where the images are assumed to be a Gaussian Markov random field. The parameters of the model are estimated, based on the Bayesian approach. The authors use arithmetic coding to store seed values and parameters of the model as it augments the compression ratio. They also have studied the use of the M-H algorithm, which updates the parameters and through which the image contents such as untexturedness are captured.

INTRODUCTION

Nowadays, storing data is a critical problem in many domains and using different formats as well as modalities in agriculture, medicine, advertising, military, education, and so on. However, a high-speed network communication technology available, it is difficult to transmit and retrieve the high volume of multimedia data

DOI: 10.4018/978-1-5225-8027-0.ch009

and Hyperspectral (HSI) images. The advances in computer vision technology such as the Internet of Things (IoTs), all sorts of digital libraries and the access to online visual data repositories, among other amenities have exponentially grown through the past decade. Multidisciplinary areas like agriculture, medicine, education, traffic management, automatic recognition, and Geographic Information Systems (GISs) effectively employ these computer vision advances (Moallem et al., 2013,2014; Razmjooy et al., 2012, 2013). Handling images entails manipulation, storage, analyzes, indexing, matching, retrieval, display, etc., which are all computationally costly tasks. Especially, handling the agricultural satellite imageries and the normal agricultural image or data, that is, storage and retrieval are complicated because of volumes of data. Thus, it requires proper visual database systems to back the previously mentioned image operations. If one conducts a literary survey, it will become apparent the amazing progress made in both theoretical and implementation levels towards effective and efficient storage and retrieval systems. Hence, in this chapter, a new compression technique is introduced, based on an Adaptive Gaussian Markov Random Field (AGMRF) model (Brezger et al., 2007), which reduces the visual data redundancy while affording speedy retrieval of the agricultural knowledge, and the images of the crops, seeds, diseased leaves, stems, fruits, and so on to help all stages of agribusiness.

BACKGROUND

The preceding decades witnessed the influx of a considerable amount of technologies have been developed for agribusiness (Fountas et al., 2015) and ranging from embedded systems to networking technologies (Munack & Speckmann, 2001), more and more advanced electronics and communication methodologies continue to be added (Pereira et al., 2010; Steinberger et al., 2009). Though the technology in communication has been developed up to some extent, it does not satisfy both data transmission and storage. Therefore, the researchers in computer vision developed the data reduction techniques based on visual data characteristics. The data reduction, that is, data compression method satisfies both rapid transmission of data and minimizing the storage space, whereas the network communication technology helps to speed up the data transmission only. In the agricultural domain, this problem was first envisioned by (Stone, 1994), who introduced the use of agrarian networks using buses of higher bandwidth or multiple of buses. However, these solutions have not addressed the problems either accordingly nor are suitable to the ISO 11783 compression standard. In this respect, an alternative option is to shrink the amount of data to be transferred, as happens with in-vehicle networks (Miucic et al., 2009; Ramteke & Mahmud, 2005), e.g., via the employment of data compression methods.

The data compression technique is more effective and efficient for data storage and transmission, which is effectively adopted in the agricultural domain

The Compression Ratio (CR) reached by any compression algorithm is contingent upon the content of the visual input or its associated data compressibility [Jain81]. In most applications, e.g., image communication, visual storage management, and image content analysis the CR has to be maintained within reasonable levels or constant,

Compression intends to diminish the redundancy of data to save storage space and minimize the transmission time. Still, getting rid of the redundancy will not heighten the CRs. A high CR can stem from the eradication of the high-frequency constituents of the signals (often using a transformation) by storing only the relevant information content. The most popular manner, Discrete Cosine Transform (DCT) algorithm, relies on this technique. There are several international image compression standards that are widespread, such as JPEG-XL, JPEG2000 (Li et al., 2016; ISO/IEC 15444, 2015; Rabbani & Joshi, 2002), MPEG 1, MPEG 2, MPEG 4, MPEG 7 (Watson, 1998; Sullivan et al., 2012), H.261, H.262, H.264, and HEVC (Flierl & Girod, 2003) relying on the DCT and some other orthonormal transforms. The fast progress of multimedia applications like text, audio, image, video and multimedia signals made image compression more imperative and demanded. These kinds of applications call for a vast amount of records. Consequently, three reasons lead to visual data compression: (i) the large multimedia information storage demands, (ii) the somewhat slow storage mechanisms that cannot handle multimedia data (particularly video) in real time, and (iii) the network bandwidth necessary for real-time data transmission.

In terms of information loss, compression can be lossless and lossy (Jain, 1981).

1. **Lossless:** Decompression renders an exact copy of the original image or video. It has various methods, such as the Variable-Length coding, that is, Huffman Coding (HC), Arithmetic Coding (AC), Lempel-Ziv-Welch (LZW), Entropy Encoding (EE), Bit-Plane Encoding (BPE), Run-Length Encoding (RLE).
2. **Lossy:** The resultant decompressed image looks very similar to the original image (given some metric or metrics), but not an exact copy. The lossy compression achieves higher CRs than the lossless, but the image quality becomes lower than in the lossless case with an inevitable degree information loss. The lossy Predictive Coding (PC), Wavelet Transform (WT) coding, and Fractal Image Compression (FIC) are lossy compression methods.

Visual data can be compressed using statistical-based and dictionary-based approaches. Statistical schemes encode the data employing the probability that a character or symbol appears. Dictionary-based modeling, to each string of characters or symbols, there is a single code. To save storage space, the dictionary is not kept with the compressed file. Instead, the dictionary is recreated when the information is decoded. The dictionary-based strategies are computationally more intensive and time-consuming than the statistical-based approaches.

Likewise, if one considers the working style, image compression practices can be twofold: (i) symmetrical and (ii) asymmetrical. In the symmetric method, the compression and the decompression procedures take coarsely the same time and effort. For example, the Lempel-Ziv-Welch (LZW) and the JPEG structures fall into a symmetric method. The LZW is a dictionary-based technique that works on an extremely textured image. The FIC and the PC techniques exemplify asymmetrical schemes, which require more compression time and effort than decompressing an image. Its basis is to perform most of the effort throughout the compression. Hence, the output can be decompressed very speedily.

In the VLE arrangement, the symbols are encoded as variable-length output codes depending on the symbols occurrence probability. Low-probability characters receive more bits while high-probability symbols employ a fewer number of bits. The HC (Jeon et al., 1998; Jagadish & Lohit, 2010) and AC (Bose & Pathak, 2006) are variable-length coding varieties.

The *HC* allocates an output code for an individual symbol, with output codes as short as *1* bit or markedly longer than the extent of the group of input symbols. This strategy strictly depends on the probabilities. If p is the probability of a certain symbol appear, then the corresponding optimal amount of bits is the $\log_2(1/p)$. This method main caveat is it has to dispense to the code either *1* or *2* bits, thus extending the compressed code.

AC replaces the input symbols stream with a single floating-point number as output, which is a single figure less than 1 as well as greater than or else zero. The AC number is individually decoded to obtain the specific stream of symbols that went into its creation. The symbols that go through coding have a group of probabilities allotted to them to permit the output number creation. The first encoded symbol is the most crucial part of an arithmetic code. AC is superior to the HC (Nelson & Gailly, 1996), has replaced the HC in JPEG-2000 and both are symmetric.

The *BPE* procedure decomposes a multilevel image into a sequence of binary images, and any binary compression method can perform compression.

RLE (Golomb, 1996) is easy to deploy and has speedy execution while dropping the size of a repeating characters string aka run, which is typically encoded using two bytes. The run count corresponds to the number of characters or symbols in the run, and it is stored in the first byte. The second byte holds the value of the character or symbol (Howard & Vitter, 1992).

Lossless PC (Chellappa et al., 1985, Motta et al., 2000; Kingston & Autrusseau, 2008) engenders an array filled in with uncorrelated random variables for a particular visual data by using an invertible transformation defined by an appropriate statistical model. By fitting this model to the input image, one produces the so-called residuals, which are less correlated than the original pixels and, therefore, the very smaller number of bits that can represent it. The original visual data can be recreated with the quantized residuals and the estimated model parameters. Agrarian image analysis employs lossy compression most of the times. The lossless compression rarely appears, e.g., Multispectral (MSI) satellite image processing.

Numerous transform coding practices appear in the literature, and they play a significant part in image coding, reconstruction as well as feature extraction because they project the signal into space spanned by orthogonal basis functions that allocate the signal energy together with a set of decorrelated components. Each kind of orthonormal transform has its specific properties.

The Transform Coding (TC) concept for coding and transmitting visual data started with the 2-D Fourier Transform (FT) of an image has become practical with the advent of a fast computational algorithm called Fast FT (FFT) (Andrews & Pratt, 1968; Anderson & Huang, 1971). Next, there was the discovery that Hadamard Transform could replace the FT with a substantial decrease in computational requirements. Investigations have begun since then the applications of Karhunen-Loeve (KL) transform aka Principal Component Analysis (PCA) analysis (Habibi & Wintz, 1971; Wintz & Tasto, 1971), Harr Transform (HT) (Andrews, 1970) and Slant Transform (ST) (Pratt et al., 1973) for image coding. Among all these transforms, the KL transform provides the minimum Mean Square Error (MSE) performance metric but regrettably does not possess a fast computational algorithm.

In contrast, the Harr transform is an advantageous computational procedure but results in a more massive coding error. Most practical systems rely on the Discrete Cosine Transform (DCT) that balances well the information packing ability in addition to the computational complexity. Hence, the DCT was incorporated in the JPEG static image compression standard.

The Wavelet Transform (WT) is a very popular low-level image processing, principally for image compression. Antonini et al. (1992) employed wavelets for image coding. Then, (Sweldens, 1996) introduced a scheme, namely, lifting, which is a general structure for constructing bi-orthogonal wavelets, which also affords a natural structure for constructing a reversible Integer WT (IWT) for lossless image compression. The Lifting scheme comprises of three stages, such as splitting, prediction, and updating. The visual data are split into two disjoint sets that will pass thru a series of prediction and updating stages to transform the information. The WT admits non-stationary signals with proper localization in both the spatial and the frequency domains that can be implemented by a fast algorithm too. In particular, these properties are appropriate for image processing. Adams & Kossentini(2000) have adopted reversible integer-to-integer WTs for image compression, performance appraisal, and analysis. Later, the projection-based and adaptive reversible IWT delivered lossless image compression (Deever & Hemami, 2003). The reversible integer WT is used in the combination of DCT and is referred to as the international standard JPEG2000.

The Fractal Image Compression (FIC) technique (Wohlberg & Jager, 1999) is lossy. Hamzaoui (1999) has introduced a fast iterative FIC method, and while (Hartension et al., 2000) have performed image compression with the region-based approach. To begin with, it localizes self-similar image sections and, then, performs the compression, and it applies a fractal algorithm to spawn the sections. This asymmetrical technique takes much effort to compress and decompress an image because it encodes a picture as a collection of transforms very similar to the photocopy machine metaphor. The supplementary details looked-for decoding at grander sizes are created automatically through the encoding transforms, which increases the CR when compared to others. The CR for the fractal structure is cumbersome to obtain since decoding an image can happen at any scale. The FIC technique is beneficial for the compression of the agrarian images, like paddy, sugarcane, groundnut, and so forth because the images of these crops are in the self-similar form with repeated patterns.

In lossy PC, the compressed images are rebuilt only with the assistance of the estimated parameters besides some initial conditions (Magli & Olmo, 2003). The PC is less time consuming when compared to other compression practices. Delp et al., (1979) have introduced a non-parametric quantizer for the Differential Pulse Code Modulation (DPCM) strategy for recreating the original image. They reported that the lossy PC technique could work with textured images, whereas the lossless PC technique can reconstruct the nontextured images without sacrificing any information. Latter, (Krishnamoorthy & Seetharaman, 2007) recommended a technique founded on PC using the autoregressive model, which gives better quality for textured and nontextured images. The lossy compression has been adopted in many places of the agricultural domain because it does not require 100% percent accuracy of the decompressed images like medical image processing.

COMPRESSION STANDARDS FOR STILL IMAGES

Firstly, the dictionary-based JPEG was established with DCT and HC. In the next version, AC replaced the HC to increase the CR and noticeably decrease the execution time. Subsequently, JPEG2000 appeared using wavelets with the AC. The test results proved that the wavelet coding boosted subjective in addition to objective image quality when compared to the DCT (Jeon et al., 1998; Rabbani & Joshi, 2002). JPEG 2000 is a current standard for still picture compression as well as decompression of continuous-tone pictures, and later after some improvements, it was devised the JPEG-XL. It targets a wide variety of images for different deployment scenarios (Rabbani and Joshi, 2002) to offer one unified standard to undertake these tasks. The main objectives of the JPEG-XL are fivefold: (a) Efficient encoding, (b) high-quality spatial features and scalability, (c) complexity reduction, (d) line and block procedures, and (e) robustness. The primary objective of the JPEG-XL is to store the video, audio files together and to reduce the file size by 60% compared to the JPEG2000.

In this Chapter, a causal model is proposed, called Adaptive Gaussian Markov Random Field (AGMRF) model, where the image is assumed a Markov Random Field (MRF). This chapter also adopts advantages of the Metropolis-Hastings (M-H) algorithm (Chib & Greenberg, 1995) to modernize the model coefficients when the image contains features such as edges and boundaries. Many different methodologies, including the Gibbs sampler, seem to be a particular instance of the M-H algorithm, which takes less than half of the time required by other iterative methods. The suggested model is fitted to obtain the image using the estimated parameters besides the seed values followed by the computation of the residuals between the original and produced images as well as the average residuals values. The AC is used to save seed values, the residual averages, and the model parameters lead to a higher CR. The reconstructed sub-images are synthesized using the seed values and the parameters to form the whole decompressed image.

PROPOSED COMPRESSION MODEL

Let X be a random variable in lieu of the pixel intensity or brightness value at position (k,l) in a given image. X may have Gaussian noise with Independent and Identically Distributed (iid) random variable belonging to a discrete time and a continuous state domains, respectively, with mean zero and variance σ^2 and is denoted by $\varepsilon(k,l)$, i.e., $\varepsilon(k,l) \sim N(0, \sigma^2)$.

The stochastic process $\{X(s);s\in S\}$, where $S=\{s:(k,l);1 \leq k,l \leq M\},\{X(s)\}$, is a Markov process following the conditional probability

$$P\{X(s_n) = i_n \mid X(s_k) = i_k \,;\, k=0,1,2,\ldots, n-1\}$$

$$= P\{X(s_n) = i_k \mid X(s_n\text{-}1) = i_n\text{-}1\}$$

$\forall\, i_k$, k=0,1,2, . . ., n-1; and s_k, which belongs to the state space S with $s_0 < s_1 < \ldots < s_n$.

Here, the 2-D images $\{X(k, l), 1 \leq k, l \leq N\}$ are modeled using discrete spatial time series of random fields. Thus, the proposed Adaptive Gaussian Markov Random Field (AGMRF) model is a family of representations by a discrete-time stochastic process {X(s)}, t = 0, ±1, ±2, ±3, ..., as follows

$$X(k,l) = \sum_{\substack{r=-1 \\ r=q\neq 0}}^{M} \sum_{q=-1}^{M} \Gamma_{rq} X(k+r,\ l+q) + \varepsilon(k,l) \tag{1}$$

where, $\Gamma_{rq} = \Gamma_{r} = \dfrac{K \sin(r\theta)\cos(r\varphi)}{\alpha^{r}},\ \forall\ q$

In the above Equation (1), $X(k+r, l+q)$ accounts for the spatial variation resulting from the image properties, and $\varepsilon(k,l)$ is the spatial variation brought about the additive noise and $\Gamma_{rq} = \Gamma_{r} = \dfrac{K \sin(r\theta)\cos(r\varphi)}{\alpha^{r}},\ \forall\ q$ is the rth coefficient of variation among the texture primitives in the small frame region. The coefficients are interrelated so that the relationship among the parameters K, α, θ, and ϕ can be established. The coefficients in Eq. (1) are the functions of the K, α, θ, and ϕ as well as *n* becomes

$$\Gamma_{r} = \Gamma_{r}(K,\ \alpha,\ \theta,\ \varphi) = \frac{K}{\alpha^{r}} \sin(r\theta)\cos(r\varphi)$$

$K \in R$, $\alpha > 1$, $\theta, \phi \in [0, 2\pi]$ and $n \in \{1,2, \ldots \}$.

The AGMRF model allows the analysis of a 2-D discrete image after partitioning it into various sub-images of size M×M, to locally depict the image nature. Using the Markovian hypothesis, the conditional probability of X(*s*) given that all other values only depends upon the nearest surrounding area values, that is, P(X(s)/X(s-j)); j=1,2, …, M.

Initially, the parameters are specified as K∈R, with α>1, and θ, where ϕ∈[0, 2π]. A further constraint on the parameter range is introduced by inspecting the model identifiability. It is worth mentioning that some of the models employed in the preceding works, e.g., white noise, finite-order Markov model as well as the infinite-order Markov model can be regarded as a particular instance of the AGMRF model. Thus,

1. If we set $\theta=0$, then the AGMRF model becomes the white noise process.
2. When α is large, the coefficients Γ_rs turn out to be negligible as r increases. Therefore, the AGMRF model reduces to order p approximately, for a suitable value of p.
3. When α is chosen less than one, then the AGMRF model becomes an explosive AGMRF model with infinite order.

The fact that $X(s)$ has conditionally dependent on its pixel vicinity originated the Markovian process terminology where the dependence of $X(s)$ on the neighborhood values may be true to some extent. The process is Gaussian under the hypothesis that the $\varepsilon(k,l)$s are Gaussian, which determines the probabilistic structure completely by its second order properties. To conclude, the range of the parameters of the model is established with the constraints, $K \in R$, $\alpha > 1$, $0 < \theta < \pi$, $0 < \phi < \pi/2$; and $\varepsilon(k,l)$ are an iid Gaussian random process with mean zero and variance σ^2.

The noncausal model given in Equation (1) states the pixel $X(s)$ as a linear combination of the nearest neighborhood values on each side as illustrated in Figure 1(a) and the influence of the horizontal and vertical pixels on the center is higher than the diagonal pixels. Figure 1(b) displays the order of influence of the neighboring pixels with respect to its center. The pixel $X(s)$ is predicted by the closest neighborhood pixels on each side.

The model considered here is very important from a theoretical point of view, since it has an infinite structure with a finite number of parameters, and hence completely avoids the problem of order determination. It can be solved systematically with many important problems, that is, testing whether an image is homogeneous or not with the choice of the relevant set of neighbors.

Figure 1. (a) Relationship among neighborhood pixels; (b) Order of influence of neighborhood pixel with respect to its center

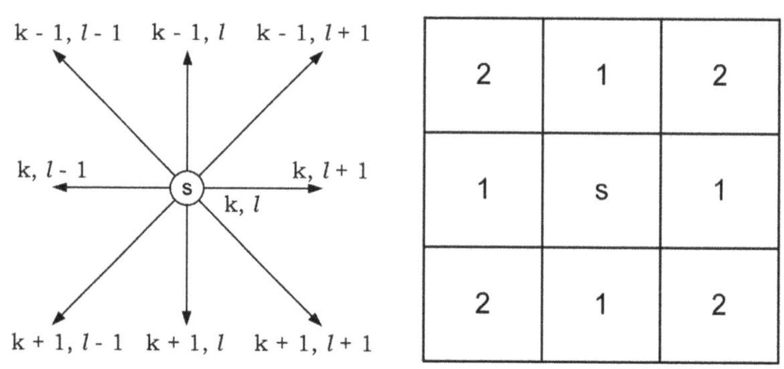

(a) (b)

PARAMETER ESTIMATION

Measure of Performance

In order to implement the AGMRF model, the parameters have to be estimated precisely. The parameters, K, α, θ, and ϕ, are estimated, by taking a suitable prior information for the hyper parameters, β, γ, and δ, based on numerical integration technique and Bayesian methodology, the posterior density of the parameters of the AGMRF model is obtained. The AGMRF model, discussed in the previous section, is defined by a discrete time stochastic process $\{X(t)\}$ as in Equation (2).

$$X(k,l) = \sum_{\substack{r=-1 \\ r=q\neq 0}}^{M} \sum_{q=-1}^{M} \Gamma_{rq} X(k+r,\ l+q) + \varepsilon(k,l) \tag{2}$$

where, $\Gamma_r = \dfrac{K \sin(r\theta)\cos(r\varphi)}{\alpha^r}$,

$\varepsilon(k,l)$ s are iid normal random variables, which have mean zero and variance $\sigma^2 > 0$, whose parameters domains are given by $K \in R$, $\alpha \in (1, \infty)$, $\theta \in (0, \pi)$ and $\phi \in (0, \pi/2)$ to ensure an identifiable model.

Presume the N observations are $X_1, X_2, \ldots, X_N$ and that the unknown parameters K, α, θ, ϕ and σ^2 will be estimated by the Bayesian methodology.

The pixel values belonging to each sub-image are stacked to form vectors X(t), t = 1,2,3, ..., N ($M \times M = M^2 = N$) for a computational purpose only. Since the error entries $\varepsilon(k, l)$ from Eq. (1) are iid and Gaussian. The joint probability density function (pdf) of $\{X(t)\}$ becomes

$$P\left(X/\Theta\right) \propto \left(\sigma^2\right)^{-N/2} \exp\left[-\frac{1}{2\sigma^2}\sum_{t=1}^{N}\left\{X_t - K\sum_{r=1}^{\infty} S_r X_{t-r}\right\}^2\right] \tag{3}$$

where,

$X = (X_1, X_2, \ldots, X_N)$,
$\Theta = (K, \alpha, \theta, \phi, \sigma^2)$, and
$S_r = \dfrac{\sin(r\theta)\cos(r\varphi)}{\alpha^r}$.

While analyzing N finite observations of the real data, the interval variation corresponding to the index r is simplified from 1 to ∞ to the integer interval $[1, N]$. Meanwhile, the joint pdf for the observations appearing in the previous expression can replace the summation $\sum_{r=1}^{\infty}$ by $\sum_{r=1}^{N}$, which yields

$$P\left(X/\Theta\right) \propto \left(\sigma^2\right)^{-N/2} \exp\left[-\frac{1}{2\sigma^2}\sum_{t=1}^{N}\left\{X_t - K\sum_{r=1}^{N} S_r X_{t-r}\right\}^2\right] \tag{4}$$

If the term inside the square in the exponent is expanded, then

$$P\left(X/\Theta\right) \propto \left(\sigma^2\right)^{-N/2} \exp\left[-\frac{1}{2\sigma^2}\left\{T_{00} + K^2\sum_{r=1}^{N} S_r^2 T_{rr} + 2K^2\sum_{\substack{r,s=1\\ r<s}}^{N} S_r S_s T_{rs} - 2K\sum_{r=1}^{N} S_r T_{0r}\right\}\right] \tag{5}$$

where $T_{rs} = \sum_{t=1}^{N} X_{t-r} X_{t-s}$, $\quad r, s = 0,1,2,\ldots,N$

Reshaping the joint pdf yields

$$P\left(X/\Theta\right) \propto \left(\sigma^2\right)^{-N/2} \exp\left[-\frac{Q}{2\sigma^2}\right] \tag{6}$$

where,

$$Q = T_{00} + K^2\sum_{r=1}^{N} S_r^2 T_{rr} + 2K^2\sum_{\substack{r,s=1\\ r<s}}^{N} S_r S_s T_{rs} - 2K\sum_{r=1}^{N} S_r T_{0r}$$

with $K \in R$, $\alpha > 1$, $0 < \theta < \pi$, $0 < \phi < \pi/2$ besides $\sigma^2 > 0$.

The prior parameters distribution is achieved as follows:

1. α has a displaced exponential pdf with parameter β, that is to say

$$P(\alpha) = \beta \exp(-\beta(\alpha - 1)) \; ; \; \alpha > 1; \; \beta > 0 \tag{7}$$

2. σ^2 is distributed according to an inverted gamma having parameters ν and δ such as

$$P(\sigma^2) \propto \exp(-\nu / \sigma^2)\,(\sigma^2)^{-(\delta+1)} \; ; \; \sigma^2 > 0, \; \nu, \delta > 0 \tag{8}$$

3. K, θ, in addition to ϕ are uniformly distributed in the domains they are defined, i.e.,

P(K, θ, ϕ) = C, where C is a constant and

$K \in R$, $0 < \theta < \pi$, $0 < \phi < \pi/2$

Consequently, the joint prior pdf of Θ turns out to be

$$\begin{aligned} P(\Theta) \propto \; & \beta \exp(-\beta(\alpha - 1) - \nu/\sigma^2)\,(\sigma^2)^{-(\delta+1)} \; ; \\ & \sigma^2 > 0, \; \alpha > 1, \; 0 < \theta < \pi, \; 0 < \varphi < \pi/2. \end{aligned} \tag{9}$$

In the previous equation, P denotes the pdf of the random variables within parentheses. With the help of (6), (9) and the Bayes theorem, allows computing the joint posterior pdf of K, α, θ, ϕ, and σ^2

$$P(\Theta | X) \propto \exp(-\beta(\alpha - 1)) \exp(-1/2\sigma^2)\,(Q + 2\nu)\,\left(\sigma^2\right)^{-\left(\frac{N}{2}+\delta+1\right)} \tag{10}$$

for $K \in R$, $\alpha > 0$, $0 < \theta < \pi$, $0 < \varphi < \pi/2$ and $\sigma^2 > 0$.

Integrating (6) with respect to σ^2 results in the posterior pdf of K, α, θ, and ϕ given by

$$P\left(K, \alpha, \theta, \varphi / X\right) \propto \exp(-\beta(\alpha - 1))(Q + 2\nu)^{-\left(\frac{N}{2}+\delta\right)} \tag{11}$$

$$K \in R,\ \alpha > 1,\ 0 < \theta < \pi,\ 0 < \varphi < \pi/2$$

where

$$\left[Q + 2\nu\right] = \left[\left(K^2 \sum_{r=1}^{N} S_r^2 T_{rr} + 2K^2 \sum_{\substack{r,s=1 \\ r<s}}^{N} S_r S_s T_{rs} - 2K \sum_{r=1}^{N} S_r T_{0r}\right)\right] + T_{00} + 2\nu$$

That is,

$$(Q+2\nu) = aK^2 - 2Kb + T_{00} + 2\nu,$$
$$= C\left[1 + a_1\left(K - b_1\right)^2\right]$$

with

$$C = T_{00} - \frac{b^2}{a} + 2\nu$$
$$a = \sum_{r=1}^{N} S_r^2 T_{rr} + 2 \sum_{\substack{r,s=1 \\ r<s}}^{N} S_r S_s T_{rs}$$

$$b = \sum_{r=1}^{N} S_r T_{0r}$$

$$a_1 = \frac{a}{C},\ b_1 = \frac{b}{C}$$

As an end result, the preceding joint posterior pdf of K, α, θ, and ϕ becomes

$$P\left(K, \alpha, \theta, \varphi / X\right) \propto \exp(-\beta(\alpha - 1))\left[C\left\{1 + a_1 (K - b_1)^2\right\}\right]^{-d} \quad (12)$$

$$K \in R, \ \alpha > 1, \ \ 0 < \theta < \pi, \ \ 0 < \varphi < \frac{\pi}{2}$$

where, $d = N/2 + \delta$.

This confirms that for known (α, θ, ϕ), then the conditional pdf of K is a *t* distribution placed at b_1 and having (2d – 1) degrees of freedom.

The Bayesian inference on K, α, θ, and ϕ can be found from their respective posterior pdfs. The joint posterior density of (α, θ, ϕ), i.e., P(α, θ, ϕ|X), arises from the integration of the Eq. (12) with respect to K. Hence, the joint posterior pdf of (α, θ, ϕ) becomes

$$P\left(\alpha, \theta, \varphi / X\right) \ \propto \ \exp(-\beta(\alpha-1))C^{-d}a_1^{-1/2} \tag{13}$$

$$\alpha > 1, \ 0 < \theta < \pi, \ \ 0 < \varphi < \frac{\pi}{2}$$

The previous posterior pdf of α, θ, and ϕ in Eq. (9) is a difficult function, and not analytically solvable. Consequently, the original posterior pdf of α, θ, and ϕ can be obtained numerically using the joint density above . That is,

$P(\alpha) \propto \iint P(\alpha, \theta, \phi|X)\, d\theta\, d\phi$.

Similarly,

$P(\theta) \propto \iint P(\alpha, \theta, \phi|X)\, d\alpha\, d\phi$, and

$P(\phi) \propto \iint P(\alpha, \theta, \phi|X)\, d\alpha\, d\theta$.

The estimates of α, θ, and ϕ might result from the means of the marginal posterior pdf or in other words, the posterior means. Initially, the posterior mean of α is computed numerically to minimize the computations. Next, keeping α fixed at the value of its posterior mean, so that the conditional means of θ and ϕ can be computed. Fixing (α, θ, ϕ) at their posterior means correspondingly, followed by the evaluation of the conditional mean of K yields to the estimates

$$\left.\begin{aligned}\hat{\alpha} &= E(\alpha)\\ (\hat{\theta},\ \hat{\varphi}) &= E(\theta,\ \varphi/\alpha\ = \hat{\alpha}),\ \text{and}\\ \hat{K} &= E(K/\hat{\alpha},\ \theta\ = \hat{\theta},\ \varphi\ =\ \hat{\varphi}).\end{aligned}\right\} \quad (14)$$

The estimated hyperparameters $\hat{K}$, $\hat{\alpha}$, $\hat{\theta}$, and $\hat{\varphi}$ help to find the coefficients Γ_rs from the AGMRF model in Eq. (1).

The advantages of the AGMRF approach are (i) the fixed number of parameters is four, and (ii) an order that does not augment the computational complexity. The second benefit occurs because the estimation of these hyperparameters is the same, regardless of the model order that increases the AGMRF model efficiency.

Metropolis-Hastings Algorithm

The Metropolis-Hastings (M-H) algorithm obtains the parameters with high precision supposing the model coefficients $\Gamma = \left(\Gamma_1,\ \Gamma_2\right)$ fulfill the stationarity conditions

$$\Gamma_1 + \Gamma_2 < 1; \quad -\Gamma_1 + \Gamma_2 < 1 \text{ and } \Gamma_2 > -1. \quad (15)$$

The next likelihood function comes from Eq. (15) for the sub-image pixel values $X_n = (x_1,\ x_2,\ \ldots,\ x_n)'$

$$L\left(\Gamma,\ \sigma^2\right) = \Psi\left(\Gamma,\ \sigma^2\right) \times \left(\sigma^2\right)^{-(n-2)/2} \times \exp\left[-\frac{1}{2\sigma^2}\sum_{t=3}^{n}\left(x_t - w_t'\Gamma\right)^2\right] \quad (16)$$

where $w_t = \left(x_{t-1},\ x_{t-2}\right)'$,

$$\psi\left(\Gamma, \sigma^2\right) = \left(\sigma^2\right)^{-1} \left|V^{-1}\right|^{1/2} \exp\left[-\frac{1}{2\sigma^2} X_2 V^{-1} X_2\right] \quad (17)$$

is the density of $X_2 = (x_1, x_2)'$,

$$V^{-1} = \begin{pmatrix} 1-\Gamma_2^2 & -\Gamma_1\left(1+\Gamma_2\right) \\ -\Gamma_1\left(1+\Gamma_2\right) & 1-\Gamma_2^2 \end{pmatrix},$$

and the third element in (16) is proportional to the pdf of the observed measurements (x_3, x_4,...,x_n) given X_2.

If the only existing prior information is the stationarity of the process, the posterior hyperparameters distribution becomes

$$\Pi\left(\Gamma,\sigma^2 \middle| X_n\right) \propto L\left(\Gamma,\sigma^2\right)\, I\left[\Gamma \in S\right],$$

where $I\left[\Gamma \in S\right]$ is 1 if $\Gamma \in S$, and 0 otherwise.

The posterior pdf can be simulated relying on two concepts,

1. Γ *and* σ^2 are considered as blocks; and
2. The application of the regression ANOVA decomposition makes the exponential term of (16) proportional to

$$\exp\left[-\frac{1}{2\sigma^2}\left(\Gamma-\hat{\Gamma}\right)' G\left(\Gamma-\hat{\Gamma}\right)\right],$$

where $\hat{\Gamma} = G^{-1}\sum_{t=3}^{n}\left(w_t x_t\right)$ and $G = \sum_{t=3}^{n}\left(w_t w_t'\right)$. This matches to the kernel of a normal pdf that have mean $\hat{\Gamma}$ and covariance matrix $\sigma^2 G^{-1}$. These observations lead to the subsequent full conditional densities for α^2 and Γ straightforwardly as

1. The density of σ^2 given Γ and X_n is an inverted gamma having parameters n/2 and this relationship holds $X_2 V^{-1} X_2 + \sum_{t=3}^{n}\left(x_t - w_t'\Gamma\right)^2$.
2. The density of Γ given σ^2 and X_n is

$$\Pi\left(\Gamma \middle| X_n,\sigma^2\right) \propto \Psi\left(\Gamma,\sigma^2\right) \times \left\{f_{nor}\left(\Gamma \middle| \hat{\Gamma},\sigma^2 G^{-1}\right) I\left[\Gamma \in S\right]\right\}, \tag{18}$$

where f_{nor} is a normal pdf.

Samples drawn from $\Pi\left(\sigma^2, \Gamma / X_n\right)$ come from consecutive samplings of the Γ *from* $\Pi\left(\Gamma \mid X_n, \sigma^2\right)$, and using this value of Γ to estimate σ^2 *from* $\Pi\left(\sigma^2 \middle| X_n, \Gamma\right)$. The latter simulation is direct. For the previous simulation, candidates were generated from the density function within the curly braces of (18) because $\left|V^{-1}\right|^{1/2}$ is bounded for all values of the Γ in the stationary region. Then, the value of Γ is simulated as follows:

At the jth iteration, draw a candidate $\Gamma^{(j+1)}$ from a normal density with mean $\hat{\Gamma}$ and covariance $\sigma^{2(j)}G^{-1}$; when it satisfies a stationarity condition, then move to this point with probability

$$\min\left\{\frac{\psi(\Gamma^{(j+1)}, \sigma^{2(j)})}{\psi(\Gamma^{(j)}, \sigma^{2(j)})}, 1\right\}$$

and, otherwise, set $\Gamma^{(j+1)} = \Gamma^{(j)}$, where $\Psi\left(\cdot, \cdot\right)$ is defined in (17).

The above mathematical concept can be cast as an algorithm to update the coefficient values (Γ_1 and Γ_2) of the model.

Step 1: Check if the stationarity conditions hold for the set of coefficient values $\Gamma = (\Gamma_1, \Gamma_2)$

i.e. $\Gamma_1 + \Gamma_2 < 1; \quad -\Gamma_1 + \Gamma_2 > -1$

Step 2: A sample of draws from the posterior pdf of the parameters, i.e., $\Pi(\Gamma, \sigma^2 / X_n)$ results from successively sampling Γ from $\Pi(\Gamma / X_n, \sigma^2)$, where $X_n = (x_1, x_2, \ldots, x_n)$.

Step 3: Using Γ simulate σ^2 from $\Pi(\sigma^2 / X_n, \Gamma)$

Step 4: Now, one can generate candidates from the normal pdf f_{nor} defined as

$$f_{nor}(\Gamma / \hat{\Gamma}, \sigma^2 G^{-1})\ I(\Gamma \in S)$$

Step 5: At the jth iteration whose current value is $\sigma^{2(j)}$, draw a candidate $\Gamma^{(j+1)}$ from a normal pdf with mean $\hat{\Gamma}$ and covariance $\sigma^{2(j+1)} G^{-1}$ so that

(i) if it satisfies the stationarity conditions, then use this point with probability

$$\min\left\{\frac{\psi(\Gamma^{(j+1)}, \sigma^{2(j)})}{\psi(\Gamma^{(j)}, \sigma^{2(j)})}, 1\right\}$$

(ii) otherwise, set $\Gamma^{(j+1)} = \Gamma^{(j)}$

Step 6: Repeat steps 2 to 5 until it satisfies the condition (i) from the Step 5.

Thus, the entire sub-image can be generated relying on the given initial values as well the estimated model parameters.

The image frame containing the residuals (changes) between the input (aka residual image $f_e(i, j)$) and the generated images is

$$f_e(i, j) = f_i(i, j) - \hat{f}_g(i, j)$$

Now, the average values for the residuals are found and saved along with the model parameters and seed values. Moreover, the AC is applied to the entire image to guarantee a higher CR.

The complete image compression algorithm enclosing the mathematical concepts above follows.

Algorithm

Step 1: Read a grayscale image $f_i(i, j)$ with size N×N to serve as input.

Step 2: Partition $f_i(i, j)$ into various non-overlapping sub-images with size M×M.

Step 3: Estimate the hyperparameters $\hat{K}, \hat{\alpha}, \hat{\theta}$ *and* $\hat{\varphi}$ using the pixel values in a sub-image.

Step 4: Calculate the Γ_1 and Γ_2 with the model parameters resulting from Eq. (1) in Step 3.

Step 5: Adjust Γ_1 and Γ_2 via the M-H algorithm.

Step 6: With the help of Eq. (1), generate an image $\hat{f}_g(i, j)$ with the corresponding coefficients and initial conditions.

Step 7: Determine $f_e(i, j)$, i.e. $f_e(i,j) = f_i(i,j) - \hat{f}_g(i,j)$
Step 8: Calculate the average of $f_e(i, j)$.
Step 9: The initial conditions, the average $f_e(i, j)$ and the model coefficients are influenced by the AC.

PERFORMANCE METRICS

The Mean Square Error (MSE) is formally stated as

$$MSE = \frac{1}{N \times N} \sum_{i=0}^{N-1} \sum_{j=0}^{N-1} \left[f(i,j) - \hat{f}(i,j) \right]^2 ,$$

where f and $\hat{f}$ are respectively the original and the reconstructed images of size N × N. The Peak Signal-to-Noise Ratio (PSNR) expressed in decibel (dB) is

$$PSNR = 10\log_{10}\left(\frac{(255)^2}{MSE} \right)$$

The expression for the Compression Ratio (CR) becomes

$$CR = \frac{\text{No. of bits required to store reconstructed image}}{\text{No. of bits required to store original image}} \text{ x } 100$$

The MSE, the PSNR in addition to the CR are the metrics used to appraise the performance of the AGMRF model compression.

EXPERIMENTS AND RESULTS

The AGMRF model has been tested with different types of agricultural images captured with the camera and the satellite as given in Figure 2. The leaves of the paddy crop, sugarcane, and cotton crop images were considered for the experiment. The input images were divided into various non-overlapping sub-images of equal size 8×8. The model parameters are estimated for each sub-image as discussed in the PARAMETER ESTIMATION section. The M-H algorithm updates the parameters estimated from the pixel values within the sub-images. A few pixel values, and the average values of the residual image between the original and generated images of each sub-image are stored with the estimated parameter set. It is then subjected to AC. The stored pixel values are taken as seed values of the AGMRF model and the model is employed to reconstruct the sub-images. Now, all the generated sub-images are synthesized to form a reconstructed image. Figure 2 presents the reconstructed images. The images in rows 1, 3, and 5 are actual images, and the images in rows 2, 4, and 6 are reconstructed (compressed) images. The first image in the first row of the Figure 2 yields a CR of 95.86, whereas the second and third images yield CRs equal to 96.52 and 96.15 respectively.

The first and second images in the second row yield CRs equal to 96.95 and 96.75 respectively, whereas the third image gives a CR up to 96.61. The first image in the first row of the Figure 2 yields 95.86 as CR, whereas the second and third images yield CRs of 96.52 and 96.15 respectively. The first, second, and third images in row 3 give CRs up to 96.01, 95.82, and 95.54 respectively. The average CRs for the three varieties of the images in Figure 2 appear in Table 1.

Furthermore, emphasis the efficacy of the AGMRF method, another set of images, such as fruits and vegetable were subjected to the experiment; and the obtained compression results were tabulated in Table 2. The images in row 1 of the Figure 3 give a CR up to 96.84, 96.03, and 95.86 respectively; and the images in row 2 of the Figure 3 yield CR up to 96.97, 96.28, and 97.53 respectively.

In addition to the images captured by the digital camera, in order to validate the efficiency and effectiveness of the AGMRF method, a number of agricultural crop imageries captured by the satellite were subjected to the experiments. Due to space constraint, a two of them were presented in Figure 4. The AGMRF method yields the average CR up to 94.56, 95.84, and 96.02 to the imageries in row 1, row 2, and row 3 of the Figure 4 respectively. The average of the CRs obtained were tabulated in Table 3.

The AGMRF image compression scheme has also been compared with JPEG XL. The compression ratios along with the PSNR values by the AGMRF scheme

Table 1. Average Compression Ratio for the images in Figure 2

Images	CR(%)		PSNR (dB)	
	AGMRF	JPEG	AGMRF	JPEG
Paddy	96.18	96.10	29.15	27.26
Sugar cane	96.77	95.97	28.94	26.86
Cotton	95.79	95.46	28.57	27.15

Figure 2. Agricultural crops: rows 1, 3, and 5 are actual images; rows 2, 4, and 6 are compressed images

Figure 3. Vegetables: rows 1 and 3 are actual images; rows 2 and 4 are compressed images

Table 2. Average Compression Ratio for the images in Figure 3

Images	CR(%)		PSNR (dB)	
	AGMRF	JPEG	AGMRF	JPEG
Fruits	96.25	96.13	28.85	27.09
Vegetables	96.93	96.48	27.94	26.94

Figure 4. Agriculture imageries captured by Satellite: First column - actual imagery; second column - compressed imagery

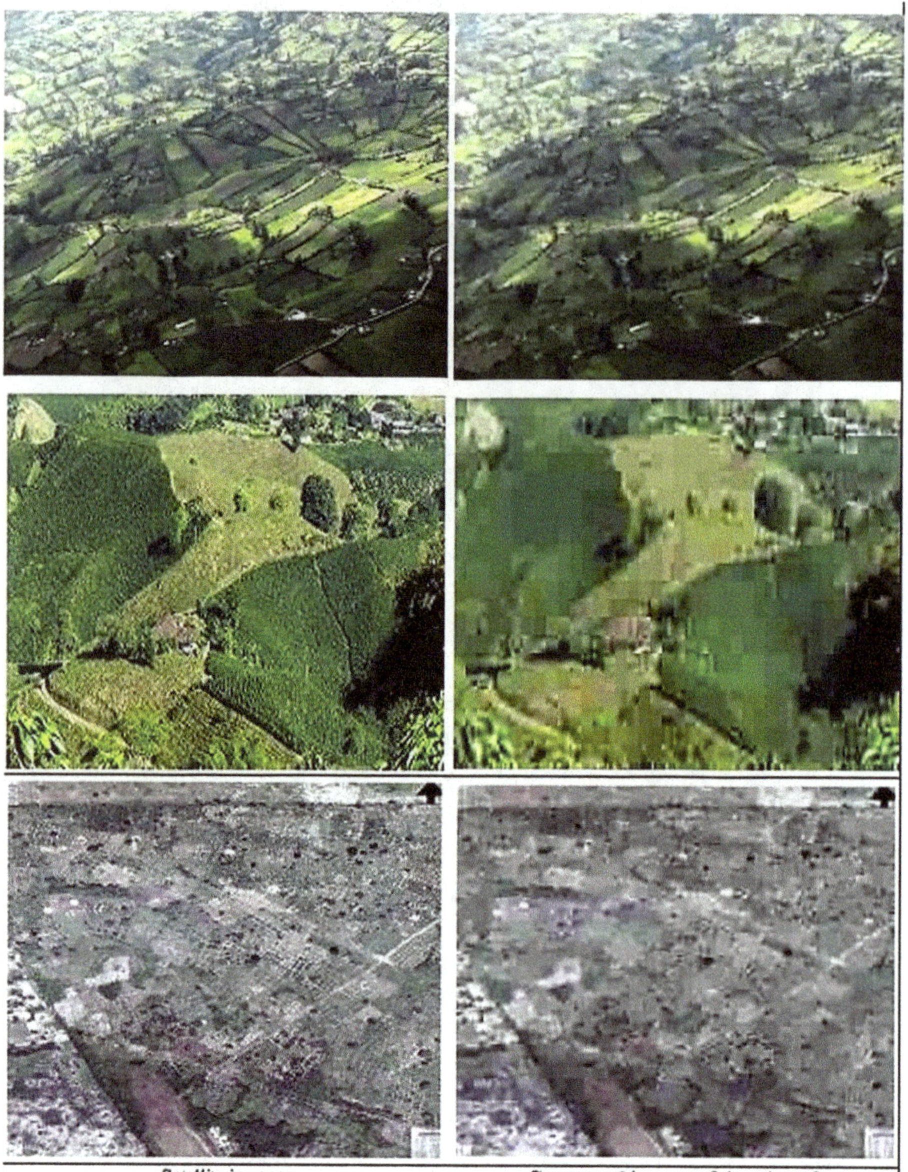

Satellite imagery | Compressed imagery of the column 1

Table 3. Average Compression Ratio for the Satellite Imageries

Images	CR(%)		PSNR (dB)	
	AGMRF	JPEG	AGMRF	JPEG
Imagery in row 1	94.56	94.32	30.28	28.61
Imagery in row 2	95.84	95.47	30.95	27.34
Imagery in row 3	96.02	95.61	31.27	28.38

and JPEG were tabulated in Tables-1 to Table-3. From the experimental results, it is observed that the AGMRF model gives a higher compression ratio and good reconstruction quality for both images captured by digital camera and the satellite.

DISCUSSION

First, the Bayesian methodology infers the estimates the hyperparameters. Later, the M-H algorithm and the AC are applied to attain the precise parameter estimate. After estimating parameters, the model generates the pixel values with the negligible period. The AGMRF parameter estimate method consumes less time than other iterative approaches such as the Expectation-Maximization (EM) (Coelho & Estrela, 2012) and Neural Network (NN) procedures (Chib & Greenberg, 1995). In addition, there is no dictionary-based data storage overhead in the AGMRF scheme as in JPEG 2000, and its sequel versions and other approaches developed based on DCT. Hence, the AGMRF scheme consumes a minimum of time and less memory when compared to others existing techniques like JPEG. The methodologies relying on the block-based DCT compression method results in the visible blocky artifacts at boundaries owing to coarse coefficients' quantization (Luo & Ward, 2003).

The AGMRF technique overcomes this drawback by predicting the sub-image pel intensities to be reinstated with the use of hyperparameters, seed values, and the average residuals' values with decent reconstruction quality. Despite that, one flaw of the AGMRF technique is the fact it smooths sharp edges.

FUTURE WORK

The MRFs accept noncausal models, with isotropic comportment and relying only on local dependencies. However, figuring out probabilities and performing parameter estimation are challenging activities because of the computational burden that comes from the iterative energy optimization schemes. Since the model builds upon the Gaussianity assumption, one has to investigate other pdfs that can give better results for plant images.

Texture analysis is paramount in agrarian imaging because it characterizes understated properties of vegetation and tissues, which are not straightforwardly quantified. As such, the most effective methods are the ones that manage intrinsic variability and noise either by redesigning the descriptors used in other CV realms or by applying sophisticated feature encoding to shrink different types of local evidence into a compact, multi-scale as well as invariant representations. All the previously cited approaches work with handcrafted texture descriptors.

Efficient IC has a significant practical and marketable interest (Vaddella, & Rama, 2010). Image compression (IC) provides a tradeoff between the quality of digital images while keeping the bandwidth and storage space to minimum levels. Error Correction Coding (ECC) aids to broadcast signals over possibly flawed communication channels by means of a skillful usage of the original visual dataredundancy.

Self-testing ECC classical methods are usually employ the Hamming distance and repeat-and-accumulate (RaA) techniques (Park, and Yang, 2002; Chandra, & Singh, 2008; Watkinsa & Sayeh, 2016) where the Hamming coding outperforms RaA coding. Three NN models are presented in (Chandra & Singh, 2008) as another solution to the ECC problem.

Deep Learning (DL) designs can learn texture models from the images straightforwardly. Although DL methods require very large image sets, it holds several promises. As DL architectures are becoming more and more accessible and rapidly growing, the amount of positive outcomes also grows.

Since most compression strategies are based either on handcrafted image descriptors or some predefined texture models (Hemanth & Estrela, 2017; Herrmann & Estrela, 2016; Hussain & Jeong, 2015). The discriminative powers of each procedure hang on (i) the use of realistic a model describing the image features undergoing examination; (ii) the descriptor efficiency regarding compactness and robustness to optical variations when it is an input to a classifier. Hence, the general research focus is the design of representations involving these two notions.

Still, two bottlenecks need to be examined. Initially, in-depth a priori knowledge about the image characteristics have to be analyzed, segmented, and classified. Second, it is strongly application- dependent, or in other words, any model has its best performance with minimal imaging conditions. Hence, a perfect descriptor for a particular class of images does not guarantee the same outcome once applied to different image varieties.

DL abandons handcrafted features, and let the texture analysis structure learn the model straight from the images. DL architectures are increasingly popular in CV and pattern recognition processing that rely on distributed data representations. The core assumption is to represent the observed data by interactions among multiple factors, organized in layers. Each layer characterizes a different abstraction level, hierarchically from the lowest to highest: the former carries more low-level information about the pixel intensities distributions, while the latter affords a more abstract input representation. Hence, the abstraction level can be simply controlled by changing the number beside the size of the layers. A raw image can be fed into a DL network as a vector containing pixel values. Each layer connects locally to the preceding one to learn and to generalize structures to describe the original image at progressively higher abstraction levels, usually using the backpropagation algorithm. The first layer usually maps the edges in the image at specific locations or orientations. A second layer accomplishes rough pattern detection, in that it detects particular edge structures. A third layer detects spatial combinations of such patterns, and so on. Hence, DL can deliver compact and non-redundant halfway representations, evading the extraction of handcrafted features.

DL algorithms can implement either supervised or unsupervised tasks that usually rely on feedforward neural networks. The NN learns to relate a fixed-size input (like an untreated image) to a fixed-size output such as a label, or the probability of finding a particular texture type.

The Autoencoder is a simple unsupervised network that transforms the unrefined input image into a satisfactory representation through a backpropagation algorithm with the least possible distortion (Dimililer, 2013; Patel &Agrawal, 2013). It has two symmetrical deep-belief networks standing as the coding and decoding system parts. The encoding layers compress the raw input image, with gradually higher reduction of the feature size. As the amount of hidden layers rise, so does the compression level. The decoding layers recreate the input with its original feature size. Therefore, the intermediate layer offers a reduced set of descriptive features aimed at texture classification. The Autoencoder can capture high-level features from raw images to distinguish textures.

The contemporary Convolutional Neural Networks (CNNs) are DL networks with convolutional layers devoted to featuring extraction. Convolutional layers are interlaced with subsampling layers (for pooling), followed by fully-connected layers dedicated to classification (Hemanth & Estrela, 2017). The success of CNNs comes from learning more complex input transformations progressively and apprehend invariances from huge labeled datasets. Moreover, CNNs delivered promising results with large image datasets to obtain the pre-trained general-purpose feature extractors that can toil with other kinds of agrarian imaging. The most important CNN caveat is the necessity for big datasets to explain representative features.

The sophisticated combination of different kinds of multi-scale descriptors through feature encoding techniques can improve knowledge handling (de Jesus et al., 2018; de Jesus et al.,2016; Fernandes et al., 2014). Nevertheless, it encounters a familiar DL caveat: its limited training set size.

A progressive although a significant topic is that a vast and complex cyber-farm built upon the Industry 4.0 standard is very likely to possess quite a few Visual Wireless Sensor and Actuator Nodes (VWSANs) to monitor vegetation from different points of the crop. According to the kinds of vegetation involved, the CV structure can profit from the cyber-physical system paradigm (Estrela et al., 2018; Estrela et al., 2017).

CONCLUSION

Handling visual and multi-dimensional information is an extremely data-intensive task, which entails outsized volumes of memory for storage and, consequently, high communication bandwidths. It is conceivable to wipe out redundant information contained in visual and multi-dimensional records utilizing Image Compression (IC) and better visual representation strategies. The IC scheme comprises two phases: compression and decompression.

In this chapter, the Adaptive Gaussian Markov Random Field (AGMRF) model was proposed for agricultural IC. The model parameters resulted from applying the Bayesian methodology. The M-H algorithm kept the model parameters up-to-date. The average of the residual between the original as well as decompressed images was calculated. The seed values, the average value of the residual, along with the model parameters underwent the Arithmetic Coding (AC) to accomplish a higher Compression Ratio (CR) of 96% and above, and the MSE and PSNR metrics evaluated the performance of the AGMRF model and compared it to the JPEG standard. The obtained results reveal that the AGMRF performs well when compared to the existing IC methods.

REFERENCES

Adams, M. D., & Kossentini, F. (2000). Reversible integer-to-integer Wavelet transforms for image compression: Performance evaluation and analysis. *IEEE Transactions on Image Processing*, *9*(6), 1010–1024. doi:10.1109/83.846244 PMID:18255472

Anderson, G. B., & Huang, T. S. (1971). Piecewise Fourier transform for picture bandwidth compression. *IEEE Transactions on Communication Technology*, *COM19*(2), 133–140. doi:10.1109/TCOM.1971.1090630

Andrews, A. C., & Pratt, W. K. (1968). Fourier Transform Coding of images. In *Proc. International Conference on Systems Sciences* (pp. 677-679). Academic Press.

Andrews, H. C. (1970). *Computer techniques in image processing*. New York: Academic Press.

Antonini, M., Barlaud, M., Mathieu, P., & Daubechies, I. (1992). Image coding using Wavelet Transform. *IEEE Transactions on Image Processing*, *I*(2), 205–220. doi:10.1109/83.136597 PMID:18296155

Bose, R., & Pathak, S. (2006). A novel compression and encryption scheme using variable model arithmetic coding and coupled chaotic system. *IEEE Transactions on Circuits and Systems—I*, *53*(4), 848–856.

Brezger, A., Fahrmeir, L., & Hennerfeind, A. (2007). *Adaptive Gaussian Markov Random Fields with Applications in Human Brain Mapping*. Academic Press.

Chandra, P., & Singh, Y. (2008). Fault tolerance of feedforward artificial neural networks- a framework of study. *Proc. of the 2008 Int'l Joint Conf. on Neural Networks*, 1, 489-494.

Chellappa, R., Chatterjee, S., & Bagdazian, R. (1985). Texture Synthesis and Compression Using Gaussian-Markov Random Field Models. *IEEE Transactions on Systems, Man, and Cybernetics*, *15*(2), 298–303. doi:10.1109/TSMC.1985.6313361

Chib, S., & Greenberg, E. (1995). Understanding the Metropolis Hastings algorithm. *American Statistical Association*, *49*(4), 327–335.

Coelho, A. M., & Estrela, V. V. (2012). EM-based mixture models applied to video event detection. In P. Sanguansat (Ed.), Principal component analysis—Engineering applications, InTech open. Academic Press. doi:10.5772/38129

de Jesus, M. A., Estrela, V. V., & Saotome, O. (2016) Super-resolution in a nutshell. In *Proceedings of the Brazilian Technology Symposium 2016 (BTSym 2016)*. Academic Press.

de Jesus, M. A., Estrela, V. V., Saotome, O., & Stutz, D. (2018). Super-resolution via particle swarm optimization variants. In J. Hemanth & V. Balas (Eds.), *Lecture Notes in Computational Vision and Biomechanics* (Vol. 25). Cham: Springer.

Deever, A. J., & Hemami, S. S. (2003). Lossless image compression with projection-based and adaptive reversible integer Wavelet transforms. *IEEE Transactions on Image Processing*, *12*(5), 489–499. doi:10.1109/TIP.2003.812374 PMID:18237926

Dimililer, K. (2013). Backpropagation neural network implementation for medical image compression. *Journal of Applied Mathematics, 2013*, 1–8. doi:10.1155/2013/453098

Estrela, V. V., Hemanth, J., Saotome, O., Grata, E. G. H., & Izario, D. R. F. (2019). Emergency response cyber-physical system for flood prevention with sustainable electronics. In *Proceedings of the 3rd Brazilian Technology Symposium. BTSym 2017*. Springer. 10.1007/978-3-319-93112-8_33

Estrela, V. V., Saotome, O., Loschi, H. J., Hemanth, J., Farfan, W. S., Aroma, J., ... Grata, E. G. H. (2018). Emergency response cyber-physical framework for landslide avoidance with sustainable electronics. *Technologies, 6*(2), 42. doi:10.3390/technologies6020042

Fernandes, S. R., Estrela, V. V., Magalhaes, H. A., & Saotome, O. (2014). On improving sub-pixel accuracy by means of B-Spline. In *Proceedings of the 2014 IEEE international conference on imaging system and techniques (IST 2014)*. Santorini, Greece: IEEE. 10.1109/IST.2014.6958448

Flierl, M., & Girod, B. (2003). Generalised B pictures and the draft H.264/AVC video compression standard. *IEEE Transactions on Circuits and Systems for Video Technology, 13*(7), 587–597. doi:10.1109/TCSVT.2003.814963

Fountas, S., Sorensen, C. G., Tsiropoulos, Z., Cavalaris, C., Liakos, V., & Gemtos, T. (2015). Farm machinery management information system. *Computers and Electronics in Agriculture, 110*, 131–138. doi:10.1016/j.compag.2014.11.011

Golomb, S. W. (1996). Run length encoding. *IEEE Transactions on Information Theory, 12*(3), 399–401. doi:10.1109/TIT.1966.1053907

Habibi, A., & Wintz, P. A. (1971). Image Coding by linear transformation and block quantization. *IEEE Transactions on Communication Technology, COM19*(1), 50–62. doi:10.1109/TCOM.1971.1090601

Hamzaoui, R. (1999). Fast iterative methods for fractal image compression. *Journal of Mathematical Imaging and Vision, 11*(2), 147–159. doi:10.1023/A:1008343530356

Hartension, H., Ruhl, M., & Saupe, D. (2000). Region based fractal image compression. *IEEE Transactions on Image Processing, 9*(7), 1171–1184. doi:10.1109/83.847831 PMID:18262956

Hemanth, D. J., & Estrela, V. V. (2017). *Deep Learning for Image Processing Applications*. IOS Press.

Herrmann, A. E., & Estrela, V. V. (2016). Content-based image retrieval (CBIR) in remote clinical diagnosis and healthcare. In M. Cruz-Cunha, I. Miranda, R. Martinho, & R. Rijo (Eds.), *Encyclopedia of E-Health and Telemedicine* (pp. 495–520). Hershey, PA: IGI Global. doi:10.4018/978-1-4666-9978-6.ch039

Howard, P., & Vitter, J. (1992). New methods for lossless image compression using arithmetic coding. *Journal of Information Processing and Management*, *28*(5), 765–779. doi:10.1016/0306-4573(92)90067-A

Hussain, F., & Jeong, J. (2015). Exploiting deep neural networks for digital image compression. *Proc. 2nd World Symp. on Web Appl. and Networking (WSWAN)*, 1-6. 10.1109/WSWAN.2015.7210294

ISO/IEC 15444 I ITU-T Recommendation T.800, Information technology - JPEG 2000 image coding system: Core coding system, November 2015.

Jagadish, H. P., & Lohit, M. K. (2010). A new lossless method of image compression and decompression using Huffman coding techniques. *Journal of Theoretical & Applied Information Technology*, *5*(2), 18–23.

Jain, A. K. (1981). Image data compression: A review. *Proceedings of the IEEE*, *69*(3), 366–406. doi:10.1109/PROC.1981.11971

Jeon, B., Park, J., & Jeong, J. (1998). Huffman coding of DCT coefficients using dynamic code word assigned and adaptive codebook selection. *Signal Processing Image Communication*, *22*(3), 253–262. doi:10.1016/S0923-5965(97)00041-6

JPEG2000 Part I Final Committee Draft Version 1.0. ISO/IECJTC1/SC29 WG1 N1646R, 2000.

Kingston, A., & Autrusseau, F. (2008). Lossless image compression via predictive coding of discrete Radon projections. *Signal Processing Image Communication*, *23*(4), 313–324. doi:10.1016/j.image.2008.03.001

Li, Z., Aaron, A., Katsavounidis, I., Moorthy, A., & Manohara, M. (2016). Toward a practical perceptual video quality metric. *The Netflix Tech Blog*. Retrieved October 13, 2018, https://medium.com/netflixtechblog/toward-a-practical-perceptual-video-quality-metric-653f208b9652

Luo, Y., & Ward, R. K. (2003). Removing the blocking artifacts of block-based DCT compressed images. *IEEE Transactions on Image Processing*, *12*(7). PMID:18237958

Magli, E., & Olmo, G. (2003). Lossy predictive coding of SAR raw data. *IEEE Transactions on Image Processing, 41*(5), 977–987.

Miucic, R., Mahmud, S., & Popovic, Z. (2009). An enhanced data-reduction algorithm for event-triggered networks. *IEEE Transactions on Vehicular Technology, 58*(6), 2663–2678. doi:10.1109/TVT.2008.2011361

Moallem, P., Razmjooy, N., & Ashourian, M. (2013). Computer vision-based potato defect detection using neural networks and support vector machine. *International Journal of Robotics and Automation, 28*(2), 137–145. doi:10.2316/Journal.206.2013.2.206-3746

Moallem, P., Razmjooy, N., & Mousavi, B. S. (2014). Robust potato color image segmentation using adaptive fuzzy inference system. *Iranian Journal of Fuzzy Systems, 11*(6), 47–65.

Motta, G., Storer, J. A., & Carpentieri, B. (2000). Lossless image coding via adaptive linear prediction and classification. *Proceedings of the IEEE, 88*(11), 1790–1796. doi:10.1109/5.892714

Nelson, M., & Gailly, J. (1996). *The data compression book* (2nd ed.). New Delhi: BPB Publications.

Park, S. I., & Yang, K. (2002). Extended Hamming accumulate codes and modified irregular repeat accumulate codes. *Electronics Letters, 38*(10), 467–468. doi:10.1049/el:20020316

Patel, B., & Agrawal, S. (2013). Image compression techniques using artificial neural network. *Int. J. Adv. Res. Comput. Eng. Technol., 2*(10), 2725–2729.

Pratt, W. K., Chen, W. H., & Welch, L. R. (1973). Slant transform image coding. *IEEE Transactions on Communications, COM22*(8), 1075–1093. doi:10.1109/TCOM.1974.1092335

Rabbani, M., & Joshi, R. (2002). An Overview of the JPEG 2000 Still Image Compression Standard. *Signal Processing Image Communication, 17*(1), 3–48. doi:10.1016/S0923-5965(01)00024-8

Ramteke, P. R., & Mahmud, S. M. (2005). *An adaptive data-reduction protocol for the future in-vehicle networks. Tech. Rep.* SAE.

Razmjooy, N. (2012). A multi layer perceptron neural network trained by invasive weed optimization for potato color image segmentation. *Trends in Applied Sciences Research, 7*(6), 445–455. doi:10.3923/tasr.2012.445.455

Razmjooy, N., Mousavi, B. S., & Soleymani, F. (2013). A hybrid neural network Imperialist Competitive Algorithm for skin color segmentation. *Mathematical and Computer Modelling*, *57*(3-4), 848–856. doi:10.1016/j.mcm.2012.09.013

Razmjooy, N., Mousavi, B. S., Soleymani, F., & Khotbesara, M. H. (2013). A computer-aided diagnosis system for malignant melanomas. *Neural Computing & Applications*, *23*(7-8), 2059–2071. doi:10.100700521-012-1149-1

Stone, M., McKee, K., Formwalt, C., & Benneweis, R. (1999). *ISO 11783: An Electronic Communications Protocol for Agricultural Equipment*. Agricultural Equipment Technology Conference, ASAE, Louisville, KY.

Sullivan, C. J., Ohm, J.-R., Han, W.-J., & Wiegand, T. (2012). Overview of the high efficiency video coding (HEVC) standard. *IEEE Transactions on Circuits and Systems for Video Technology*, *22*(12), 1649–1668. doi:10.1109/TCSVT.2012.2221191

Sweldens, W. (1996). The lifting scheme: A custom-design construction of biorthogonal wavelets. *Applied and Computational Harmonic Analysis*, *3*(2), 186–200. doi:10.1006/acha.1996.0015

Vaddella, V., & Rama, K. (2010). Artificial neural networks for compression of digital images: A review. *Int. J. Rev. Comput*, *3*, 75–82.

Watkinsa, Y. Z., & Sayeh, M. R. (2016). Image data compression and noisy channel error correction using deep neural network. *Procedia Computer Science*, *95*, 145–152. doi:10.1016/j.procs.2016.09.305

Watson, A. B. (1998). Toward a perceptual video quality metric. *Human Vision, Visual Processing, and Digital Display. VIII*, *3299*, 139–147.

Wintz, P. A., & Tasto, M. (1971). Image coding by adaptive block quantization. *IEEE Transactions on Communication Technology*, *COM19*, 957–972.

Wohlberg, B., & Jager, G. D. (1999). A review of the fractal image coding literature. *IEEE Transactions on Image Processing*, *8*(12), 1716–1729. doi:10.1109/83.806618 PMID:18267449

ADDITIONAL READING

Elvira, C., Chainais, P., & Dobigeon, N. (2017). Bayesian antisparse coding. *IEEE Transactions on Signal Processing*, *65*(7), 1660–1672. doi:10.1109/TSP.2016.2645543

Estrela, V. V., & Galatsanos, N. P. (2000). Spatially adaptive regularized pel-recursive motion estimation based on the EM algorithm. In B. Vasudev, T. R. Hsing, A. G. Tescher, & R. L. Stevenson (Eds.), *In Proc. Image and Video Communications and Processing 2000* (Vol. 3974, pp. 372–383). San Jose, CA, USA: SPIE; doi:10.1117/12.382969

Estrela, V. V., Rivera, L. A., & Bassani, M. H. (2004). Pel-recursive motion estimation using the expectation-maximization technique and spatial adaptation. In *Proc. 2004*. WSCG.

Jing, J., Huang, M., Li, P., & Ning, X. (2018). Automatic measurement of yarn hairiness based on the improved MRMRF segmentation algorithm. *Journal of the Textile Institute*, *109*(6), 740–749. doi:10.1080/00405000.2017.1368106

Kamilaris, A., & Prenafeta-Boldú, F. X. (2018, April 1). Deep learning in agriculture: A survey. *Comp. and Electronics in Agriculture*, Elsevier B.V. doi:10.1016/j.compag.2018.02.016

Krishnamachari, S., & Chellappa, R. (1997). Multiresolution Gauss-Markov random field models for texture segmentation. *IEEE Transactions on Image Processing*, *6*(2), 251–267. doi:10.1109/83.551696 PMID:18282921

Liu, Z., Li, X., Luo, P., Loy, C. C., & Tang, X. (2018). Deep learning Markov random field for semantic segmentation. *IEEE Transactions on Pattern Analysis and Machine Intelligence*, *40*(8), 1814–1828. doi:10.1109/TPAMI.2017.2737535 PMID:28796610

Mariño, I. P., Zaikin, A., & Míguez, J. (2017). A comparison of Monte Carlo-based Bayesian parameter estimation methods for stochastic models of genetic networks. *PLoS One*, *12*(8), e0182015. doi:10.1371/journal.pone.0182015 PMID:28797087

Wang, C., Komodakis, N., & Paragios, N. (2013). Markov Random field modeling, inference & learning in computer vision & image understanding: A survey. *Computer Vision and Image Understanding*, *117*(11), 1610–1627. doi:10.1016/j.cviu.2013.07.004

Wu, Z., Lin, D., & Tang, X. (2016). Deep Markov random field for image modeling. In B. Leibe, J. Matas, N. Sebe, & M. Welling (Eds.), Lecture Notes in Computer Science: Vol. 9912. *Proc. Computer Vision – ECCV 2016. ECCV 2016*. Cham: Springer. doi:10.1007/978-3-319-46484-8_18

KEY TERMS AND DEFINITIONS

Compression Ratio: It measures the relation between the number of bits for an image after decompression and its original size.

Handcrafted Image Descriptor: An image descriptor where the set of features is handpicked by the designer of the algorithm.

Image Compression: It is the process of reducing the size of an image of video by means of redundant information.

Image Descriptor: Set of features used to describe a scene or object.

Markov Random Field: Undirected graphical model containing a set of random variables with the Markov property that can be represented by an undirected graph.

Metropolis-Hastings (M-H) Algorithm: Markov Chain Monte Carlo (MCMC) scheme for attaining a sequence of random samples following a given probability distribution whose direct sampling is problematic.

Texture: It is the nature of a surface defined by the small, local deviations from the perfectly flat ideal. It provides information about the spatial organization of color or brightness patterns in an image or particular region of an image.

Chapter 10
Deep Learning-Based Mobile Application for Plant Disease Diagnosis:
A Proof of Concept With a Case Study on Tomato Plant

Shradha Verma
https://orcid.org/0000-0002-7053-1654
GGSIP University, India

Amit Prakash Singh
https://orcid.org/0000-0002-8675-6903
GGSIP University, India

Anuradha Chug
GGSIP University, India

Shubham Sharma
GGSIP University, India

Puranjay Rajvanshi
VIT University, India

ABSTRACT

With the increasing computational power, areas such as machine learning, image processing, deep learning, etc. have been extensively applied in agriculture. This chapter investigates the applications of the said areas and various prediction models in plant pathology for accurate classification, identification, and quantification of plant diseases. The authors aim to automate the plant disease identification process. To accomplish this objective, CNN has been utilized for image classification. Research shows that deep learning architectures outperform other machine learning tools significantly. To this effect, the authors have implemented and trained five CNN models, namely Inception ResNet v2, VGG16, VGG19, ResNet50, and Xception, on PlantVillage dataset for tomato leaf images. The authors analyzed 18,160 tomato leaf images spread across 10 class labels. After comparing their performance measures, ResNet50 proved to be the most accurate prediction tool. It was employed to create a mobile application to classify and identify tomato plant diseases successfully.

DOI: 10.4018/978-1-5225-8027-0.ch010

INTRODUCTION

Advanced research in agriculture has led to the concept of smart farming and development of novel techniques, to improve the crop yield and provide assistance to the farmers, but pests and plant diseases are still a dominant threat to the overall plant health and food production.

With the current population growth rate, it is imperative that researchers investigate every potential solution that would facilitate the protection of crop yield. For improving the sustainability, we require to minimize the economic and production losses, along with the degrading impact of fertilizers and pesticides on the environment. Several researchers have consecrated their valuable efforts in developing new and innovative practices for disease diagnosis in plants (Razmjooy et al, 2012; Moallem et al, 2013, 2014). Previously, this domain has remained highly dependent on the conventional and long-established approaches.

Traditional methods encompass visual inspections by experienced individuals, be it the farmer or a botanist. Detecting plant diseases in a lab are one other option. It comprises bringing back samples from the field and conducting a microscopic evaluation and diagnostic experiments, such as ELISA, PCR etc. While this is an accurate procedure, it is time-consuming, not to mention costly, requiring laboratory equipment set-up and highly labor-intensive.

Given the restricted access to resources and limited expertise in plant pathology, there is a vital requirement for processes automation. Farmers are struggling throughout the world to protect their crop from the onslaught of several harmful microorganisms or pathogens such as a virus, bacteria, fungus, nematodes, protozoa (Shankar, Harsha, & Bhandary, 2014) as well as feeding off the insects.

A susceptible host and favorable environmental conditions are all these pathogens require, in order to infect the plants and eventually degrade their growth, sometimes resulting in high mortality. They are ubiquitous and stay inactive in the soil, air and/ or water, even in the crop debris for multiple seasons, until they encounter suitable conditions to infest the plant.

In the event of infestation of pests in the crop, it undergoes biological and chemical stress that initially fails to provide any visual cues. It is only after the disease has manifested to a greater severity level, that visual symptoms appear and demand immediate action to control the outbreak.

The visual symptoms may cause some changes in different parts of the plant like shape, size, and color.

Also, to detect any such symptoms, constant monitoring of the crop is required, which can prove to be a tiresome task.

For Agronomists and researchers, plant disease diagnosis has become a greater concern over time and has led them to devise a multitude of techniques for early detection and prediction of diseases that acutely affect the yield and quality of the grain.

Recent research in the sensor based techniques has led to the identification and development of imaging technologies that are extensively being utilized for detection as well as identification of plant diseases, in addition to being non-destructive in nature.

Images capture any anomalies in the occurrence of a plant, which might indicate the attack of a pathogen. Optical properties of a plant are exploited to detect plant stress levels and disease severity. Lately, the emphasis has been towards early disease detection, which is possible due to the recent advances in the remote sensing and imaging technologies (Mahlein, 2016) such as multispectral imaging, hyperspectral imaging, thermal imaging etc. As in humans, early diagnosis in plants can help eliminate and cure the disease, with minimum cost, effort and use of pesticide, hence preserving the ecosystem from its damaging effects.

For more than last four decades, researchers have contributed significantly towards achieving this goal by utilizing the computational power of advanced technologies such as machine learning, deep learning, digital image processing, computer vision, genetic algorithms, big data, internet of things etc. These areas of computing have led to various suitable solutions in agriculture, from crop protection and monitoring to crop quality and its management.

With their combined application, the process of data collection and analysis can be largely automated, leading to the timely prediction of most probable diseases in crops. It would also eliminate the dependency on manual tasks and subject matter experts. Table 1 gives the nomenclature of the general terms used throughout the text.

Following sections in this chapter highlights numerous image processing, machine learning and deep learning based plant disease diagnosis techniques devised in the last few decades, along with the implementation of five CNN models and a mobile application to predict tomato plant diseases.

BACKGROUND

Pathogen outbreak is highly dependent on environmental conditions. Early prediction approaches employing forecasting technologies, based on weather, can prove to be highly beneficial in minimizing the pesticide usage for disease control in plants; thereby reducing the subsequent cost and environmental damage. Such systems can act as a warning system and heed in the timely control of the disease.

Kaundal et al. (2006) have designed cross-year and cross-location models for the duration of five consecutive years from 2000-2004 and 5 distant locations, for rice blast. The coefficient correlation, the coefficient of determination and %MAE

Table 1. Nomenclature

ANN	Artificial Neural Networks
CNN	Convolutional Neural Networks
REG	Regression
SVM	Support Vector Machines
BPNN	Back Propagation Neural Networks
GRNN	Generalized Regression Neural Networks
MAE	Mean Absolute Error
VI	Vegetation Indices
MDC	Minimum Distance Criterion
FFBPNN	Feed Forward Back Propagation Neural Networks
GA	Genetic Algorithms
ROC	Receiver Operator Characteristic
PCA	Principal Components Analysis
RBF	Radial Basis Function
MLP	Multi-Layer Perceptron
FCM	Fuzzy C Means
LVQ	Learning Vector Quantization
SIMCA	Soft Independent Modeling of Class Analogy
DPLS	Discriminant Partial Least Squares
MD	Mahalanobis Distance
MSE	Mean Squared Error

were computed and validated by implementing and comparing prediction accuracies of SVM, REG, GRNN, and BPNN.

Six environmental predictor variables were recorded: temperature (min/max), relative humidity (min/max), rainy days in a week and rainfall, where rainfall was found to be the most effective among the other variables. In response to the attack of several micro-organisms, plants have developed certain vital self-defense mechanisms viz. genes for resisting diseases, also known as R genes. Pal et al. (2016) have investigated an SVM-based machine learning tool, trained for fast and accurate identification of R Proteins (112 in total), with 91.11% accuracy.

These R Proteins can be exploited for the breeding of novel plant species that are inherently disease resistant. SVM has been the most efficient machine learning technique employed in various researches. Pattern recognition techniques are also majorly applied for disease identification in plants, whereby numerous features pertaining to color, shape, and texture are extracted (Camargo & Smith, 2009).

However, not all features are helpful for training the machine learning tool, they may provide superficial, redundant or no information at all. Hence, such features can be discarded after a thorough analysis. Petrellis (2015) proposed a cost-effective, low complexity approach for a mobile application that focuses on visible lesions, their coverage and counts, rather than extracting features, which as mentioned, is a common practice in approaches based on Image Processing.

For the majority of Image processing based techniques, plant leaf is the most utilized part of a plant for disease identification, as it shows clear visual symptoms of changes in color, shape or size. Recent advancements and research in sensor-based techniques have resulted in various non-destructive approaches, applicable to disease diagnosis and plant pathology.

Measurements of disease severity level and stress in a plant can be captured by observing its optical properties. RGB imaging, multi- and hyperspectral imaging, thermal imaging, fluorescence spectroscopy are amongst the few non-invasive methods for prediction and prevention of plant diseases (Mahlein, 2016). As mentioned previously, visual symptoms indicate the manifestation of the disease to a greater severity level, hence modifying the appearance of the plant.

In such cases, RGB images of the diseased plant are captured via a digital camera deployed in the field. These images are preprocessed and transformed into some suitable color model for feature extraction and finally, computed features are used for training a machine learning based tool for classification of the disease. Al-Bashish et al. (2011) have converted the RGB images into HSI color model, employed k-means clustering for segmentation, co-occurrence method for texture-based feature extraction (eleven features in total) and trained FFBPNN with ten hidden layers for classification that resulted in 93% accuracy.

Similarly, Sabrol and Kumar (2016) have suggested extracting ten statistical features such as mean, skewness, standard deviation and correlation, and exercising a decision tree with 80 nodes for classification of tomato diseases.

These four steps (color space conversion, segmentation, feature extraction, and classification) have been the most commonly used approach in Image Processing field concerning RGB imaging for disease identification in plants.

With minor modifications, researchers have investigated the variations of this process to better the results such as using mean, normalized mean and green ratios of R, G, B values, for computing features and accuracy (Liu, Zhang, Shu, & Jin, 2013).

Another promising imaging technique is Hyperspectral imaging that has emerged as an effective method for early prediction, without the necessity of appearance of the visual symptoms. It captures spatial as well as spectral data, conforming to the subtle modifications in the biological and chemical properties of a plant in the form of reflectance values in various wavelengths.

These reflectance values are then employed to compute predefined VIs such as Normalized Difference Vegetation Index (NDVI), Simple Ratio Index (SRI), Photochemical Reflectance Index (PRI), Plant Senescence Reflectance Index (PSRI) etc., whose values indicate the biotic and abiotic stress in the plants such as water content, chlorophyll content, nutrients deficiency etc., hence, initiating the necessary course of action.

Multispectral Imaging is similar to hyperspectral, as in both capture spectral data, but multispectral imaging records data for 3-10 broad bands while hyperspectral imaging captures data in thousands of narrow bands, hence proves to be more accurate, while posing difficulty in identifying significant bands for data analysis (Zhao, Li, Yu, Cheng, & He, 2016), not to mention expensive and sophisticated hardware setup. Increased temperature of the plant surface can also indicate an infestation of pathogens, where different levels of infection have distinct heat signatures.

For such cases, thermal imaging can be used to capture the reflected infrared radiations for disease diagnosis.

Table 2 and Table 3 summarizes the numerous machine learning based statistical methods and imaging techniques successfully applied for identification, quantification, and classification of plant diseases. Genetic Algorithms based segmentation of colored images of the cotton crop has been proposed by Singh and Misra (2016), whereby after preprocessing and image enhancement, green colored pixels are masked followed by the segmentation.

Features such as energy, contrast, shade etc. are computed via the color co-occurrence method, and SVM and MDC are trained for classification. SVM outperforms MDC and gives a prediction accuracy of 95.71%. The combination of GA and SVM has been successfully applied for examining the quality of rice seeds at an early stage to differentiate between healthy and diseased food grains, infected with a seed-borne disease called Bakanae disease (Chung et al., 2016).

DEEP LEARNING: APPLICATIONS IN AGRICULTURE

Deep learning has been extensively applied to various computational fields viz. computer vision, speech recognition, natural language processing, image classification, but first it was utilized for detecting handwritten digits in documents.

While machine learning requires pre-computed features to be fed to the classifier for training, deep learning streamlines the process and elicits these features directly from the data; hence, it encapsulates the feature extraction as well as classification processes.

Its successful application in classifying images was exploited to distinguish among healthy and diseased plants, based on leaf images. But its performance is

Table 2. Summary of statistical approaches for plant disease identification and classification

Publication and Year	Statistical Approach	Crop	Disease
Ferentinos (2018)	Deep Learning (AlexNetOWTBn, VGG)	25 Plant Species	58 different plant-disease pair
Wang, Sun, & Wang (2017)	Deep Learning (VGG16, VGG19, Inception-v3, ResNet50)	Apple	Apple Black Rot
Brahimi, Boukhalfa, & Moussaoui (2017)	Deep Learning (AlexNet, GoogleNet)	Tomato	Spider Mites, Septoria Spot, Late Blight, Early Blight, Bacterial Spot, Tomato Yellow Leaf Curl Virus, Leaf Mold, Tomato Mosaic Virus, Target Spot
Durmus, Günes, & Kirci (2017)	Deep Learning (AlexNet, SqueezeNet)	Tomato	Spider Mites, Septoria Spot, Late Blight, Early Blight, Bacterial Spot, Tomato Yellow Leaf Curl Virus, Leaf Mold, Tomato Mosaic Virus, Target Spot
Lu, Zhou, Gao, & Jiang (2017)	ROC curve analysis	Tomato	Yellow Leaf Curl Disease
Mohanty, Hughes, & Salathé (2016)	Deep Learning (GoogleNet, AlexNet)	14 crops	26 diseases
Sabrol & Kumar (2016)	Decision Trees	Tomato	Late Blight, Bacterial Leaf Spot, Septoria Leaf, Leaf Curl, Bacterial Canker
Zhao et al. (2016)	Partial Least Square Regression (PLSR)	Cucumber	Angular Leaf Spot (ALS) disease
Singh & Misra (2016)	SVM classifier + Minimum Distance Criterion+ K-means	Rose, beans leaf, lemon leaf, banana leaf	Banana leaf: Early scorch disease; Lemon leaf: Sun burn disease; Beans & Rose: Bacterial disease; Beans leaf: Fungal disease
Chung et al. (2016)	SVM	Wheat	Bakanae disease
Sabrol & Kumar (2016)	Fuzzy Inference System (FIS), Adaptive neuro-fuzzy inference system, Multi-layer FFBPNN	Tomato	Late Blight, Bacterial Leaf Spot, Tomato Leaf Curl, Septoria Leaf Spot, Bacterial Canker
Rupanagudi, Ranjani, Nagaraj, Bhat, & Thippeswamy (2015)	K-means Clustering and PCA	Tomato	Borer Insects
Raza, Prince, Clarkson, & Rajpoot (2015)	SVM classifier with a linear kernel	Tomato	Powdery Mildew
Xie, Shao, Li, & He (2015)	Extreme Learning Machine (ELM) classifier model	Tomato	Early Blight, Late Blight

continued on following page

Table 2. Continued

Publication and Year	Statistical Approach	Crop	Disease
P Moallem, N Razmjooy, BS Mousavi (2014)	Adaptive Fuzzy method	Potato	Grain Borer, discordant shape
Liu, Zhang, Shu, & Jin (2013)	RBF SVM	Wheat	Stripe Rust, Powdery Mildew, Leaf Rust, Leaf Blight
Arivazhagan, Shebiah, Ananthi, & Varthini (2013)	MDC, SVM	Tomato	Bacterial Leaf Spot, Leaf Spot
Navid Razmjooy, BS Mousavi, & F Soleymani (2012)	MLP, RBF, SVM	Potato	Grain Borer, discordant shape
Sankaran, Mishra, Ehsani, & Davis (2011)	LCA, QDA, K-nearest neighbor, SIMCA	Citrus Orchards	Huanglongbing (HLB)
Al-Bashish, Braik, & Bani-Ahmad (2011)	K-means clustering and FFBPNN	General leaf images	Late Scorch, Tiny Whiteness, Cottony Mold, Ashen Mold, Early Scorch
Rumpf et al. (2010)	Decision Trees, ANN, SVM+RBF	Sugar Beet	Powdery Mildew, Sugar Beet Rust, Cercospora Leaf Spot
Singh, Jayasa, & Paliwala (2010)	Linear, Quadratic, Mahalanobis and a BPNN classifier	Wheat Kernels	Grain Borer, Rusty Grain Beetle, Rice Weevil, Red Flour Beetle
Ghaffari et al. (2010)	**Clustering:** K-Means clustering, PCA and FCM **For classification:** MLP, LVQ, and RBF based ANNs	Tomato	Spider Mite infected plants and Powdery Mildew (Oidium lycopersicum)
Camargo & Smith (2009)	SVM	Cotton	Southern Green Stink Bug, Bacterial Angular, Ascochyta Blight
Wang, Zhang, Zhu, & Geng (2008)	BPNN with gradient descent	Tomato	Late Blight
Kuo-Yi Huang (2007)	BPNN	Phalaenopsis	Phalaenopsis Seedling disease
Xu, Zhu, Ying, & Jiang (2006)	DPLS, MD	Tomato	Tomato Mosaic Virus

greatly dependent on two factors: the size of the data and diversity of the data, both must be extensive for the deep learning model to present exceptional results.

There are essentially four elementary operations performed by the CNN architectures: Convolution, which utilizes kernels or filters of varied sizes to generate multiple and distinct feature maps; Pooling (most commonly max pooling); ReLu

Table 3. Imaging Techniques successfully investigated for several crops

Imaging Technique	Crop	Publication and Year
RGB Imaging	Apple, Tomato, Plantvillage Dataset (Blueberry, Potato, Raspberry, Strawberry, Cherry, Corn, Grape, Orange, Peach, Pepper, Soybean), Rose, Beans Leaf, Lemon Leaf, Banana Leaf, Wheat, Cotton, Phalaenopsis, Wheat Kernels, Watermelon, Pumpkin, Bell, Peach	Ferentinos (2018), Wang, Sun, & Wang (2017), Brahimi, Boukhalfa, & Moussaoui (2017), Durmus, Günes, & Kirci (2017), Mohanty, Hughes, & Salathé (2016), Sabrol & Kumar (2016), Singh & Misra (2016), Chung et al. (2016), Raza, Prince, Clarkson, & Rajpoot (2015), Liu, Zhang, Shu, & Jin (2013), Al-Bashish, Braik, & Bani-Ahmad (2011), Camargo & Smith (2009), Kuo-Yi Huang (2007), Singh, Jayasa, & Paliwala (2010)
Hyperspectral Reflectance	Cucumber, Tomato, Sugar Beet, Wheat Kernels	Zhao et al. (2016), Xie, Shao, Li, & He (2015), Rumpf et al. (2010), Singh, Jayasa, & Paliwala (2010), Wang, Zhang, Zhu, & Geng (2008), Zhang, Qin, & Liu (2005), Zhang, Qin, Liu, & Ustin (2003)
Thermal & Stereo Visible Light Imaging	Tomato	Raza, Prince, Clarkson, & Rajpoot (2015)
Visible-near Infrared Spectroscopy	Citrus Orchards	Sankaran, Mishra, Ehsani, & Davis (2011), Xu, Ying, Fu, & Zhu (2007), Xu, Zhu, Ying, & Jiang (2006)

(Rectified Linear Units) for introducing non-linearity; and Fully Connected Layer. Research shows that deep learning architectures outperform other machine learning tools significantly.

In recent years, deep learning architectures have been largely evaluated with digital/RGB images and hyperspectral data as well. GoogleNet and AlexNet have been previously implemented on all plant images across each crop & disease pair of PlantVillage dataset (Mohanty, Hughes, & Salathé, 2016). This dataset is an open access depository of 54,306 images covering 14 crops and corresponding 26 diseases. This dataset has been utilized by many authors for implementing deep learning architectures. Another implementation includes evaluation of AlexNet and SqueezeNet models on PlantVillage dataset (Durmus, Günes, & Kirci, 2017).

In contrast to other image processing techniques, where a few hundred images suffice for image classification, deep learning requires thousands of images for training, as less number of training images might lead to overfitting. Apart from disease classification, deep learning has also been effectively applied to determine the severity levels of a disease ranging from healthy, early, to the middle, and finally, end-stage (Wang, Sun, & Wang, 2017).

Authors trained the deep learning architectures on Apple black rot images from the PlantVillage dataset and achieved a test accuracy of 90.4%. Ferentinos (2018) has implemented AlexNetOWTBn and VGG on 87,848 images of 25 distinct plant species with 58 plants & disease pair, with 99.53% accuracy.

These experiments have been performed on images collected/downloaded from the Internet, not digital images captured on the field in real environmental conditions. This can be considered as a shortcoming; hence, these applications must reach the farmers for actual evaluation of the technique.

SOLUTIONS AND RECOMMENDATIONS

Tomato plant was selected as the crop under consideration, with India being the second largest producer of tomatoes worldwide, as per the Food and Agriculture Organization of the United Nations (FAOSTAT). The authors have analyzed 18,160 images of tomato plant leaves, spread across 10 class labels assigned to them.

Each class label is a predefined crop & disease pair. With the image of the plant leaf as input, the authors attempt to predict the correct plant-disease pair. Table 4 lists a few common tomato plant diseases along with their scientific names and category of pathogens.

This application can be effectively utilized by farmers to meet their needs and protect the tomato yield by timely application of control measures.

Table 4. Common tomato plant diseases

Disease Name	Pathogen Scientific Name	Type
Early Blight	Alternaria solani	Fungal
Late Blight	Phytophthora infestans	Oomycete
Buck Eye Rot	Phytophthora parasitica	Oomycete
Fusarium Wilt	Fusarium oxysporum f. sp. lycopersici	Hyphomycete
Septoria Leaf Spot	Septoria lycopersici	Fungal
Powdery Mildew	Leveillula taurica	Fungal
Bacterial Wilt	Pseudomonas solanacearum	Bacterial
Bacterial Leaf Spot	Xanthomonas campestris pv. Vesicatoria	Bacterial
Bacterial Canker	Clavibacter michiganensis pv. michiganensis	Bacterial
Tomato Mosaic Virus	(ToMV)	Viral
Tomato yellow leaf curl	Tomato yellow leaf curl virus	Viral
Anthracnose	Colletotrichum gloeosporioides	Fungal
Cercospora leaf mold	Pseudocercospora fuligena	Fungal

Source: *(Shankar, Harsha, & Bhandary, 2014)*

Through this chapter, the authors promote the practice of smart agriculture via automation. This application aims to make the existing methods reach the growers, as well as, be user-friendly and accurate, with minimal human intervention. Once the disease has been diagnosed, treatment can be suggested through any communication medium.

Figure 1 depicts images of healthy as well as diseased tomato leaves from the PlantVillage dataset. In every approach described in this chapter, the authors have resized the given images to 256 × 256 pixels, and performed model optimization along with predictions, on all downscaled images. Figure 2 demonstrates the resultant images after intermediate convolution, pooling and activation operations performed on the input images.

Due to hardware limitations, the authors decided to only use bottlenecking. The original approach was to take a pre-trained network, slice off the end (the part that classifies inputs), stick a new classification network on to replace it, freeze the convolutional layers, and then train the new part of the network. Easier said than done, also, much more expensive to do. By creating bottlenecks, one can simply record the feature map outputs as they are, before they would have been fed into the classification layers that were removed.

Figure 1. Tomato healthy and diseased leaf images form PlantVillage Dataset (Top row left to right) (a) Healthy (b) Tomato Mosaic Virus (c) Early Blight (d) Bacterial Spot (Bottom row left to right) (e) Late Blight (f) Septoria Leaf Spot (g) Yellow Leaf Curl (h) Leaf Mold

Figure 2. Intermediate resultant images after image processing by CNN

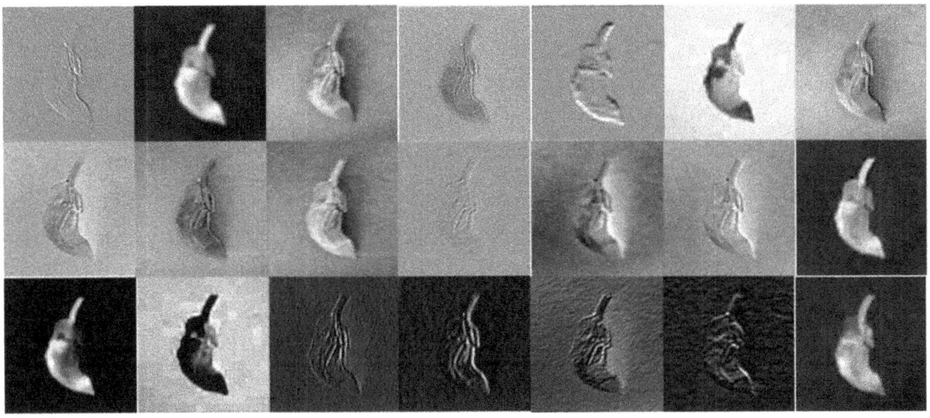

In other words, bottlenecks require zero training. They are produced by the model predicting on your new data set. It is exponentially quicker to use a model to predict that it is to train it (For example, the computation that goes into stochastic/ mini-batch gradient descent). Additionally, since bottlenecking does not require parallel computing, it can be done on a CPU. The authors have a train-test split of 70%-30%. And the performance measures include accuracy, loss, mean precision, mean recall, confusion matrix, and mean F1-micro score (as the proportion of images per class is not balanced).

The authors assess the relevance of deep convolutional neural networks for the classification problem mentioned above. The authors target the five popular architectures, namely Inception_ResNet_V2, ResNet50, VGG16, VGG19 and Xception for image classification. Table 5 provides some differentiating details of all five models.

Table 5. Few details of the implemented CNN Architectures

Model	Size (in MB)	Parameters	Depth
Xception	88	22,910,480	126
VGG16	528	138,357,544	23
VGG19	549	143,667,240	26
ResNet50	99	25,636,712	168
InceptionResNetV2	215	55,873,736	572

Source: https://github.com/GeorgeSeif/Transfer-Learning-Suite

Inception_ResNet_V2

Inception-ResNet-v2 is a modified version of Inception V3 model, roughly based on Microsoft's ResNet. It is a CNN that delivered better accuracy on the ILSVRC image classification benchmark. With residual connections, it leads to decreased training times.

- Basic structure of Inception-ResNet-v2 (layers, dimensions) are as follows:
    - Image → Stem → 5x Module A → Reduction A → 10x Module B → Reduction B → 5x Module C → AveragePooling → Droput 20% → Linear, Softmax
    - 299x299x3 → 35x35x256 → 35x35x256 → 17x17x896 → 17x17x896 → 8x8x1792 → 8x8x1792 → 1792 → 1792 → 1000
- Modules A, B and C are similar.
- They contain 2 (B, C) or 3 (A) branches.
- Each branch starts with a 1x1 convolution on the input.
- All branches merge into one 1x1 convolution (which is then added to the original input, as usually in residual architectures).
- Module A uses 3x3 convolutions, B uses 7x1 and 1x7, and C uses 3x1 and 1x3.
- The reduction modules also contain multiple branches. One has max pooling (3x3 stride 2), the other branches end in convolutions with stride 2.

VGG16 and VGG19

Simonyan & Zisserman introduced VGG network in 2014. VGG is individualized by its uncomplicated structure, utilizing 3×3 convolutional layers, each placed on top of the other in increasing depth. Max Pooling takes care of the reduced volume size.

Two fully-connected layers with each having 4,096 nodes, is followed by a softmax classifier. The numbers '16' and '19' that are included in the names represent the number of weight layers in the network.

Deployment of VGG networks is tedious due to inherently slow training speeds and large architecture weights. It can be observed in Table 5 that VGG takes up more space as compared to other networks.

ResNet50

It was introduced in 2015. Unlike ordinary architectures like VGG & OverFeat, AlexNet, and ResNet with 50 weight layers, is build on micro-architecture modules also known as network-in-network architectures. Micro-architecture can also be viewed as the set of "building blocks" utilized for construction of a network.

Updating the residual module for user identity mappings can help in obtaining better accuracy. ResNet being deeper than VGG16 & VGG19, the model size is remarkably smaller, as fully-connected layers have been replaced by the utilization of global average pooling— this significantly reduces the network size to 99MB for ResNet50.

Xception

Xception was first proposed by François Chollet, the creator of the Keras library. Xception is the extended version of the original Inception architecture, where the standard Inception modules are replaced with depth-wise separable convolutions. Xception exercises the smallest weight serialization at only 88 MB. Xception, when trained on ImageNet dataset for image classification significantly outperformed Inception V3, due to the efficient usage of model parameters and incorporation of depth-wise separable convolution operation instead of Inception modules.

The authors exercised the following hyper-parameters in all of the implementations:

- **Solver type**: Stochastic Gradient Descent,
- **Base learning rate**: 0.001,
- **Decay**: 1e-6,
- **Momentum**: 0.9,
- **Gamma**: 0.1,
- **Batch size**: 128 (in case of Inception ResNet V2, ResNet50, Xception), 100 (in case of VGG16, VGG19).

ResNet50 had the highest train-test accuracies (99.12% - 96.14%), the highest F1 score (0.9612) and also the best *Confusion matrix*. This can be observed in Figures 3-7, depicting the resultant graphs of all five CNN architectures for performance measures accuracy, loss, and F1 mean score. Similarly, Table 6 and Table 7 present the confusion matrix and the summaries of the performance measures, respectively.

ResNet50 was utilized to create a mobile application for the identification and classification of Tomato plant diseases. The Tomato disease detection app is an android based application. The purpose of this application is to help the growers detect diseases in the tomato plant, using Image-based detection techniques.

In the current scenario, because of their simplicity and user-friendly interface, android phones are widely used, by the majority of the people in India. Moreover, due

Figure 3. Model accuracies against epoch of all five CNN architectures

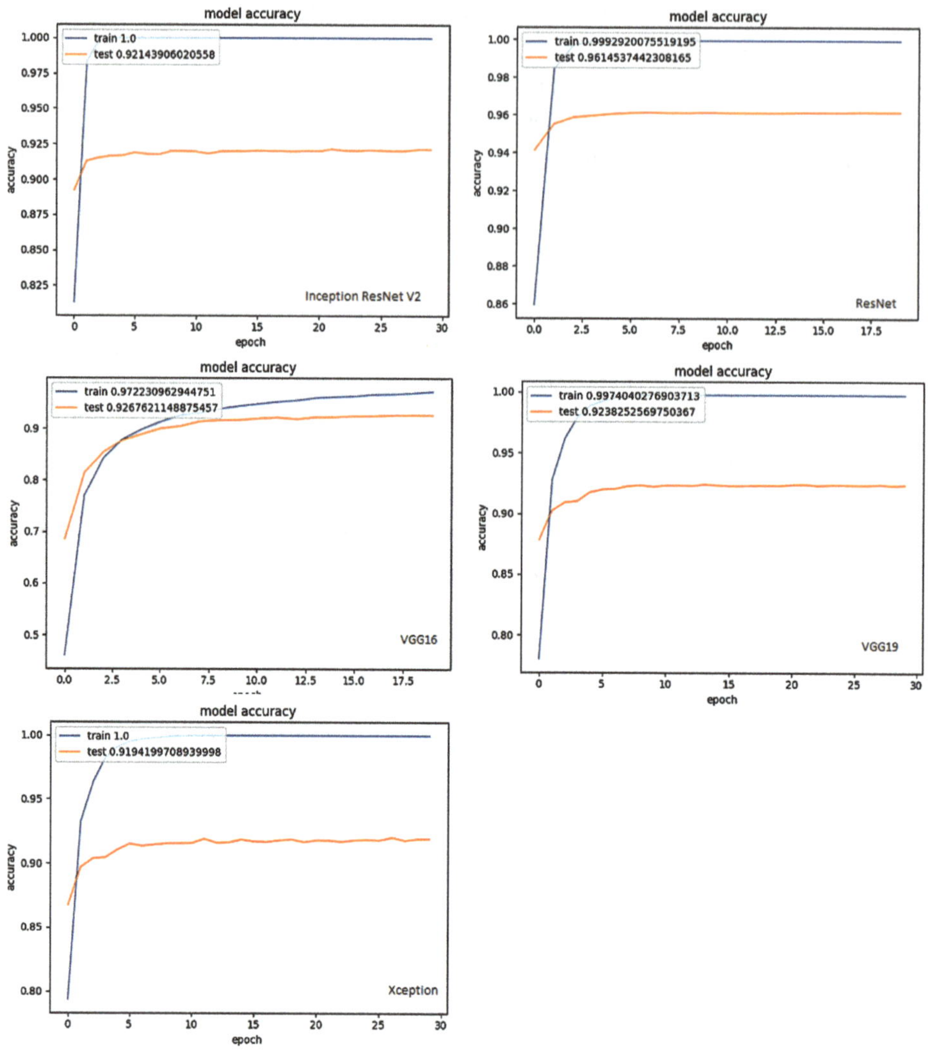

to a highly competitive market, there is a drop in their prices, hence, it has become easy for all sections of society to own an android phone and reap its benefits along with a lot of applications present on this platform.

This app can be utilized in the agricultural sector to a great extent for disease detection. It could prove to be of great assistance, particularly for farmers, as it would eliminate the dependency on expensive machinery, high-end setup, and experts, for diagnosing diseases in the plant.

Figure 4. F1 Mean Score against epoch of all five CNN architectures

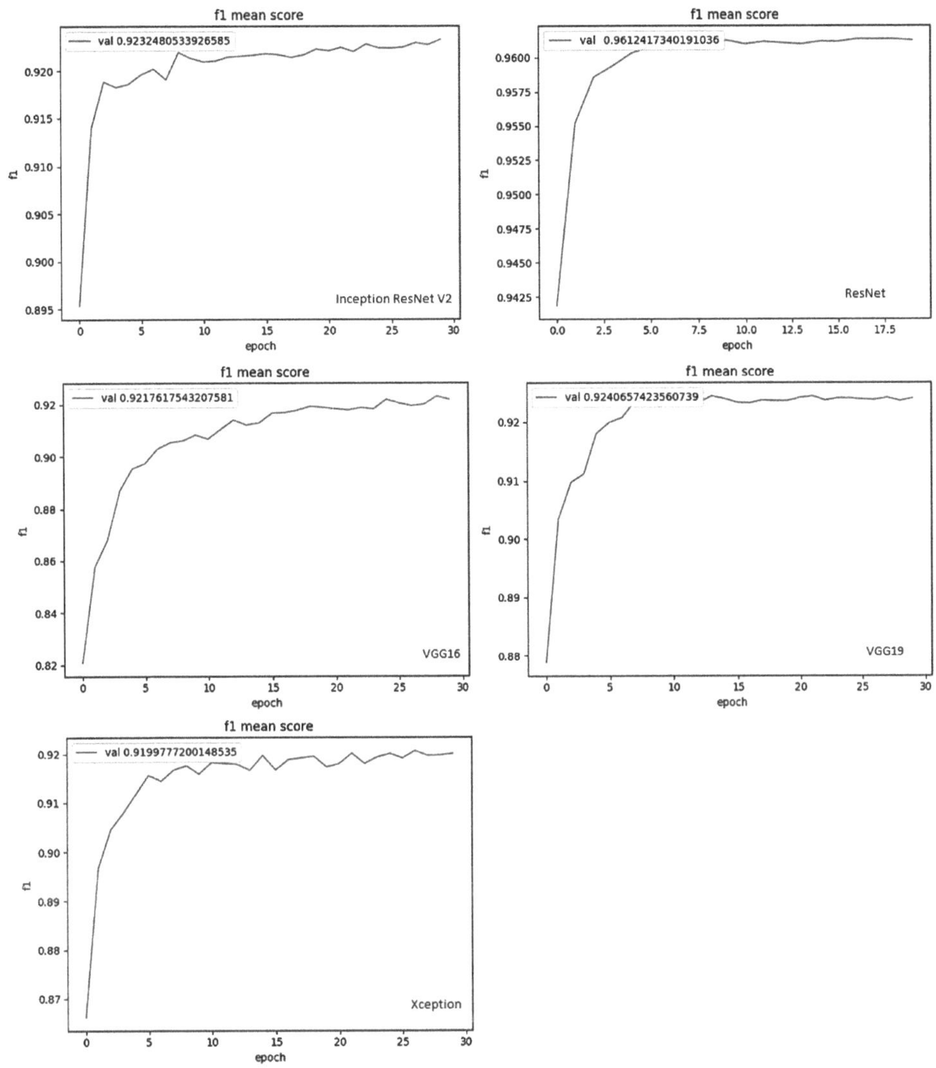

This app is easy to use and can be easily habituated by anyone because of its simple design and highly accurate results. After installing the application, it will appear as shown in Figure 8. The figure demonstrates the complete flow of the application.

It has two buttons and an image view, where the images are displayed.

Figure 5. Model Loss against epoch of all five CNN architectures

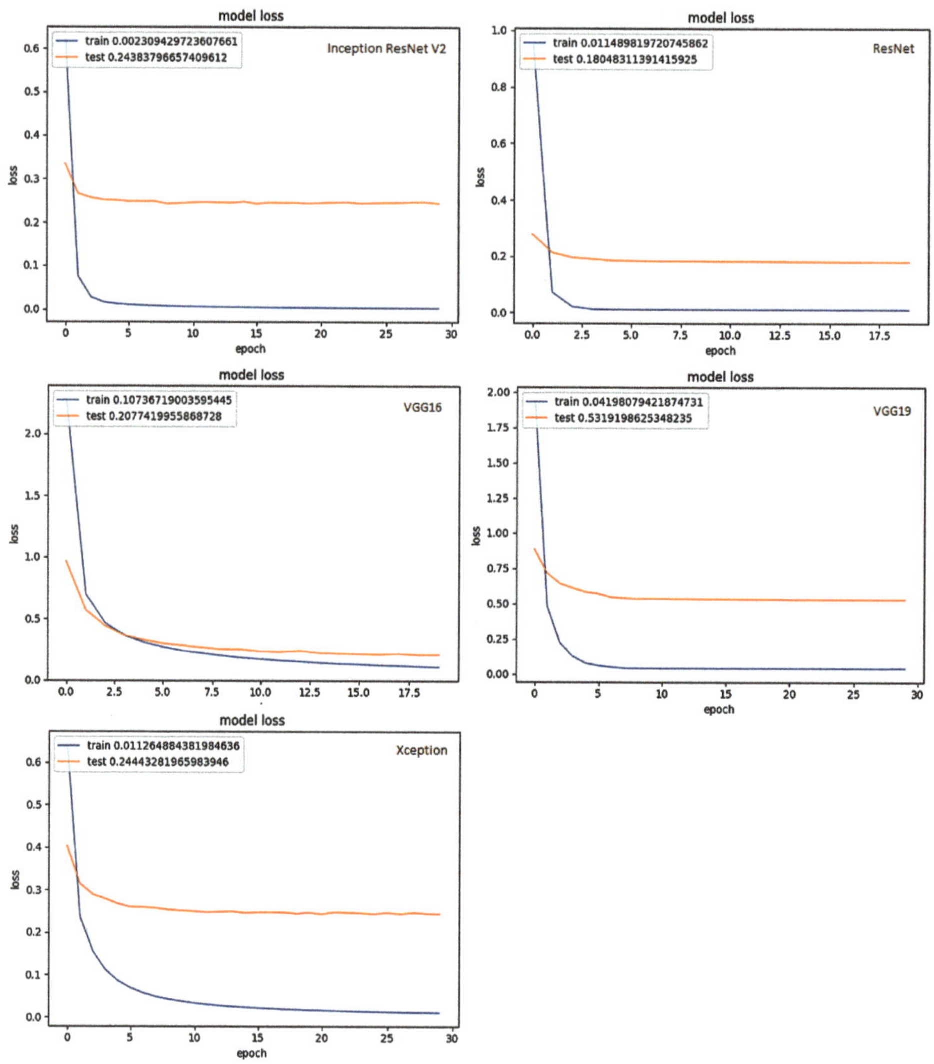

- **Load Image:** This button is used to load an image from the phone's gallery. Choose a leaf image from the gallery, for which you want to test for a disease. This image is then displayed on the imageView.
- **Upload:** This button sends the loaded image to the Flask-based server, which returns the result and displays it in the result space. Click *Upload* to send the selected image to the server for further analysis and classification results. Figure 9 depicts snapshots of classification results for healthy as well as Tomato mosaic virus infected leaf.

Figure 6. Train-Test dataset Comparison of all five CNN architectures

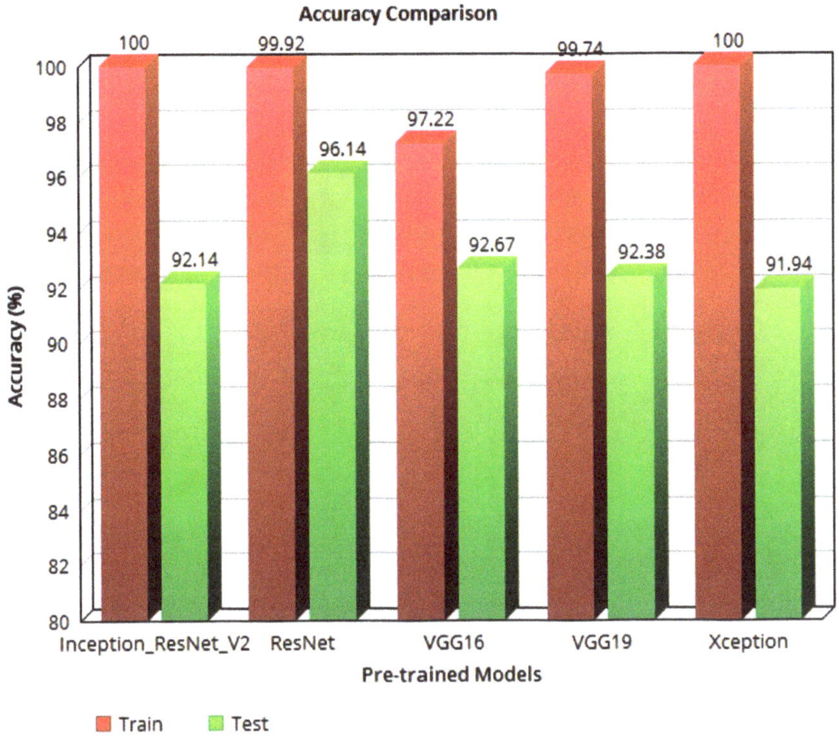

Tech-Stack Used

1. **Android Studio:** The android application was made from scratch in Android Studio. This app is compatible with nearly 100% of the android devices. The primary programming language used was JAVA and XML.
2. **Volley:** It was used in this application for sending the image to our Flask-based server in base64 String format as a request, and receiving a response, which consists of predictions made after processing of the input image by the server.
3. **Flask Application:** For machine learning model an API is utilized based on Python software along with the lightweight framework Flask which will perform as an access point for the model across Flask, used to deploy the Resnet-50 Model, trained on images of tomato leaves, as an API service for handling the requests to the server. It receives a POST request with an image in base64 string format, sent via Volley from the android application and in turn, sends a response back to the Android application in JSON format, which is then displayed on the app as the final classification result.

Figure 7. Performance Comparison of all five CNN architectures

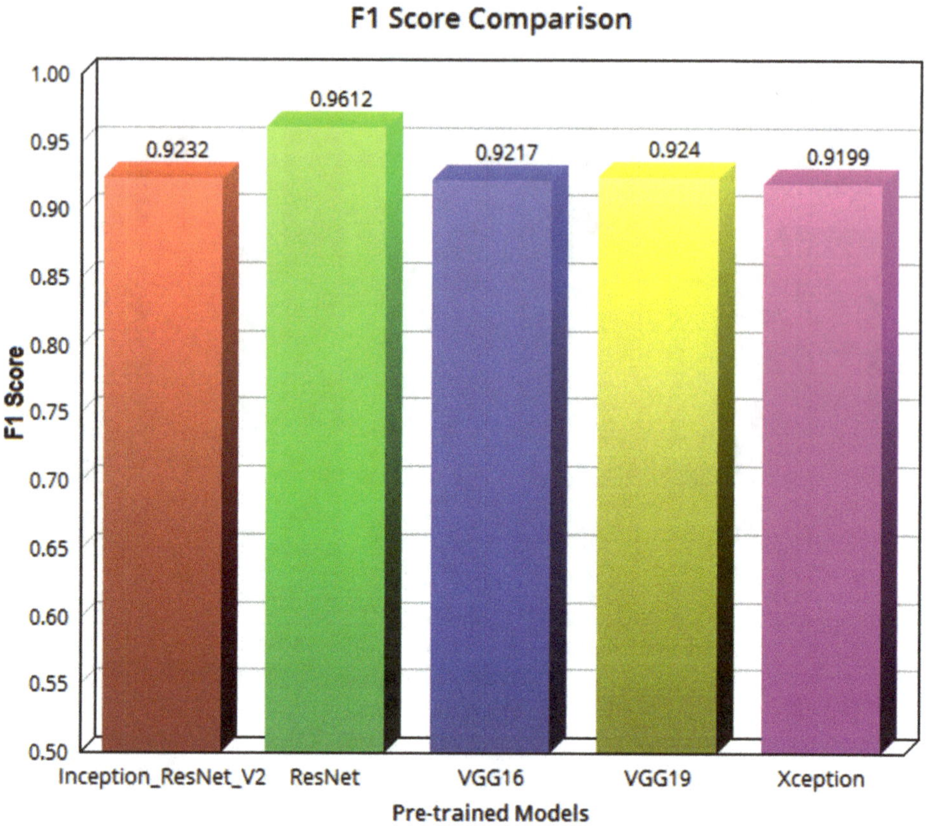

Final Note

1. This app can be later launched on PlayStore so that anyone can use it and get its benefits.
2. The Flask API is currently deployed on the local server for testing purposes but can be scaled for heavier usage at later stages, by deploying on servers like AWS, Google Cloud Services, Heroku, etc. so that it can handle a large number of requests at the same time.

FUTURE RESEARCH DIRECTIONS

The mobile application is in its initial stages of development. The authors aim to replicate it for all crucial crops in the Indian agricultural domain and distribute it

Table 6. Comparison Matrices of all five CNN architectures

Inception_ResNet_V2										
	Target Spot	**Late Blight**	**Tomato Mosaic Virus**	**Leaf Mold**	**Bacterial Spot**	**Early Blight**	**Healthy**	**Tomato Yellow Leaf Curl Virus**	**Two Spotted Spider Mite**	**Septoria Leaf Spot**
Target Spot	**361**	0	1	2	4	6	5	1	22	8
Late Blight	12	**523**	0	9	3	13	1	5	3	7
Tomato Mosaic Virus	5	0	**81**	3	0	1	0	2	2	6
Leaf Mold	14	2	2	**234**	2	3	0	1	4	6
Bacterial Spot	8	3	0	0	**598**	4	0	11	0	3
Early Blight	26	28	1	6	13	**230**	2	7	5	21
Healthy	15	1	0	0	0	1	**445**	0	0	2
Tomato Yellow Leaf Curl Virus	8	1	0	2	3	3	0	**1596**	8	0
Two Spotted Spider Mite	26	2	0	2	1	2	5	9	**473**	1
Septoria Leaf Spot	25	5	5	11	20	3	0	1	1	**451**
ResNet50										
	Target Spot	**Late Blight**	**Tomato Mosaic Virus**	**Leaf Mold**	**Bacterial Spot**	**Early Blight**	**Healthy**	**Tomato Yellow Leaf Curl Virus**	**Two Spotted Spider Mite**	**Septoria Leaf Spot**
Target Spot	**377**	3	1	2	0	4	5	1	13	4
Late Blight	3	**549**	0	7	1	10	1	3	0	2
Tomato Mosaic Virus	1	0	**92**	1	0	0	0	2	1	3
Leaf Mold	3	2	0	**250**	1	2	0	1	3	6
Bacterial Spot	1	3	0	0	**617**	1	0	4	0	1
Early Blight	13	25	2	2	7	**277**	0	6	3	4
Healthy	5	1	0	0	0	0	**457**	0	1	0
Tomato Yellow Leaf Curl Virus	3	2	1	1	4	0	0	**1605**	4	1
Two Spotted Spider Mite	22	0	2	0	1	1	1	6	**487**	1
Septoria Leaf Spot	7	2	3	1	4	4	0	1	1	**499**

continued on following page

Table 6. Continued

VGG16										
	Target Spot	**Late Blight**	**Tomato Mosaic Virus**	**Leaf Mold**	**Bacterial Spot**	**Early Blight**	**Healthy**	**Tomato Yellow Leaf Curl Virus**	**Two Spotted Spider Mite**	**Septoria Leaf Spot**
Target Spot	**376**	2	1	1	1	1	4	0	20	4
Late Blight	21	**521**	0	3	1	20	1	0	2	7
Tomato Mosaic Virus	4	0	**93**	0	0	0	0	0	1	2
Leaf Mold	20	7	4	**212**	0	4	0	2	6	13
Bacterial Spot	14	5	0	0	**585**	8	0	7	1	7
Early Blight	31	31	0	0	9	**252**	0	2	2	12
Healthy	21	2	1	0	0	0	**438**	0	2	0
Tomato Yellow Leaf Curl Virus	13	0	1	1	15	4	0	**1573**	14	0
Two Spotted Spider Mite	44	3	2	1	0	2	2	8	**458**	1
Septoria Leaf Spot	28	4	3	0	4	3	2	0	7	**471**
VGG19										
	Target Spot	**Late Blight**	**Tomato Mosaic Virus**	**Leaf Mold**	**Bacterial Spot**	**Early Blight**	**Healthy**	**Tomato Yellow Leaf Curl Virus**	**Two Spotted Spider Mite**	**Septoria Leaf Spot**
Target Spot	**346**	3	1	3	3	11	9	0	25	9
Late Blight	7	**535**	0	11	0	14	0	0	1	8
Tomato Mosaic Virus	1	0	**91**	3	0	0	1	1	0	3
Leaf Mold	2	8	2	**234**	1	5	0	2	10	4
Bacterial Spot	1	8	0	0	**595**	7	0	10	1	5
Early Blight	9	31	1	6	12	**257**	0	7	3	13
Healthy	11	2	1	0	0	1	**445**	0	4	0
Tomato Yellow Leaf Curl Virus	0	4	3	4	8	3	0	**1581**	18	0
Two Spotted Spider Mite	21	2	1	2	1	8	3	7	**472**	4
Septoria Leaf Spot	5	10	3	11	3	6	2	2	2	**478**

continued on following page

Table 6. Continued

Xception										
	Target Spot	**Late Blight**	**Tomato Mosaic Virus**	**Leaf Mold**	**Bacterial Spot**	**Early Blight**	**Healthy**	**Tomato Yellow Leaf Curl Virus**	**Two Spotted Spider Mite**	**Septoria Leaf Spot**
Target Spot	**367**	2	0	0	1	5	7	2	17	9
Late Blight	20	**518**	0	9	1	13	0	7	2	6
Tomato Mosaic Virus	7	0	**81**	4	0	1	0	0	4	3
Leaf Mold	11	9	1	**224**	2	4	0	2	8	7
Bacterial Spot	7	5	0	0	**602**	3	0	4	0	6
Early Blight	30	35	0	4	10	**237**	2	2	5	14
Healthy	14	3	0	0	0	2	**442**	0	2	1
Tomato Yellow Leaf Curl Virus	6	4	0	1	7	4	0	**1585**	11	3
Two Spotted Spider Mite	42	1	0	1	2	0	5	5	**462**	3
Septoria Leaf Spot	23	4	5	5	8	11	2	4	3	**457**

Table 7. Resultant Performance Measures

Model	**Accuracy (%)**		**Loss**		**F1 Score**
	Train	**Test**	**Train**	**Test**	
VGG16	97.22	92.67	0.1073	0.2077	0.9217
VGG19	99.74	92.38	0.0419	0.5319	0.9240
ResNet50	99.92	**96.14**	0.0114	**0.1804**	**0.9612**
Inception_ResNet_V2	**100**	92.14	**0.0023**	0.2438	0.9232
Xception	**100**	91.94	0.0112	0.2444	0.9199

for everyday usage. As mentioned earlier, tomato leaf images from the Plant Village dataset have been utilized for training purposes, as the dataset is openly available and that too in large numbers, which is essential for training a deep learning architecture. The authors aim to build a similar database that would be open access as well, and include other parts of a plant in addition to leaves viz. fruit, stem etc.

Figure 8. Complete view and flow of the Mobile Application

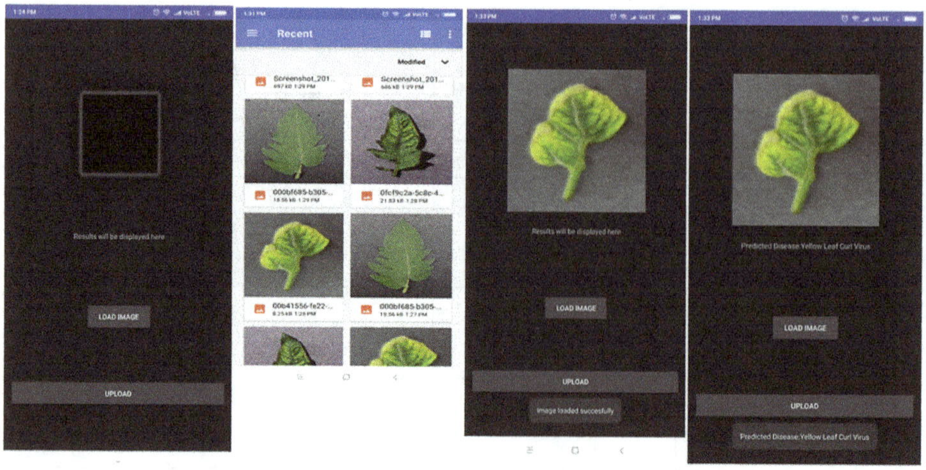

Figure 9. Few snapshots of Classification results

Work should be performed to analyze the visual symptoms appearing on different parts of a plant, along with the challenge of capturing images of hidden plants and their diseased elements, with the consideration of several sunlight conditions.

Among other imaging sensors, hyperspectral data and its applications in agriculture are being researched extensively and combined with deep learning it can prove to be monumental in differentiating among healthy and diseased plants. Spectral data measurements obtained from the hyperspectral sensors are utilized for computation of vegetation indices (VIs), which hint at the biophysical and biochemical properties of a plant, hence, enabling early disease prediction.

These values are used to determine the stress levels and/or disease severity level in a plant. Internet of Things (IoT) based sensors can be supplemented and deployed along with imaging sensors for data collection and subsequent analysis.

Implementation of a model, where the inputs are provided by the sensor nodes and intelligent decision-based applications are deployed using machine learning algorithms, a well designed "Smart Plant Disease Surveillance System" could be demonstrated on accurate real-time field data.

The authors intend to develop a smart system in which various environmental parameters such as temperature, humidity, pressure, CO_2, water level etc. along with images, can be observed and timely action can be taken to prevent any hazards. Images could be captured via the Digital, Infrared and Hyperspectral cameras, and various sensors deployed in the field would monitor the environmental factors. The study would include different parts of the plant, where the symptoms might occur.

Once the disease has been diagnosed, the concerned would be notified and treatment can be suggested via any communication medium, also basic treatment steps can be incorporated into the mobile application itself.

CONCLUSION

Due to the rapid growth in the overall computational power and researchers constantly striving for faster and accurate problem-solving approaches, there are a plethora of non-destructive, non-invasive techniques for plant disease detection in agriculture.

The authors aim to bring these technologies to the field and bridge the gap between experimentation and real-life application.

This chapter showcased the recent advancements in applications of statistical methods and sensor-based technologies to aid in plant disease diagnosis, followed by the implementation details and comparative results of CNN architectures namely Inception ResNet50, VGG16, ResNet V2, VGG19 & Xception, along with demonstrating a mobile application that can help in timely prediction of diseases in Tomato plant.

Precision agriculture and smart farming technologies can assist and provide guidance and support to farmers with the aim to minimize losses, economically as well as ecologically. Hence, this requires for the existing methods to reach the farmers effectively, to be user-friendly, secure and accurate.

REFERENCES

Al Bashish, D., Braik, M., & Bani-Ahmad, S. (2011). Detection and classification of leaf diseases using K means-based segmentation and Neural networks-based classification. *Information Technology Journal, 10*(2), 267–275. doi:10.3923/itj.2011.267.275

Arivazhagan, S., Shebiah, R. N., Ananthi, S., & Varthini, S. V. (2013). Detection of the unhealthy region of plant leaves and classification of plant leaf diseases using texture features. *Agricultural Engineering International: CIGR Journal, 15*(1), 211–217.

Brahimi, M., Boukhalfa, K., & Moussaoui, A. (2017). Deep Learning for Tomato Diseases: Classification and Symptoms Visualization. *Applied Artificial Intelligence, 31*(4), 299–315. doi:10.1080/08839514.2017.1315516

Camargo, A., & Smith, J. S. (2009). An image-processing based algorithm to automatically identify plant disease visual symptoms. *Biosystems Engineering, 102*(1), 9–12. doi:10.1016/j.biosystemseng.2008.09.030

Camargo, A., & Smith, J. S. (2009). Image pattern classification for the identification of disease causing agents in plants. *Computers and Electronics in Agriculture, 66*(2), 121–125. doi:10.1016/j.compag.2009.01.003

Chung, C. L., Huang, K. J., Chen, S. Y., Lai, M. H., Chen, Y. C., & Kuo, Y. F. (2016). Detecting Bakanae disease in rice seedlings by machine vision. *Computers and Electronics in Agriculture, 121*, 404–411. doi:10.1016/j.compag.2016.01.008

Durmus, H., Günes, E. O., & Kirci, M. (2017, August). Disease Detection on the Leaves of the Tomato Plants by Using Deep Learning. In *Agro-Geoinformatics, 2017 6th International Conference on* (pp. 1-5). IEEE. 10.1109/Agro-Geoinformatics.2017.8047016

Ferentinos, K. P. (2018). Deep learning models for plant disease detection and diagnosis. *Computers and Electronics in Agriculture, 145*, 311–318. doi:10.1016/j.compag.2018.01.009

Food and Agriculture Organization of the United Nations. (n.d.). Retrieved from http://www.fao.org/faostat/en/#country/100

Ghaffari, R., Zhang, F., Iliescu, D., Hines, E., Leeson, M., Napier, R., & Clarkson, J. (2010, July). Early Detection of Diseases in Tomato Crops: An Electronic Nose and Intelligent Systems Approach. In *Neural Networks (IJCNN), The 2010 International Joint Conference on* (pp. 1-6). IEEE.

Heckmann, D., Schluter, U., & Weber, A. P. (2017). Machine Learning Techniques for Predicting Crop Photosynthetic Capacity from Leaf Reflectance Spectra. *Molecular Plant. Cell Press*, *10*(6), 878–890. PMID:28461269

Huang, K. Y. (2007). Application of artificial neural network for detecting Phalaenopsis seedling diseases using color and texture features. *Computers and Electronics in Agriculture*, *57*(1), 3–11. doi:10.1016/j.compag.2007.01.015

Kaundal, R., Kapoor, A. S., & Raghava, G. P. (2006). Machine learning techniques in disease forecasting: A case study on rice blast prediction. *BMC Bioinformatics*, *7*(1), 485. doi:10.1186/1471-2105-7-485 PMID:17083731

Liu, L., Zhang, W., Shu, S., & Jin, X. (2013, August). Image Recognition of Wheat Disease Based on RBF Support Vector Machine. In *Proceedings of the International Conference on Advanced Computer Science and Electronics Information* (pp 307-310). Atlantis Press. 10.2991/icacsei.2013.77

Lowe, A., Harrison, N., & French, A. P. (2017). Hyperspectral image analysis techniques for the detection and classification of the early onset of plant disease and stress. *Plant Methods*, *13*(1), 80. doi:10.118613007-017-0233-z PMID:29051772

Lu, J., Zhou, M., Gao, Y., & Jiang, H. (2017). Using hyperspectral imaging to discriminate yellow leaf curl disease in tomato leaves. *Precision Agriculture*, 1–16.

Mahlein, A. K. (2016). Plant Disease Detection by Imaging Sensors – Parallels and Specific Demands for Precision Agriculture and Plant Phenotyping. *Plant Disease. APS Journals*, *100*(2), 241–251.

Moallem, P., Razmjooy, N., & Ashourian, M. (2013). Computer vision-based potato defect detection using neural networks and support vector machine. *International Journal of Robotics and Automation*, *28*(2), 137–145. doi:10.2316/Journal.206.2013.2.206-3746

Moallem, P., Razmjooy, N., & Mousavi, B. S. (2014). Robust potato color image segmentation using adaptive fuzzy inference system. *Iranian Journal of Fuzzy Systems*, *11*(6), 47–65.

Mohanty, S. P., Hughes, D. P., & Salathé, M. (2016). Using deep learning for image-based plant disease detection. *Frontiers in Plant Science*, *7*, 1419. doi:10.3389/fpls.2016.01419 PMID:27713752

Osroosh, Y., Khot, L. R., & Peters, R. T. (2018). Economical thermal-RGB imaging system for monitoring agricultural crops. *Computers and Electronics in Agriculture*, *147*, 34–43. doi:10.1016/j.compag.2018.02.018

Pal, T., Jaiswal, V., & Chauhan, R. S. (2016). DRPPP: A machine learning based tool for prediction of disease resistance proteins in plants. *Computers in Biology and Medicine*, *78*, 42–48. doi:10.1016/j.compbiomed.2016.09.008 PMID:27658260

Petrellis, N. (2015, September). Plant Disease Diagnosis Based on Image Processing, Appropriate for Mobile Phone Implementation. In *Proceedings of 7th International Conference on Information and Communication Technologies in Agriculture, Food and Environment* (pp. 238-246). Academic Press.

PlantVillage. (n.d.). Retrieved from https://plantvillage.org/

Raza, S. A., Prince, G., Clarkson, J. P., & Rajpoot, N. M. (2015). Automatic detection of diseased tomato plants using thermal and stereo visible light images. *PLoS One*, *10*(4), e0123262. doi:10.1371/journal.pone.0123262 PMID:25861025

Razmjooy, N. (2012). A multi layer perceptron neural network trained by invasive weed optimization for potato color image segmentation. *Trends in Applied Sciences Research*, *7*(6), 445–455. doi:10.3923/tasr.2012.445.455

Razmjooy, N., Mousavi, B. S., & Soleymani, F. (2012). A real-time mathematical computer method for potato inspection using machine vision. *Computers & Mathematics with Applications (Oxford, England)*, *63*(1), 268–279. doi:10.1016/j.camwa.2011.11.019

Rumpf, T., Mahlein, A. K., Steiner, U., Oerke, E. C., Dehne, H. W., & Plümer, L. (2010). Early detection and classification of plant diseases with Support Vector Machines based on hyperspectral reflectance. *Computers and Electronics in Agriculture*, *74*(1), 91–99. doi:10.1016/j.compag.2010.06.009

Rupanagudi, S. R., Ranjani, B. S., Nagaraj, P., Bhat, V. G., & Thippeswamy, G. (2015, January). A Novel Cloud Computing based Smart Farming System for Early Detection of Borer Insects in Tomatoes. In *Communication, Information & Computing Technology (ICCICT), 2015 International Conference on* (pp. 1-6). IEEE. 10.1109/ICCICT.2015.7045722

Sabrol, H., & Kumar, S. (2016). Fuzzy and Neural Network based Tomato Plant Disease Classification using Natural Outdoor Images. *Indian Journal of Science and Technology*, *9*(44), 1–8. doi:10.17485/ijst/2016/v9i44/92825

Sabrol, H., & Kumar, S. (2016). Intensity based feature extraction for tomato plant disease recognition by classification using decision tree. *International Journal of Computer Science and Information Security*, *14*(9), 622–626.

Sankaran, S., Mishra, A., Ehsani, R., & Davis, C. (2010). A review of advanced techniques for detecting plant diseases. *Computers and Electronics in Agriculture*, *72*(1), 1–13. doi:10.1016/j.compag.2010.02.007

Sankaran, S., Mishra, A., Maja, J. M., & Ehsani, R. (2011). Visible-near infrared spectroscopy for detection of Huanglongbing in citrus orchards. *Computers and Electronics in Agriculture*, *77*(2), 127–134. doi:10.1016/j.compag.2011.03.004

Shankar, R., Harsha, S., & Bhandary, R. (2014). *A Practical Guide to Identification and Control of Tomato Diseases*. Tropica Seeds Pvt.Ltd.

Singh, C. B., Jayasa, D. S., Paliwala, J., & White, N. D. G. (2010). Identification of insect-damaged wheat kernels using short-wave near-infrared hyperspectral and digital colour imaging. *Computers and Electronics in Agriculture*, *73*(2), 118–125. doi:10.1016/j.compag.2010.06.001

Singh, V., & Misra, A. K. (2017). Detection of Plant Leaf Diseases Using Image Segmentation and Soft Computing Techniques. *Information Processing in Agriculture*, *4*(1), 41–49. doi:10.1016/j.inpa.2016.10.005

Wang, G., Sun, Y., & Wang, J. (2017). Automatic Image-Based Plant Disease Severity Estimation Using Deep Learning. *Hindawi Computational Intelligence and Neuroscience*, 1–8.

Wang, X., Zhang, M., Zhu, J., & Geng, S. (2008). Spectral prediction of Phytophthora infestans infection on tomatoes using artificial neural network (ANN). *International Journal of Remote Sensing*, *29*(6), 1693–1706. doi:10.1080/01431160701281007

Xie, C., Shao, Y., Li, X., & He, Y. (2015). Detection of early blight and late blight diseases on tomato leaves using hyperspectral imaging. *Scientific Reports*, *5*(16564), 1–11. PMID:26572857

Xu, H., Zhu, S., Ying, Y., & Jiang, H. (2006, October). Application of multispectral reflectance for early detection of tomato disease. In Optics for Natural Resources, Agriculture, and Foods (vol. 6381, p. 63810R). International Society for Optics and Photonics. doi:10.1117/12.685531

Xu, H. R., Ying, Y. B., Fu, X. P., & Zhu, S. P. (2007). Near-infrared Spectroscopy in detecting Leaf Miner Damage on Tomato Leaf. *Biosystems Engineering*, *96*(4), 447–454. doi:10.1016/j.biosystemseng.2007.01.008

Ye, X., Sakai, K., Okamoto, H., & Garciano, L. O. (2008). A ground-based hyperspectral imaging system for characterizing vegetation spectral features. *Computers and Electronics in Agriculture*, *63*(1), 13–21. doi:10.1016/j.compag.2008.01.011

Zhang, M., Qin, Z., & Liu, X. (2005). Remote Sensed Spectral Imagery to Detect Late Blight in Field Tomatoes. *Precision Agriculture*, *6*(6), 489–508. doi:10.100711119-005-5640-x

Zhang, M., Qin, Z., Liu, X., & Ustin, S. L. (2003). Detection of stress in tomatoes induced by late blight disease in California, USA, using hyperspectral remote sensing. *International Journal of Applied Earth Observation and Geoinformation*, *4*(4), 295–310. doi:10.1016/S0303-2434(03)00008-4

Zhao, Y. R., Li, X., Yu, K. Q., Cheng, F., & He, Y. (2016). Hyperspectral Imaging for Determining Pigment Contents in Cucumber Leaves in Response to Angular Leaf Spot Disease. *Scientific Reports*, *6*(27790), 1–9. PMID:27283050

ADDITIONAL READING

Chollet, F. (2017). Xception: Deep learning with depthwise separable convolutions. *arXiv preprint*, 1610-02357.

Golhani, K., Balasundram, S. K., Vadamalai, G., & Pradhan, B. (2018). A review of neural networks in plant disease detection using hyperspectral data. *Information Processing in Agriculture*, *5*(3), 354–371. doi:10.1016/j.inpa.2018.05.002

He, K., Zhang, X., Ren, S., & Sun, J. (2016). Deep residual learning for image recognition. In *Proceedings of the IEEE conference on computer vision and pattern recognition* (pp. 770-778).

Lee, S. H., Chan, C. S., Mayo, S. J., & Remagnino, P. (2017). How deep learning extracts and learns leaf features for plant classification. *Pattern Recognition*, *71*, 1–13. doi:10.1016/j.patcog.2017.05.015

Rangaswami, G., & Mahadevan, A. (2016). *Diseases of Crop Plants in India* (4th ed.). PHI Learning Pvt. Ltd.

Simonyan, K., & Zisserman, A. (2014). Two-stream convolutional networks for action recognition in videos. In Advances in neural information processing systems (pp. 568-576).

Simonyan, K., & Zisserman, A. (2014). Very deep convolutional networks for large-scale image recognition. *arXiv preprint arXiv:1409.1556.*

Szegedy, C., Ioffe, S., Vanhoucke, V., & Alemi, A. A. (2017, February). *Inception-v4, inception-resnet and the impact of residual connections on learning* (Vol. 4). AAAI.

KEY TERMS AND DEFINITIONS

Artificial Neural Networks (ANN): It is a computational technique inspired by the human brain. It consists of nodes (neurons) and connections (also known as synapses) between them, to exchange and transfer data. The network learns automatically according to the flow of the data.

Convolutional Neural Networks (CNN): It is a subtype of ANNs with a collection of deep FFNNs utilized mainly for image analysis and classification.

F1 Score: Given precision and recall values, F1 score is computed as the harmonic mean of both.

Image Processing: Image processing is the area of computer science that deals with the analysis, enhancement and manipulation of digital images for feature extraction, recognition, and classification purposes.

Plant Pathology: It is the study of diseases in plants and their causal factors such as environmental conditions, pathogens, etc. that affect the overall plant growth.

Precision: It is a statistical measure of random errors. It is the ratio of valid outputs (also known as true positives) to retrieved samples only.

Recall: Similar to precision, it is also a statistical measure. It is the ratio of valid outputs to total number of relevant samples.

Support Vector Machine (SVM): It is a supervised machine learning tool utilized for data analysis, regression, and classification.

Chapter 11
Intelligent Farming With Surveillance Agribot

Tripty Singh
Amrita Vishwa Vidhyapeetham, India

Dasari Naga Vinod
Amrita Vishwa Vidhyapeetham, India

ABSTRACT

This chapter endeavors to execute the rural cultivating robot which can move naturally, instinctually, automatically to furrow, seeding, and water system in shut field. This examination additionally adds to upgrade usefulness of agri-bot in the field for spying. The agri-robot is outfitted with a camera and sends the information to pc through wi-fi arrange. The robot has furrowing cutting edges and servo engine for dispersing the seeds and water system into the field. It works with ultrasonic sensor and IR sensor. Ultrasonic sensor is for dodging deterrents in the field; IR sensor is for detecting felid limit. Arduino controller goes about as heart and cerebrum of the framework; it makes quick, precise, self-governing development. This exploration includes an agri-bot that utilization Wi-Fi 802.11G alongside TCP/IP convention. IP address and video are gotten in workstation for further handling. This examination means to decrease human endeavors and give a clever guide to the farmers.

INTRODUCTION

Today, the horticulture area is more consolidate on worst culture creation. By the numerous provincial areas in India does not have the essential aptitudes, procedures and difficult work in cultivating that declines the advancement of creating nations. In this manner, the farmers are essentially required to move towards the innovation for

DOI: 10.4018/978-1-5225-8027-0.ch011

sophisticating the field for furrowing, seed appropriating, and burrowing (Moallem et al., 2013;2014;Razmjooy et al., 2012). By the Furrowing and seed conveying machines are worked physically, there is no robotization for both the exertion. The customary strategy incorporates disseminating the seeds by hand and furthermore dibbling (burrowing the dirt and putting the seeds by hand) is performed. The fundamental thought of this paper is to do the activity of furrowing the field and seed sowing with the assistance of mechanization. By utilizing this procedure, craftsmanship can be diminished up to 95 percent. The utilization of this robot is chiefly for horticultural purposes. The robot can likewise catch the minute-by-minute work by rehearsing the camera. The information can be checked in the PC/PC by rehearsing the IP Webcam through Wi-Fi organizes. Along these lines, observation can be executed. The robot can see and break the deterrents like trees and stones in the cultivating field by utilizing the proximity sensor. The impediments of the proposed technique can be considered as the following:

- The robot will actualize work inside limit and will likewise escape impediments.
- The robot will be actualized in cultivating field without including any handicraft in the base measure of time.
- To actualize robotized furrowing and seed circulation utilizing Arduino controller.

Robots in Agriculture-Agri-Bots

Mechanical technology in agribusiness in India has been a pushed prerequisite for our farmers from a decade ago. The principal report was distributed from the February 1934 Current Mechanix diary. This idea strives to build up emergency that an agriculturist could be isolated from a machine's field activity where at first there was just little advancement in the robotization of Agribot or in fact a car division. Agribusiness is a key development in the rising of inactive human and by helpful in business exercises like giving the sustenance and vitality for our reality. There is diminishing in the rural robot chart, to build the horticultural development diagram we need to create more machines that abatement the general expense to the farmer. The sprouting of the cultivating robots, which can move consequently in the field to ploughing the cultivating field with mechanized furrowing and can appropriate the seeds into the cultivating field inside the limit of the cultivating field.

The Figure 1 shows how the robot can accomplish in the agricultural fields. The merits of the proposed method distinguished to the existing approach of the robot are as follows:

Figure 1. Architecture of movement of autonomous robot in an agricultural field

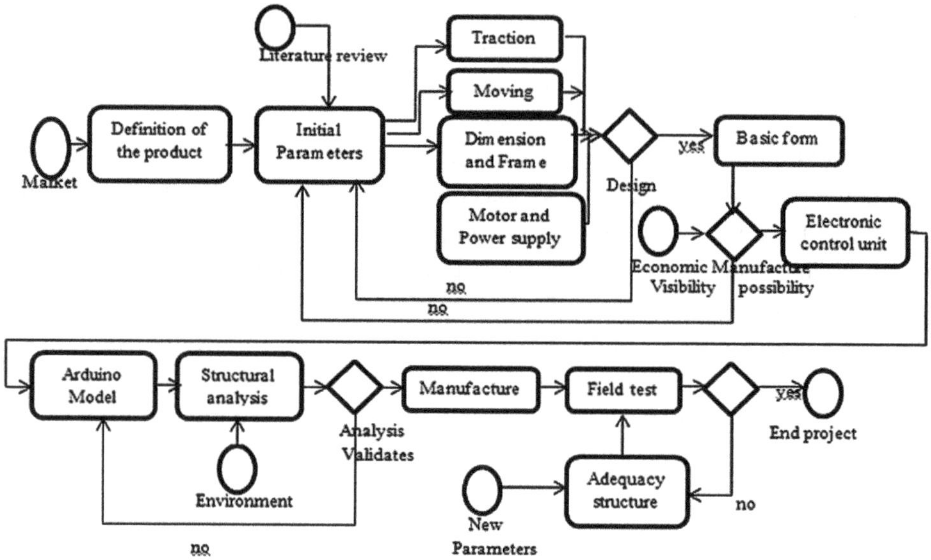

- Robots can perform the operation without break and even in hazardous situations like rainy season etc.
- The robot can sense and escape the obstacles effectively without re-organizing the main system.
- The robot can be implemented in ploughing the field in different depths and in the distribution of seeds within the adjacent boundary.
- Monitoring the work in laptop/pc by using a mobile camera with IP Webcamandroid application through a Wi-Fi network.

Farming is basic to the proceeded presence of human life as they straightforwardly rely upon it for the creation of sustenance. The exponential ascent in populace requires a fast increment in sustenance with the utilization of innovation to lessen the work and augment creation.

Figure 2 shows the functions performed in an agricultural field.

STATE OF THE ART: LITERATURE SURVEY

The Robot ideology is used to establish the mechanism of cultivating agricultural land by utilizing the Micro Controller. The agricultural robot will be utilizing a chassis, as a base to associate and assemble the whole system on it will be consisting of

Figure 2. Functions performed in an agricultural field

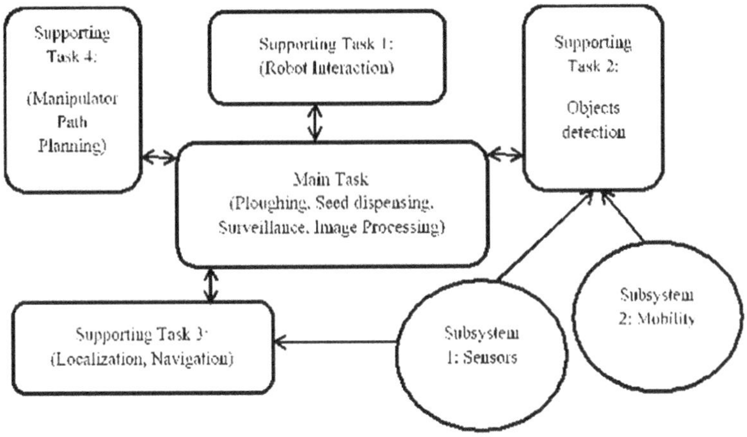

four motors. If an Ultrasonic Sensor exposes any obstacle, it will terminate the DC motor and Seed sowing operation. In preceding, the robot can be enforced the seed dispensing, fruit picking, automatic ploughing, and pesticide spraying activities in manual restraint with the assistance of AVR At mega microcontroller it containments the whole mechanism. In this approach, having profit and efficiency is less. In the Seed sowing robotic mechanism, acquire the two elements: PC element, Robot element. In PC, element keyboard is used to contribute the user input for the motion of the robot in the appropriate command. ZigBee is used to obtain the indication form the PC elements & send the signals to the ARM. Conferring to the directions given by the user, the robot moves in forward, reverse, left and right direction to place the seeds in a described position. Two of which are toy motors and the other existence gear motors. The robot is accomplished of doing three independent objectives are Ploughing, seeding; Harvesting will be functioning in distinct modes. Estimating of distinct modes will be done separately the different modes. It is manually accomplished, profit and accuracy are less. Currently, various analyses have been undertaken that deal precise with things of innovative geometry. A new limited fundamental model (Movahhedy 2002) of the machining technique has been utilized to an established result of a blunt or chamfered edge on cutting exaction.

Accuracy Farming is believed to be arrangement required to accomplish the needed creation rate. A critical change, in the region of picture preparing and information handling, has been elected a noteworthy test already in the act of exactness Farming. A database of pictures is gathered through remote detecting, broke down and a model is produced to decide the correct treatment anticipates distinctive yield composes and unique areas. Highlights of pictures from vegetation's should be removed, characterized, and divided lastly nourished into the model. Distinctive procedures have been connected to the procedures from the utilization of the neural system, bolster vector machine, fluffy rationale approach and as of late, the best methodology producing quick also, amazing outcomes utilizing the profound learning approach of convolution neural system for picture groupings. The fundamental thought is to join the ideas of Web of things alongside the procedures of picture handling to confer at exact outcomes. It is realized that temperature, dampness, soil dampness and light force prompt unpretentious to intense changes in the soundness of the plant. The morphological changes that a plant involvement is caught and examined on the MATLAB programming utilizing the calculations to confer at the outcome. In remote detecting information improvement, it is vital to feature such picture includes as edges, points of interest and surfaces. In this paper, a strong reverse dispersion condition is introduced, which utilize picture slope size and the capacity to stay away from the commotion amplification, what's more, to dispense with viably overshoots innate in the Laplacian. Trial results exhibit that this calculation can improve imperative subtle elements of picture information viably without overshoots, managing the open door for a decent picture elucidation and resulting preparing.

IMAGE PROCESSING IN AGRICULTURE

Field of View Process

Using the basic camera is embedded in the robot to capture the sight of the field. The acquired view will be passed through various processing methods like noise, removed contrast enhancement edge detection, filtrating the unnecessary components and obstacle detection (Moallem, 2012; Razmjooy et al., 2011;2014; Makaremi et al., 2018). The processed image will be used for extracting the important image parameter matrix. Parameter matrix will be the quantifiers for the robot for assisting and aiding for taking the appropriate steps about navigation and ploughing. The image processing can be utilized in agricultural relevance for the specified objectives:

1. To disclose diseased stem, leaf, and fruit
2. To determine the model of the deformed area.
3. As a disease quantification system for the affected areas.
4. For determining the size and the format of the fruits.
5. To determine the shade of the reformed area

Image Acquisition

The data assortment is the dominant aspect of the data assortment in the field. The same assortment has been utilizing in other studies of robotic examine images segmentation. Distinct image databases' feasible worldwide along with their name, explanation, and applications. The image acquisition is enforced to assemble the actual source image. An image must be changed to numerical form before processing. This alteration process is called digitization. This mechanism is performed by a CCD (charge-coupled device) which is embedded in the digital camera.

Image Preprocessing

The main purpose of the image enhancement is to mechanism an image for a precise task so that the refined image is better explores toward the original image. Image enhancement technique performs in both spatial and Frequency domains. Spatial domain: as the name exhorts, this domain uses a different approach, which will affect the manipulation of pixel values of an image.

Figure 3 shows the steps of the Image processing how it will be classified in the systematic process.

Image Segmentation

In image processing, the analysis decline within the group of a various image feature of an original image. Segmentation segregation an image into essential scope or objects. The equivalent to which that class carried out is a problem specific. The essential approach among all analysis channels is threshold-based form, whose volume utilizes either a physical or automated accomplish threshold principles for analysis.

Image Representation and Description

Representation and description approximately succeed in the output of the determined phase. The first conclusion must be built in case the data should be described as a boundary or exhaustive region. Boundary description is appropriate when the target is on external form attributes whereas regional exhibition settles on internal possessions, such as texture and skeletal form.

Image Recognition

Recognition is the method that entitles a label to an entity based on awareness maintain by its descriptors. Categorization is a usual method utilized to recognize

Figure 3. Steps of image processing

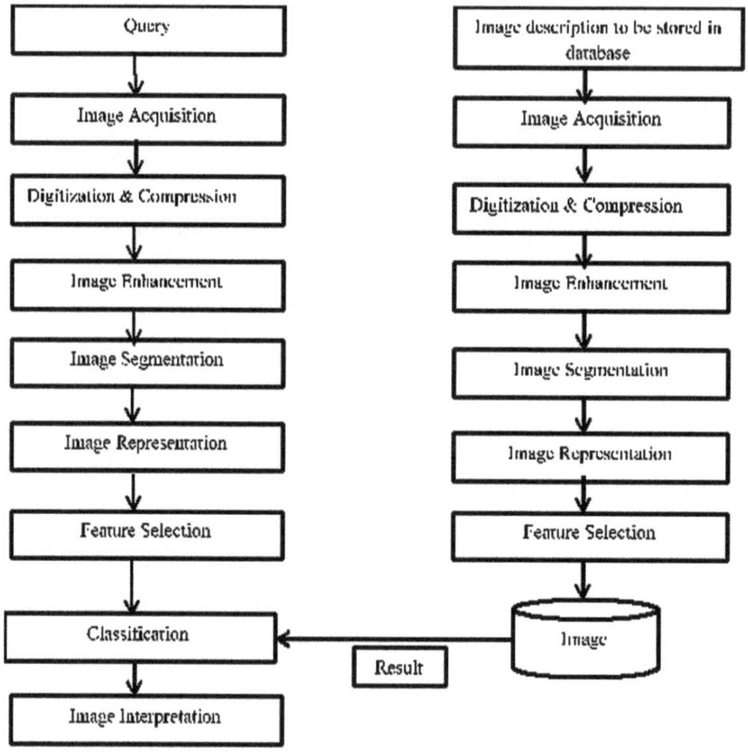

the image. Categorization is essential to differentiate a plant species with other species established on the data attained from feature preference. The descriptions from the image information reserved in the database are related with the descriptors from the suspicion image.

The Table 1 show the different parameters such as original, Binary gradient mask, Binary Image with filled holes, Binary Image with filled holes, Dilated Gradient Mask, Segmented Image, Cleared Border Image, Complement, Outlined Original Image, After finding small structures, Extracting Large Objects and After applying Tophat on the different soils.

Picture preparing has been so far being connected for different applications freely. Their individual application in the field of farming exists and has accomplished a secure level of achievement; anyway, the blend of both these innovations so far is non-existent. This paper depicts a way to deal with join and picture preparing to decide the ecological factor or fabricated factor (pesticides/manures) which is particularly obstructing the development of the plant. Utilizing a MATLAB Software detecting system, which takes the readings of the critical natural components and

Table 1. Image processing techniques on different parameters showing in the different soils

PARAMETERS	SOIL 1	SOIL 2	SOIL 3	SOIL 4
Original		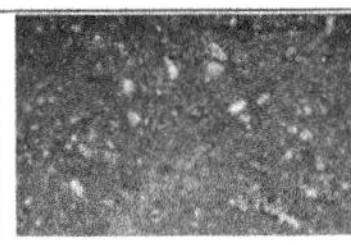	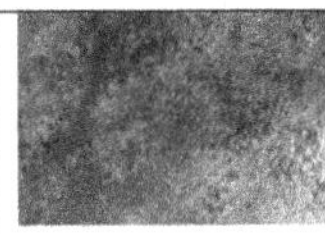	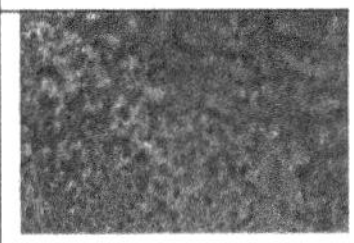
Binary Gradient Mask			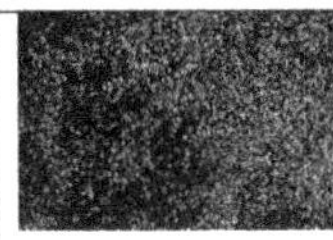	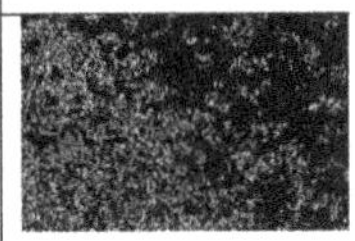
Binary Image with filled holes		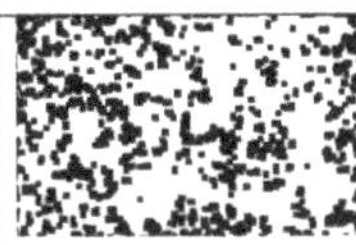		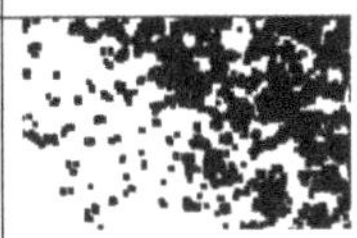
Dilated Gradient Mask			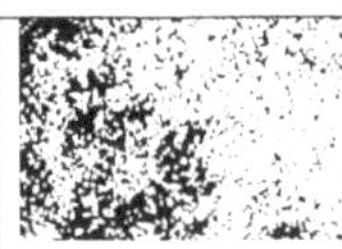	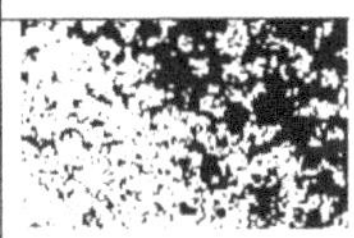
Binary Image with Filled Holes		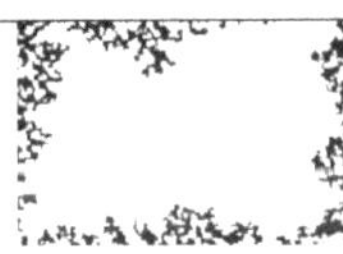	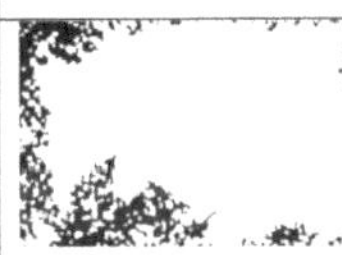	
Cleared Border Image	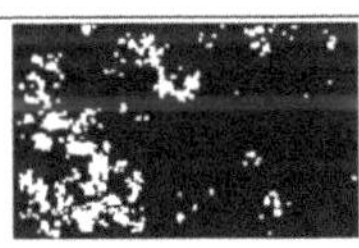	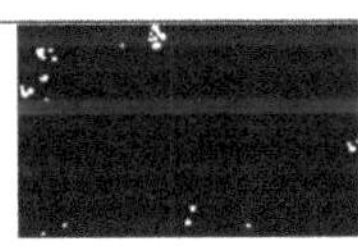	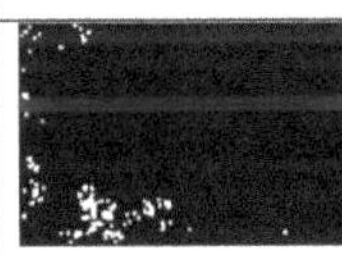	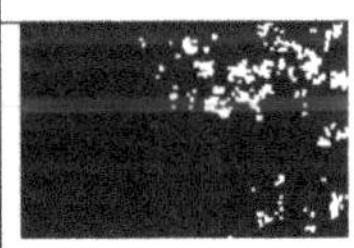
Segmented Image	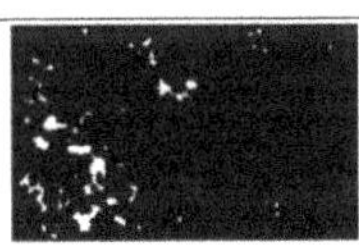			

continued on following page

Table 1. Continued

Outlined Original Image		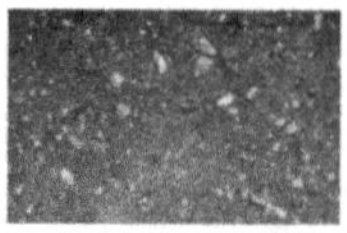		
Complement		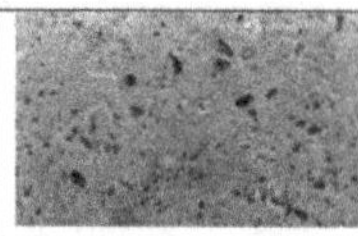		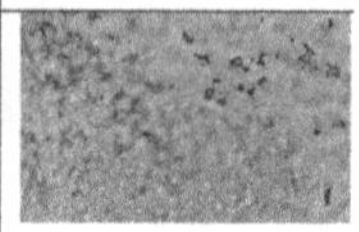
After finding small structures	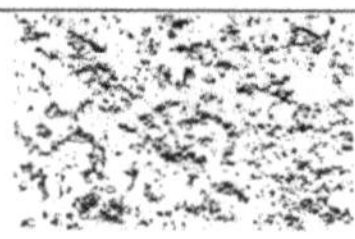	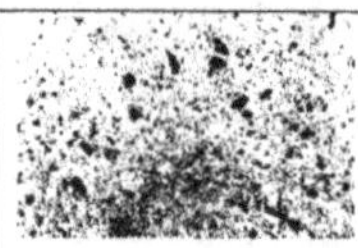	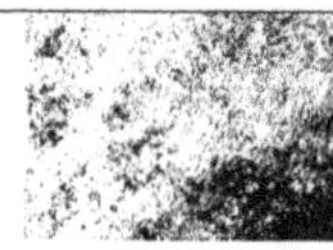	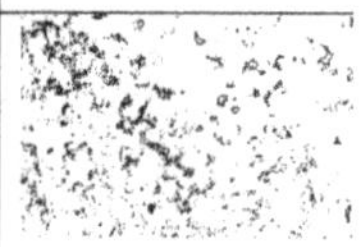
Extracting Large Objects	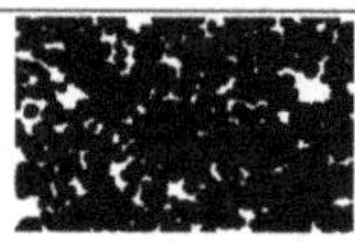	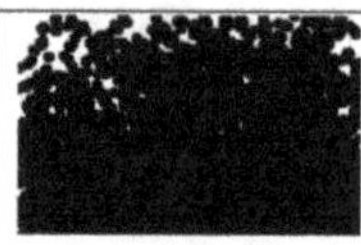	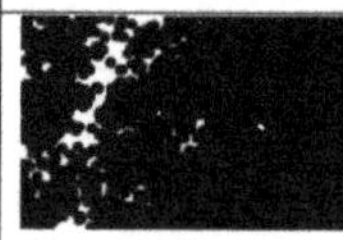	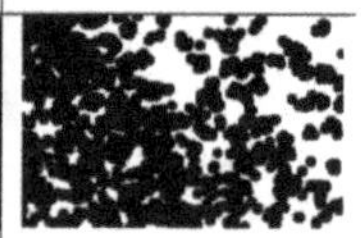
After Applying Top hat	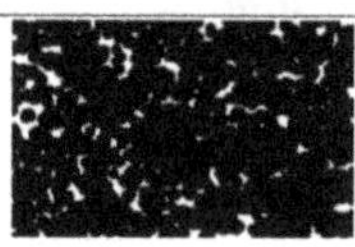	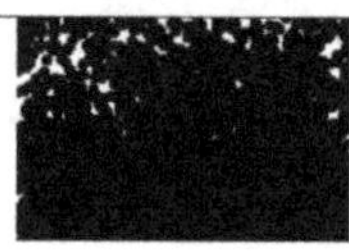	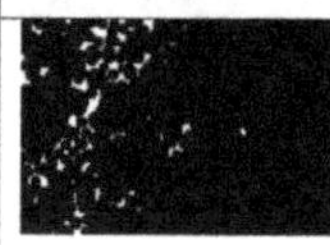	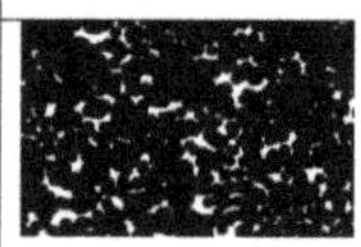

the picture of the field exploited, it is handled under MATLAB programming by the assistance of histogram investigation to confer at definitive outcomes (*Moallem, Razmjooy and Mousavi, 2014*).

Table 2 shows the different parameters such as original, binary gradient mask, Binary Image with filled holes, Dilated Gradient Mask, Binary Image with filled holes, Cleared Border Image, Segmented Image, Outlined Original Image, Complement, after finding small structures, Extracting Large Objects and After applying Top hat on the Black, Dry and Moist soils.

In the field of agribusiness designing, farming picture is examined and prepared by expelling of clamor, upgrade, pressure and edge recognition of two dimensional discrete wavelet change, also, the likelihood of discrete wavelet change in agribusiness building is seen keeping in mind the end goal to enhance the explore in the field (Sneha et al. 2015).

Table 2. Image Processing Techniques on different parameters showing in the Black, Dry and Moist soils

PARAMETERS	SOIL 1	BLACK	DRY	MOIST
Original			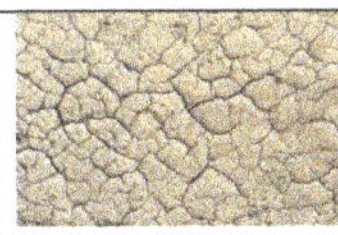	
Binary Gradient Mask			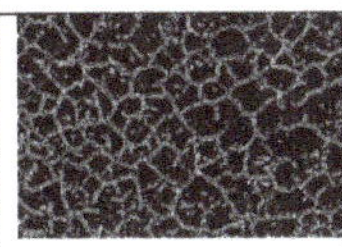	
Binary Image with filled holes	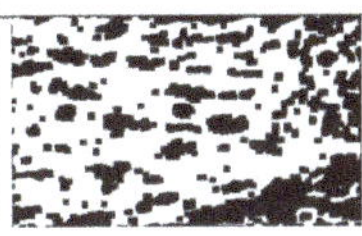	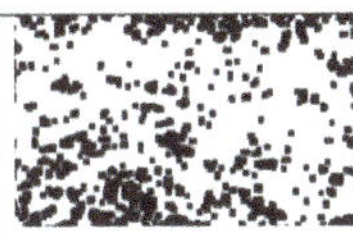	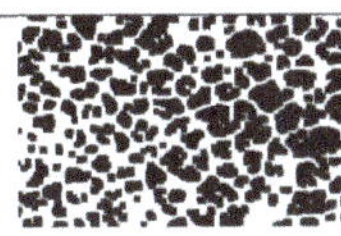	
Dilated Gradient Mask	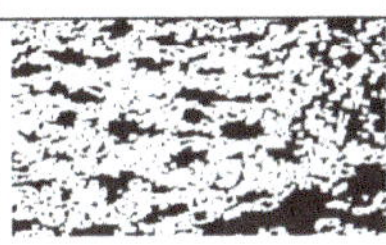	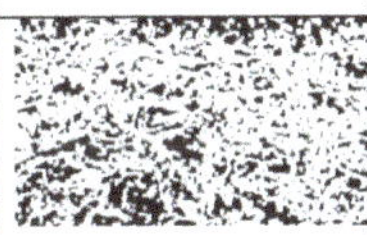	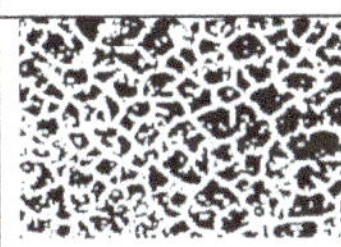	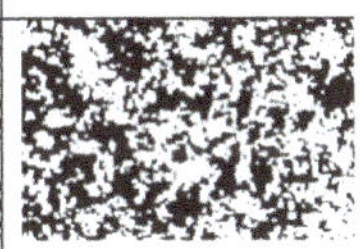
Binary Image with Filled Holes	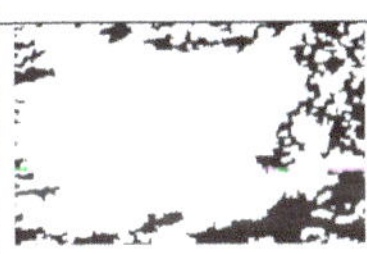		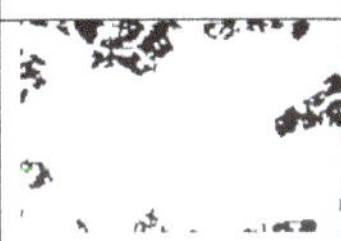	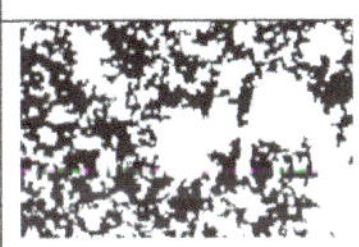
Cleared Border Image	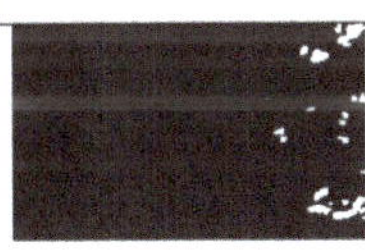			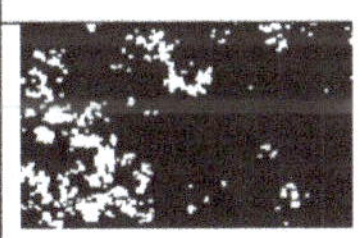
Segmented Image				

continued on following page

Table 2. Continued

Outlined Original Image				
Complement	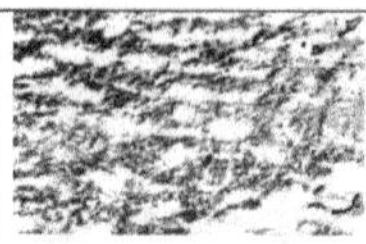		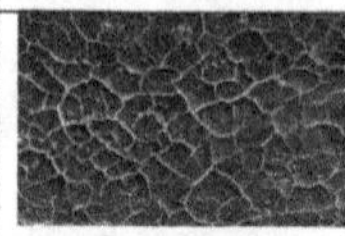	
After finding small structures		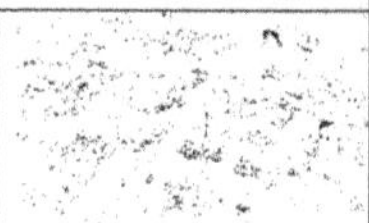	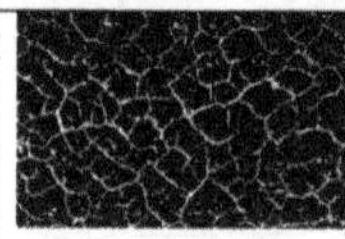	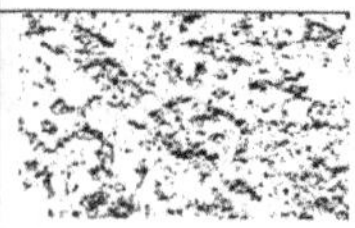
Extracting Large Objects	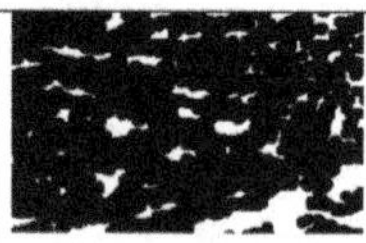	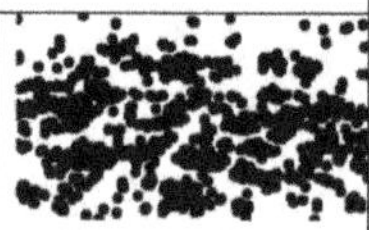		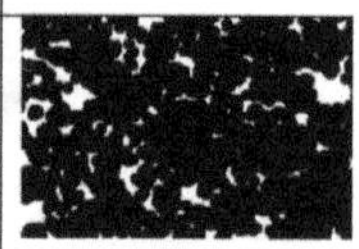
After Applying Tophat	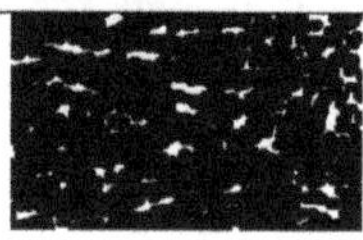	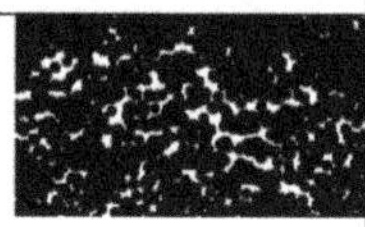		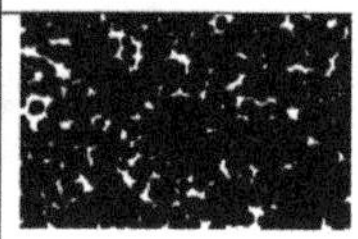

SOFT COMPUTING IN AGRICULTURE

Soft computing is a confirmed of "imprecise" computing capabilities, which are adequate to model and evaluate very complex issues. For these complex issues, more approaches that are conventional have not been adequate to yield cost-effective, analytical, or exhaustive solutions. Soft computing has been broadly studied and enforced in the last thirty-five years for scientific research and engineering computing. In agricultural and biological systemizing, researchers and organizers have advanced techniques of fuzzy logic, artificial neural networks, genetic innovations, arbitration trees, and support vector machines to consideration soil and water management associated to crop growth, evaluate the activity of food processing, and support decision fabricating in precision farming.

Image processing handling has been ended up being a feasible device for examination in different fields and applications. Agribusiness division where the parameters like covering, yield, nature of item were the vital measures from the farmers' perspective. Ordinarily, master exhortation may not be moderate, dominant part times the accessibility of master and their administrations may expend time. Image processing preparing alongside accessibility of correspondence system can change the circumstance of getting the master counsel well inside time and at reasonable expense since the picture, preparing was the compelling apparatus for examination of parameters. This paper plans to center around the study of the benefit of Image handling in the agricultural field, for example, imaging methods, land, and natural product reviewing. The examination of the parameters has ended up being precise and less tedious when contrasted with conventional strategies. Use of Image process preparing can enhance basic leadership for vegetation estimation, land, natural product arranging, and so forth (Umarkar et al. 2016).

PROPOSED SYSTEM

The block diagram of Agribot consists of Arduino Uno controller gives the commands to the robot as shown in Figure 4 and battery is append to the robot for giving the power supply which is providing +5v for Arduino Uno and +12v supply for moving DC motors. The Ultrasonic Sensor is used to detect and escape the obstacles in the field. The servomotor is used to distributing the seeds into the field. The camera is fixed in front of the robot by using the camera the work can be monitored in Pc/ Laptop through Wi-Fi network with "IP webcam" Android application in this way surveillance can be performed.

Figure 4. Functional block diagram

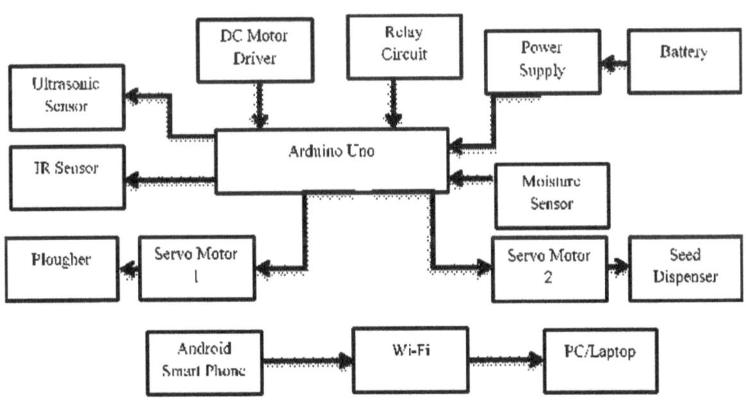

The robot is started through the initialization of IP WebCam using android application with a Wi-Fi module for transmitting the data to the Pc/Laptop for surveillance. After that robot is going further with navigating the field and also accomplish the different actions such as ploughing by using the sharp blades is equipped with the robot, seed sowing and supporter are fixed at the back so that removed soil is covered. If the ultrasonic sensor HC-SR04 detects any obstacles like stones, trees, it will detect and avoid the robot within a specific time without disturbing the main system and changing the path. Otherwise, it performs the normal operation like ploughing, seed sowing, area of field detection and the following path.

Ploughing Mechanism

The essential objective of ploughing is to revolution over the upper bed of the soil, gather original minerals to the exterior, although burying weeds and the residue of preceding crops and concede them to resolve. As the plough is starved concluded the soil it establish long ditch of fertile soil termed as furrows. In modern employment, a ploughed field is generally left to alleviate and is then demolished previously planting. Ploughing and farming a soil accommodate and customizes the above 10 to 30 cm of the soil to mode a plough soil into the field. In an abundant soil, the preponderance of expensive plant feeder roots can be settled in the topsoil or plough the soil into the field.

When the robot follows the path simultaneously, the Agribot has to do ploughing the soil with the assistance of plougher. The Agribot has to attach with the servomotor; the servo motor shaft can be attached with the plougher. The plougher can be loosening the soil in different depths. There are different categories of soils are there, on that seeds also different. Particular seeds having different depths for farming the field, based on that the servomotor can be given the movement to the shaft. The shaft can be rotated with the certain angle the plougher can loosen the soil normally. Suppose we need to plough the soil heavily for another crop, we need to give the more than the sixty degrees angle to the servomotor, based on that the shaft can rotate that much and it is suitable to alleviate the soil and distributing the seeds and farming the field. Variables to assess techniques: Quantity of fertilizer is used.

- Number of ploughing operation
- The decade of first ploughing
- Ploughing Direction
- Ploughing Depth
- Cover cropping
- A number of clod-smothering operation.
- Number of ripping intervention

- Depth of ripping
- Number of the milling operation
- Milling Depth
- Number of uprooting operation
- Uprooting Depth
- Number of other "light" operation
- Number of weeding interventions
- Chemical utilized to weed
- Harrowing
- Subsoiling Depth
- Crowning
- Mulching
- EU 2078/92 (as a proxy of rotation)

Seed Sowing Mechanism

In the extension with seed sowing, multipurpose activities such as Levelling & Plugging are also required. However, farmers sustain various issues until seed sowing proceeding, corresponding appropriate improvement of range at intervals two crops, the distance between two successions of lines. Seed sowing is an appropriate essential and paramount process in the agricultural field. Also, a couple of bullocks is needed to fetch the heavy appliances of identical and seed dropping.

Another process of seed sowing is to be helped tractor in farms. The excessive appliances of seed repository and dropping tool are appended to the tractor to globule the seeds. A ground wheel is connected at the contemptible of the seed-sowing tool (Fu, Ruan and Wang, 2008). The competence transmission ideology is used to convey the action of the turn to the metering appliance. The metering mechanism bounded number of spades to globule exhausted the seeds from the hopper.

The poor state of farming in the nation is the purpose of worry for Indians. The country farmers in India experience the ill effects of destitution and the vast majority of them are unskilled so there is an absence of good expansion administrations. Another significant issue looked by Indian farmers is their reliance on nature and inadequately kept up water system frameworks. Current agrarian practices are neither financially nor naturally maintainable and India's yields for some, horticultural wares are low. Ineffectively kept up water system frameworks and relatively widespread absence of good augmentation administrations are among the variables dependable. Poor streets to display from town, simple market foundation, and exorbitant direction are few of the other concerned focuses the farming segment in India, these are the problems faced by the agriculture. The producing crops ideology consists of ploughing; cultivating, weeding, harvesting, etc. is the various methods. All the methods are leading to customizing the appliance in cultivating which works consequently without the human competence concerned. The small tool would be assembled from extant mass formed elements without the extremity of specialized idea and tooling. Also, exertion desire to this tool is less as related to tractors or any agricultural tool. Seeding formation is the day-to-day life we utilize tractor in farms. However, it utilizes extra time and man-made deficiency is opposed continuously. When the robot follows the path, simultaneously it has distributed the seeds into the field through servomotor. The servomotor rotates the shaft with some delay, on that delay the shaft rotates, while attached to the hopper container the seed has to fall to the soil by moving the shaft.

IMPLEMENTATION

The robot is started through the initialization of IP WebCam using android application with a Wi-Fi module for transmitting the data to the pc/laptop for surveillance.

Figure 5 shows the flow of enforcement of the whole system. After that robot is going further with navigating the field and also accomplish the different actions such as ploughing by using the sharp blades is equipped with the robot, seed sowing and supporter are fixed at the back so that removed soil is covered.

The final expected accuracy of the plough-guiding robot depends not only on the control algorithm but also on registration, which relates the coordinate frame 1 of the robot to the frame 2 of the images captured by video camera. If the ultrasonic sensor HC-SR04 detects any obstacles like stones, trees, it will avoid the robot within specific time without disturbing the main system and changing the path. Otherwise, it performs the normal operation like ploughing, seed sowing, the area of field detection and following the line autonomously.

Table 3. Comparison between the seed sowing method and utilities

Si.no	Specifications	Manual	Tractor	Seeding appliance
1.	Human Power	High	Limited	No
2.	Time Required	High	Limited	Low
3.	Seeding Mechanism	Manually	Manually	Automatically
4.	Required Power	More	Very high	Less
5.	Labor Expenses	More	More	Initial cost only
6.	Pollution	No	More	No

Figure 5. Flowchart of execution

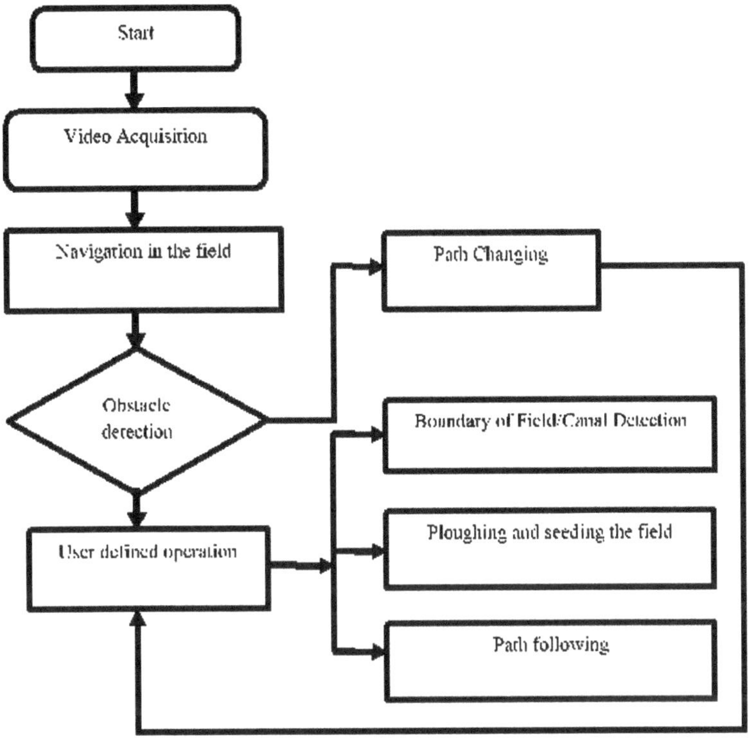

The ploughing appliance has been deliberate for around fifty decades in an effort to illustrate distinct cutting experience such as the intensity effect of chip thickness on extortion machined Surface effects, burr composition, and cutting mechanism damping. Ploughing has also been inspected for its own improvement to cutting exaction and its alliance to worn tool flank efforts (Usui 1984; Elanayar 1994). Currently, various analyses have been undertaken that deal precise with things of innovative geometry. A new limited fundamental model (Movahhedy, 2002) of the machining technique has been utilized to the result of a blunt or chamfered edge on cutting exaction (Jingfu and Zhijun, 2010). The approach uses a linked Eulerian-Lagrangian idea and a mechanism-work contact innovation for sliding-sticking friction environment based on an alliance such as:

$$\varphi_k = \varphi[1 - f^{-\frac{\tau\sigma}{\varphi}}] \tag{1}$$

Where φK is the frictional Stress, σ is the Normal stress, φ is the shear abundance emphasis, $\tau = 1$ based on experimental force measurements. A slip-line field exhibiting the distortion below a spherical cutting border in orthogonally accomplished. The model was established on a dead-metal zone established at the rounded edge, a raised prow of substantial ahead of cutting, and a wedge of distortion below the dead-metal zone inferior on divided stresses there. Assuming a negligible prow (ρ=0o) and a friction stress on the dead-metal region identical to the machining shear stress of the assignment objective (i.e., φk = φ from Equation (1)). Considering the angle that slip-lines expedient the dead-metal zone turns on this friction circumstances, the latter description implies that η=0o and the slip-line "field" destruction to a single slip line (i.e., δ=0). The cutting and thrust power on the lower horizon of the dead-metal zone can then be written as:

$$M_k = \varphi.s.t_f.\tan(\frac{\pi}{2} + \frac{\beta}{2})$$
$$M_L = \varphi.s.(1 + \frac{\pi}{2}).t_f.\tan(\frac{\pi}{2} + \frac{\beta}{2}) \quad (2)$$

Where s is the cutting breadth, t_f is the area radius, β is the rake angle, φ is the fabricate shear stress. Resolved from deliberate cutting power during orthogonal.

The orthogonal width of cut s in Equation (2) can be retrieved with the innovative length.

$$s = f_g[\frac{\pi}{2} - \alpha_n + \sin^{-1}(\frac{g}{2f_g})] + \frac{e - f_g(1 - \sin(\alpha_n))}{\cos(\alpha_n)} \quad (3)$$

For a typical turning effort, Where fg is the tool nose radius, α_n is the side cutting border angle, g is the nominal feed per rev and e is the nominal depth of cut.

$$M_L = V_Z.S \quad (4)$$

$$V_Z = \varphi.(1 + \frac{\pi}{2}).t_f.\tan(\frac{\pi}{4} + \frac{\beta}{2}) \quad (5)$$

The longitudinal ploughing action is then

$$M_Z = \int_0^S V_Z \cos\alpha(n)\, dn \quad (6)$$

Where n exhibits the length of the cutting border from 0 to s as Defined Equation (3) and α (n) is the deviating side cutting border of Angle along the curving cutting border.

$$\gamma = \tan^{-1}[(\tan(\gamma_a)\cos(\alpha_t) + (\tan(\gamma_b)\sin(\alpha_t)\cos(j)] \quad (7)$$

Where j is the inclination angle given by

$$j = \tan^{-1}[(\tan(\gamma_a)\cos(\alpha_t) - (\tan(\gamma_a)\sin(\alpha_t)] \quad (8)$$

where, γ_a and γ_b are the side and back rake angles and α_t is the effective lead angle given by

$$\alpha_t = \tan^{-1}[(\tan(\alpha_t) + \frac{(f_g + \frac{\upsilon}{2})}{e}(\frac{1-\sin(\alpha_t)}{\cos(\alpha_t)})] \quad (8)$$

A chamfered border can be contained in Equation (2) by simply using an equivalent edge radius. For a chamfered of width w at an angle δ, the correspondent edge radius would average the two "legs" of chamfer i.e. w cos(δ) and w sin(δ).

RESULTS

The recommend structure gives a compact, less power and inexpensive system with the coherent output. Therefore, the Agribot gives near about 95% precision concerning ploughing the field, dropping of seeds, avoiding the obstacles and following the path with surveillance the surrounding area.

An agricultural mobile robot has been considered to be employed as an experimental platform for the enlargement of navigation, control, and data acquisition technologies in the agricultural area.

The main application of the proposed robot is to utilize it as a remote sensing system of the agronomic parameters for the most important Indian culture in large areas.

There is no need for actions that demand high power, as in agricultural operations, but only moving efficiently in this environment. The main structure of the system is shown in Figure 6, which was designed by the studying of work conditions required in an area and the desired characteristics of the project.

The Figure 7 shows the hardware component of the project and the robot can perform the work like path following and the Agricultural Mobile Robot is designed to sensing agronomic parameters of most important Indian culture (maize, sugar cane, soybeans, and orange) during almost the entire cycle of growth and post-harvest in large areas. It does not require actions that demand high power, as in agricultural operations, but only moving efficiently in this environment.

The Figure 8 shows the monitoring the area in pc through Wi-Fi network and simultaneously ploughing if any obstacles are present it can detect and avoid the obstacles along with that surveillance can be performed by using the mobile camera and monitor in pc through IP webcam with Wi-Fi network. The Camera is used for display monitor as well as it will take the pictures of the land for processing and other techniques like Binary gradient mask, Binary Image with filled holes, Dilated

Figure 6. Front view of the robot with following the path and top view of the robot

Figure 7. Right and left view of the robot

Figure 8. Monitoring area in PC

Gradient Mask, Binary Image with filled holes, Cleared Border Image, Segmented Image, Outlined Original Image, Complement, After finding small structures, Extracting Large Objects and After applying Tophat on the different soils by using the MATLAB Software (*N, 2017*).

Figure 9 shows the seed-dispensing tub, this robot can also be integrated with Seed Sowing process such as we will attach the seeding tub to the robot while moving the robot the seed can be dropped to the field by the help of servomotor. The proposed approach is the robot can be implemented in ploughing the field, distribution of seeds within the boundary.

The Figure 10 shows the student working with the Agribot in the sample field with ploughing and seed sowing mechanism. The process in farming where seeds are adequately spaced apart during planting to allow for optimal crop growth. It

Figure 9. Seed dispensing tub

Figure 10. Working with the agribot

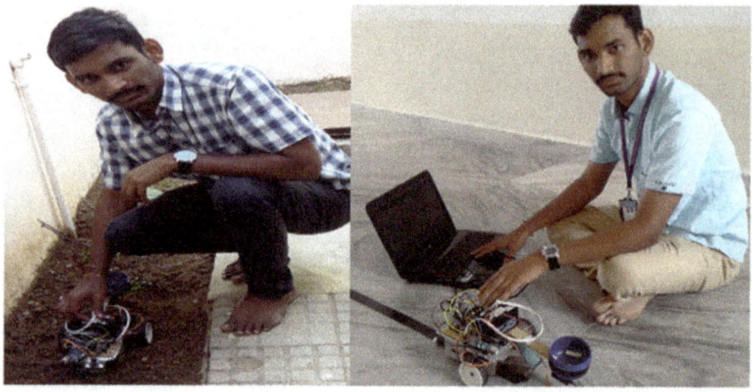

can also be a time-consuming process. There are many applications for robots in agriculture. For instance, Merlin Robot Milker, Rosphere, Harvest Automation, Orange Harvester, lettuce bot, and weeder are some kinds of these robots. One case of a large scale use of robots in farming is the milk bot. This technology is widely increasing among British dairy farms due to its efficiency and non-requirement to move. Need of precise evaluating, arranging of leafy foods and fruit or farming items emerges on account of expanded desires in quality sustenance and security guidelines. It causes expanded handling and work. PC vision and picture handling are not ruinous, precise and dependable techniques to accomplish focus of reviewing. Picture preparing in agribusiness, what's more, nourishment businesses has been connected in the regions of arranging, reviewing of new items, recognition of imperfections, for example, dim spots, breaks, and wounds on crisp foods grown from the ground, and so on. Same sorts of ideas were investigated by numerous analysts with distinctive picture handling approaches. In addition, the robot will disclose the diseased leaf, stem, fruit, and quantify damaged area by disease.

Figure 11 shows the data analysis on overlap, error score, volume, and diff.score in red grain soil. On X-axis data set is represented and on Y-axis efficiency of the red grain soil is represented. Red soils are broadly derived from crystalline rock. They are frequently poor flourishing soils, low in minerals, humus, and demanding to manage because of its low water inheritance quantity.

The data analysis on overlap, error score, volume, and diff.score in Black Sticky soil is as shown in Figure 12. On X-axis data set is represented and on Y-axis Efficiency of the Black Sticky soil is represented.

Figure 13 shows the data analysis on overlap, error score, volume, and diff.score in Sandy soil. On X-axis data set is represented and on Y-axis Efficiency of the Sandy soil is represented.

Figure 11. Data analysis on overlap, error score, volume and diff. score in red grain soil

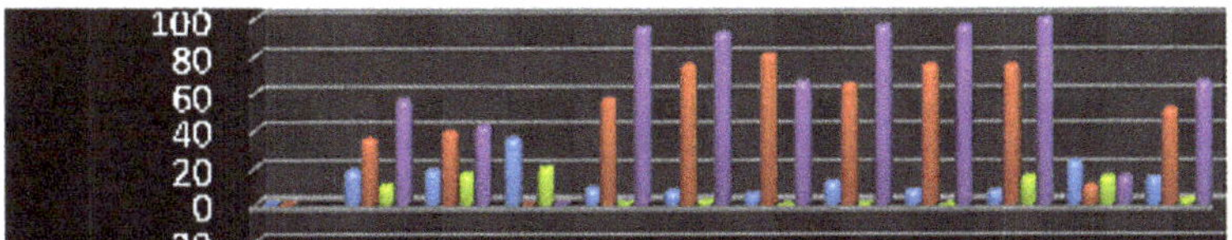

	set	1	2	3	4	5	6	7	8	9	10	Average
Overlap	0	17.9	18.2	35.2	8.4	7	5.7	12.2	7.7	7.5	23	14.3
Error	0	35	39	0	57	75	81	65	75	75	10	51.2
Volume [%]		10.7	16.5	20.3	1.3	1.8	-1.4	0.7	-1.6	15	15	3.1
Diff. Score		56	42	0	95	92	66	96	96	100	15	65.8

Figure 12. Data analysis on overlap, error score, volume and diff. score in black sticky soil

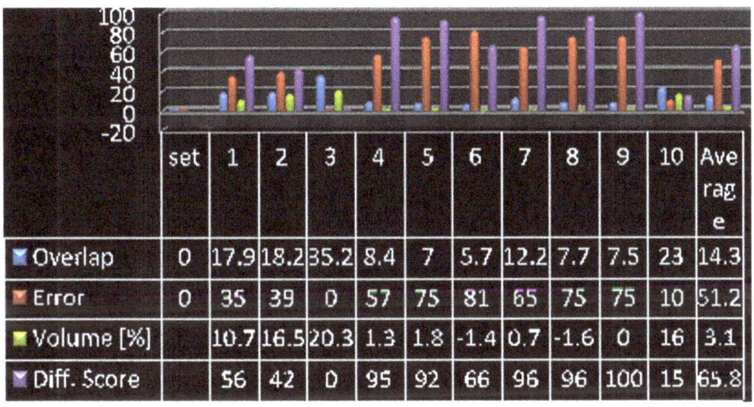

	set	1	2	3	4	5	6	7	8	9	10	Average
Overlap	0	17.9	18.2	35.2	8.4	7	5.7	12.2	7.7	7.5	23	14.3
Error	0	35	39	0	57	75	81	65	75	75	10	51.2
Volume [%]		10.7	16.5	20.3	1.3	1.8	-1.4	0.7	-1.6	0	16	3.1
Diff. Score		56	42	0	95	92	66	96	96	100	15	65.8

The trends in the agriculture Achievement of picture preparing and its development to various fields of uses like medicinal, designing and remote detecting has cleared its approach to an application in horticulture. In the agricultural area, the exhortation of topic authorities is constantly expected to tackle the greater part of the field issues. It requires additional time, costlier and is excessively expensive. Because of the higher achievement proportion in different divisions, horticulture area has started to work in add with the inventive data innovation utilizing systems like

Figure 13. Data analysis on overlap, error score, volume and diff. score in sandy soil

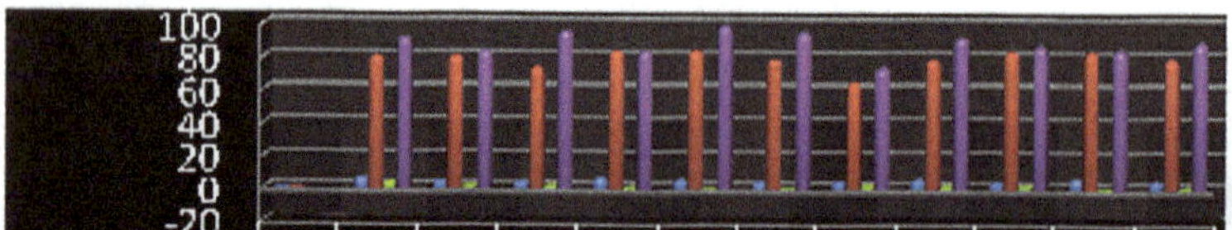

	set	1	2	3	4	5	6	7	8	9	10	Average
Overlap	0	6.5	4.5	4.1	6.4	5.4	4.1	4.4	5.7	4.2	6.6	5.2
Error	0	81	81	74	84	84	79	65	79	84	84	79.5
Volume [%]		4.2	2.2	1.5	1	0.1	-0.9	2.5	1.7	2.4	-2.8	1
Diff. Score		92	84	96	84	99	95	74	92	87	85	89

picture handling, PC vision and machine vision which could mirror the choices of topic pro. These methods create a yield for the info picture procured through any of the imaging processes.

CONCLUSION

The growth of multipurpose robot vehicles as platforms for a robot farming network. The vehicles involve an electronic robot vehicle that can assign crop enlightenment using sensors and two-robot tractors ability to conclude different implement purposes. The approach of inexpensive navigation sensors to the robot farming approach makes the system inexpensively flexible with the environment. In this way, robots can perform operations like moving, avoiding obstacles, monitoring, farming the field, etc. It allows the farmers to just manage the robots without the required to operate them. With the expansion of robot farming network, food fabrication can be increased considerably and economically.

This robot can also be integrated with Seed Sowing process such as we will attach the seeding tub to the robot. The proposed approach is the robot can be implemented in ploughing the field, distribution of seeds within the boundary. The robot can detect and avoid small canals and monitoring the work in laptop/pc by using a mobile camera with IP Webcam through a Wi-Fi network. In the way both processes can be performed in the same robot, such that the cost will be reduced, time also consumed and profit will be increased to the farmers.

The future work will include the implementation of Robot on the field in real time to evaluate and improving the Ploughing and Seed Sowing for accurate Farming.

REFERENCES

Abdullahi, H. S., Sheriff, R. E., & Mahieddine, F. (2017). Convolution neural network in precision agriculture for plant image recognition and classification. *2017 Seventh International Conference on Innovative Computing Technology (INTECH)*. 10.1109/INTECH.2017.8102436

Aminzadeh, R., & Fotouhi, R. (2014). Novel Design of a Precision Planter for a Robotic Assistant Farmer. *Volume 5B: 38th Mechanisms and Robotics Conference*. doi:10.1115/detc2014-35402

Fu, S., Ruan, Q., & Wang, W. (2008). Remote sensing image data enhancement based on robust inverse diffusion equation for agriculture applications. *2008 9th International Conference on Signal Processing*. doi:10.1109/icosp.2008.4697353

Gollakota, A., & Srinivas, M. B. (2011). Agribot — A multipurpose agricultural robot. *2011 Annual IEEE India Conference*. 10.1109/INDCON.2011.6139624

Jingfu, L., & Zhijun, L. (2010). Study on agriculture image processing based on discrete wavelet transform. *2010 Second IITA International Conference on Geoscience and Remote Sensing*. 10.1109/IITA-GRS.2010.5603095

Kapoor, A., Bhat, S. I., Shidnal, S., & Mehra, A. (2016). Implementation of IoT (Internet of Things) and Image processing in smart agriculture. *2016 International Conference on Computation System and Information Technology for Sustainable Solutions (CSITSS)*. 10.1109/CSITSS.2016.7779434

Makaremi, M., Razmjooy, N., & Ramezani, M. (2018). A new method for detecting texture defects based on modified local binary pattern. *Signal, Image and Video Processing*, 1–7.

Moallem, P., & Razmjooy, N. (2012). Optimal threshold computing in automatic image thresholding using adaptive particle swarm optimization. *Journal of Applied Research and Technology*, *10*(5), 703–712.

Moallem, P., Razmjooy, N., & Ashourian, M. (2013). Computer vision-based potato defect detection using neural networks and support vector machine. *International Journal of Robotics and Automation*, *28*(2), 137–145. doi:10.2316/Journal.206.2013.2.206-3746

Moallem, P., Razmjooy, N., & Mousavi, B. S. (2014). Robust potato color image segmentation using adaptive fuzzy inference system. *Iranian Journal of Fuzzy Systems*, *11*(6), 47–65.

N, M. M. (2017). Solar Powered Digging and Seed Sowing Machine. *International Journal for Research in Applied Science and Engineering Technology, 5*(3), 446-450. doi:10.22214/ijraset.2017.3081

Patil, D. A., Upadhye, M. Y., Kazi, F. S., & Singh, N. M. (2015). Multi Robot Communication and Target Tracking system with controller design and implementation of SWARM robot using Arduino. *2015 International Conference on Industrial Instrumentation and Control (ICIC)*. 10.1109/IIC.2015.7150777

Razmjooy, N. (2012). A multi layer perceptron neural network trained by invasive weed optimization for potato color image segmentation. *Trends in Applied Sciences Research, 7*(6), 445–455. doi:10.3923/tasr.2012.445.455

Razmjooy, N., Mousavi, B. S., Khalilpour, M., & Hosseini, H. (2014). Automatic selection and fusion of color spaces for image thresholding. *Signal, Image and Video Processing, 8*(4), 603–614. doi:10.100711760-012-0303-7

Razmjooy, N., Mousavi, B. S., Sargolzaei, P., & Soleymani, F. (2011). Image thresholding based on evolutionary algorithms. *International Journal of Physical Sciences, 6*(31), 7203–7211.

Razmjooy, N., Mousavi, B. S., & Soleymani, F. (2012). A real-time mathematical computer method for potato inspection using machine vision. *Computers & Mathematics with Applications (Oxford, England), 63*(1), 268–279. doi:10.1016/j.camwa.2011.11.019

Sneha, Abirami, Ankita, Praveena, & Srimeena. (2015). Agricultural Robot for automatic ploughing and seeding. *2015 IEEE Technological Innovation in ICT for Agriculture and Rural Development (TIAR)*. doi:10.1109/tiar.2015.7358525

Srivastva, Neda, & Kriti. (2017). Robotic Farmer. *International Journal of Engineering Development and Research, 5*(2).

Umarkar, S., & Karwankar, A. (2016). Automated seed sowing agribot using arduino. *2016 International Conference on Communication and Signal Processing (ICCSP)*. 10.1109/ICCSP.2016.7754380

Vibhute, A., & Bodhe, S. K. (2012). Applications of Image Processing in Agriculture: A Survey. *International Journal of Computers and Applications, 52*(2), 34–40. doi:10.5120/8176-1495

Compilation of References

Abbas, A. K., Lichtman, A. H., & Pillai, S. (2012). *Cellular & Molecular Immunology*. Elsevier.

Abdullahi, H. S., Sheriff, R. E., & Mahieddine, F. (2017). Convolution neural network in precision agriculture for plant image recognition and classification. *2017 Seventh International Conference on Innovative Computing Technology (INTECH)*. 10.1109/INTECH.2017.8102436

Abushmmala, F. F., & Abushmmala, F. F. (2014). Processing overlapped cells using k-means and watershed. *International Journal of Intelligent Information Systems*, *3*(1), 8–12. doi:10.11648/j.ijiis.20140301.12

Adamala, S., Raghuwanshi, N. S., Mishra, A., & Singh, R. (2017). Generalized wavelet neural networks for evapotranspiration modeling in India. *ISH Journal of Hydraulic Engineering*, *29*(1), 1–13. doi:10.1080/09715010.2017.1327825

Adams, M. D., & Kossentini, F. (2000). Reversible integer-to-integer Wavelet transforms for image compression: Performance evaluation and analysis. *IEEE Transactions on Image Processing*, *9*(6), 1010–1024. doi:10.1109/83.846244 PMID:18255472

Adapa, P., Tabil, L., & Schoenau, G. (2011). Grinding performance and physical properties of non-treated and steam exploded barley, canola, oat and wheat straw. *Biomass and Bioenergy*, *35*(1), 549–561. doi:10.1016/j.biombioe.2010.10.004

Adeleke, O. O., Makinde, V., Ayobami, O. E., Oluwaseun, F. D., Akintayo, O. O., & Taiwo, J. A. (2015). Estimation of Groundwater Recharges Using Empirical Formulae in Odeda Local Government Area, Ogun State, Nigeria. *Challenges*, *6*, 271–281. doi:10.3390/challe6020271

Ahmadvand, G., Mondani, F., & Golzardi, F. (2009). Effect of crop plant density on critical period of weed competition in potato. *Scientia Horticulturae*, *121*(3), 249–254. doi:10.1016/j.scienta.2009.02.008

Aidi, B., Philen, M. K., & Case, S. W. (2015). Progressive damage assessment of centrally notched composite specimens in fatigue. *Composites. Part A, Applied Science and Manufacturing*, *74*, 47–59. doi:10.1016/j.compositesa.2015.03.022

Aitkenhead, M. J., Dalgetty, I. A., Mullins, C. E., McDonald, A. J. S., & Strachan, N. J. C. (2003). Weed and crop discrimination using image analysis and artificial intelligence methods. *Computers and Electronics in Agriculture*, *39*(3), 157–171. doi:10.1016/S0168-1699(03)00076-0

Akaike, H. (1976). An information criterion (AIC). *Math. Sci*, *14*, 1–5.

Al Bashish, D., Braik, M., & Bani-Ahmad, S. (2011). Detection and classification of leaf diseases using K means-based segmentation and Neural networks-based classification. *Information Technology Journal*, *10*(2), 267–275. doi:10.3923/itj.2011.267.275

Alizadeh, M. J., Shahheydari, H., Kavianpour, M. R., Shamloo, H., & Barati, R. (2017). Prediction of longitudinal dispersion coefficient in natural rivers using a cluster-based Bayesian network. *Environmental Earth Sciences*, *76*(2), 86. doi:10.100712665-016-6379-6

Alizamir, M., Kisi, O., & Zounemat-Kermani, M. (2018). Modelling long-term groundwater fluctuations by extreme learning machine using hydro-climatic data. *Hydrological Sciences Journal*, *63*(1), 63–73. doi:10.1080/02626667.2017.1410891

Alves, F. D. S., Estrela, V. V., & Matos, L. J. (2011). *Hyperspectral Analysis of Remotely Sensed Images, Sustainable water management in the tropics and subtropics - and case studies in Brazil* (Vol. 2). University of Kassel.

Ambuhl, M. E., Brepsant, C., Meister, J. J., Verkhovsky, A. B., & Sbalzarini, I. F. (2012). High-resolution cell outline segmentation and tracking from phase-contrast microscopy images. *Journal of Microscopy*, *245*(2), 161–170. doi:10.1111/j.1365-2818.2011.03558.x PMID:21999192

Aminzadeh, R., & Fotouhi, R. (2014). Novel Design of a Precision Planter for a Robotic Assistant Farmer. *Volume 5B: 38th Mechanisms and Robotics Conference*. doi:10.1115/detc2014-35402

Anderson, G. B., & Huang, T. S. (1971). Piecewise Fourier transform for picture bandwidth compression. *IEEE Transactions on Communication Technology*, *COM19*(2), 133–140. doi:10.1109/TCOM.1971.1090630

Andrade, E. L., Khan, E., Woods, J. C., & Ghanbari, M. (2003). Segmentation and tracking using region adjacency graphs, picture trees and prior information. In *Proc. IEEE Visual Information Engineering (VIE'03)* (pp. 45–48). IEEE. 10.1049/cp:20030483

Andreo, B., Vías, J., Durán, J. J., Jiménez, P., López-Geta, J. A., & Carrasco, F. (2008). Methodology for groundwater recharge assessment in carbonate aquifers: Application to pilot sites in southern Spain. *Hydrogeology Journal*, *16*(5), 911–925. doi:10.100710040-008-0274-5

Andrews, A. C., & Pratt, W. K. (1968). Fourier Transform Coding of images. In *Proc. International Conference on Systems Sciences* (pp. 677-679). Academic Press.

Andrews, H. C. (1970). *Computer techniques in image processing*. New York: Academic Press.

Antonini, M., Barlaud, M., Mathieu, P., & Daubechies, I. (1992). Image coding using Wavelet Transform. *IEEE Transactions on Image Processing*, *1*(2), 205–220. doi:10.1109/83.136597 PMID:18296155

Arivazhagan, S., Shebiah, R. N., Ananthi, S., & Varthini, S. V. (2013). Detection of the unhealthy region of plant leaves and classification of plant leaf diseases using texture features. *Agricultural Engineering International: CIGR Journal*, *15*(1), 211–217.

Arivu, S. K., & Sathiya, M. (2012). Analyzing blood cell images to differentiate WBC and counting of linear & non-linear overlapping RBC based on morphological features. *Elixir Comp. Sci. & Engg*, *48*, 9410–9413.

Arthur, J. F., Kepner, R. A., Dobie, J. B., Miller, G. E., & Parsons, P. S. (1982). Tub grinder performance with crop and forest residues. *Transactions of the ASAE. American Society of Agricultural Engineers*, *25*(6), 1488–1494. doi:10.13031/2013.33751

Arumugam, V., Kumar, C. S., Santulli, C., Sarasini, F., & Stanley, A. J. (2011). A Global Method for the Identification of Failure Modes in Fiberglass Using Acoustic Emission. *Journal of Testing and Evaluation*, *39*(5).

Babak, S., & Ahmed, E. (2015). Large-scale Classification of Fine-Art Paintings: Learning The Right Metric on The Right Feature. *Computer Vision and Pattern Recognition, 21.*

Baboo, S. S., & Shereef, I. K. (2010). An efficient weather forecasting system using artificial neural network. *International Journal of Environmental Sciences and Development*, *1*(4), 321–326. doi:10.7763/IJESD.2010.V1.63

Baggett, D., Nakaya, M., Mcauliffe, M., Yamaguchi, T. P., & Lockett, S. (2005). Whole cell segmentation in solid tissue sections. *Cytometry. Part A*, *67*(2), 137–143. doi:10.1002/cyto.a.20162 PMID:16163696

Bahadure, N. B., Ray, A. K., & Thethi, H. P. (2016). Performance analysis of image segmentation using watershed algorithm, fuzzy C-means of clustering algorithm and Simulink design. *Proc. 2016 3rd International Conference on Computing for Sustainable Global Development (INDIACom)*, 1160-1164.

Bakhshipour, A., Jafari, A., Nassiri, S. M., & Zare, D. (2017). Weed segmentation using texture features extracted from wavelet sub-images. *Biosystems Engineering*, *157*, 1–12. doi:10.1016/j.biosystemseng.2017.02.002

Bandyopadhyay, S. K. (2011). Method for blood cell segmentation. *Journal of Global Research in Computer Science*, *2*(4).

Bao, W., Chen, Y., & Wang, D. (2014). Prediction of protein structure classes with flexible neural tree. In Bio-Medical Materials and Engineering (Vol. 24, pp. 3797–3806). Academic Press. doi:10.3233/BME-141209

Bao, W., Huang, Z., Yuan, C. A., & Huang, D. S. (2017). Pupylation sites prediction with ensemble classification model. *International Journal of Data Mining and Bioinformatics*, *18*(2), 91. doi:10.1504/IJDMB.2017.086441

Bao, W., Wang, D., & Chen, Y. (2017). Classification of Protein Structure Classes on Flexible Neutral Tree. *IEEE/ACM Transactions on Computational Biology and Bioinformatics*, *14*(5), 1122–1133. doi:10.1109/TCBB.2016.2610967 PMID:28113983

Bapat, D., Kulkarni, S., & Bhandarkar, V. (1997). Design and operating experience on fluidised bed boiler burning biomass fuels with high alkali ash. In *The 15th International Conference on Fluidized bed combustion*. Academic Press.

Barati, R., Neyshabouri, S. S., & Ahmadi, G. (2014). Sphere drag revisited using shuffled complex evolution algorithm. River Flow, 345–353. doi:10.1201/b17133-51

Barati, R. (2017). Comment on "Evaluation of a physically based quasilinear and a conceptually based nonlinear Muskingum methods. *Journal of Hydrology (Amsterdam)*, *550*, 218–219. doi:10.1016/j.jhydrol.2017.05.006

Barnard, K., Duygulu, P., Freitas, N., Forsyth, D., Blei, D., & Jordan, M. I. (2003). Matching words and pictures. *JMLR*, *3*, 1107–1135.

Bar-Shalom, Y., & Blair, W. D. (Eds.). (2000). *Multitarget/multisensor tracking: Applications and advances* (Vol. 3). Artech.

Barzegar, R., Moghaddam, A. A., Deo, R., Fijani, E., & Tziritis, E. (2018). Mapping groundwater contamination risk of multiple aquifers using multi-model ensemble of machine learning algorithms. *The Science of the Total Environment*, *621*, 697–712. doi:10.1016/j.scitotenv.2017.11.185 PMID:29197289

Bay, H., Ess, A., Tuytelaars, T., & Van Gool, L. (2008). Speeded-up robust features (SURF). *Computer Vision and Image Understanding*, *110*(3), 346–359. doi:10.1016/j.cviu.2007.09.014

Beucher, S., & Meyer, F. (1993). The morphological approach to segmentation: The watershed transformation. In E. Dougherty (Ed.), *Mathematical Morphology in Image Processing* (pp. 433–481). Marcel Dekker.

Bhagavathi, S. L., & Niba, T. S. (2016). An automatic system for detecting and counting RBC and WBC using fuzzy logic. *J. of Eng. and Appl. Sc.*, *11*(11), 2006–2016.

Bieniek, A., & Moga, A. (2000). An efficient watershed algorithm based on connected components. *Pattern Recognition*, *6*(33), 907–916. doi:10.1016/S0031-3203(99)00154-5

Bishop, C. M. (1995). Neural networks for pattern recognition. *Journal of the American Statistical Association*, *92*, 482. doi:10.2307/2965437

Bitra, V. S. P., Womac, A. R., Chevanan, N., Miu, P. I., Igathinathane, C., Sokhansanj, S., & Smith, D. R. (2009). Direct mechanical energy measures of hammer mill comminution of switchgrass, wheat straw, and corn stover and analysis of their particle size distributions. *Powder Technology*, *193*(1), 32–45. doi:10.1016/j.powtec.2009.02.010

Bose, R., & Pathak, S. (2006). A novel compression and encryption scheme using variable model arithmetic coding and coupled chaotic system. *IEEE Transactions on Circuits and Systems—I, 53*(4), 848–856.

Boyat, A. K., & Joshi, B. K. (2015). *A review paper: noise models in digital image processing.* arXiv preprint arXiv:1505.03489

Brahimi, M., Boukhalfa, K., & Moussaoui, A. (2017). Deep Learning for Tomato Diseases: Classification and Symptoms Visualization. *Applied Artificial Intelligence, 31*(4), 299–315. doi:10.1080/08839514.2017.1315516

Brezger, A., Fahrmeir, L., & Hennerfeind, A. (2007). *Adaptive Gaussian Markov Random Fields with Applications in Human Brain Mapping*. Academic Press.

Brian, F. T., Ali, B., & James, S. F. (2016). Precipitation Intensity Effects on Groundwater Recharge in the Southwestern United States. *Water (Basel), 8*(3), 90–105. doi:10.3390/w8030090

Brown, G. (2011). Ensemble learning. In *Encyclopedia of Machine Learning*. Springer.

Burgos-Artizzu, X. P., Ribeiro, A., Guijarro, M., & Pajares, G. (2011). Real-time image processing for crop/weed discrimination in maize fields. *Computers and Electronics in Agriculture, 75*(2), 337–346. doi:10.1016/j.compag.2010.12.011

Buttarello, M., & Plebani, M. (2008). Automated blood cell counts: State-of-the-art. *American Journal of Clinical Pathology, 130*(1), 104–116. doi:10.1309/EK3C7CTDKNVPXVTN PMID:18550479

Camargo, A., & Smith, J. S. (2009). An image-processing based algorithm to automatically identify plant disease visual symptoms. *Biosystems Engineering, 102*(1), 9–21. doi:10.1016/j.biosystemseng.2008.09.030

Camargo, A., & Smith, J. S. (2009). Image pattern classification for the identification of disease causing agents in plants. *Computers and Electronics in Agriculture, 66*(2), 121–125. doi:10.1016/j.compag.2009.01.003

Chandra, P., & Singh, Y. (2008). Fault tolerance of feedforward artificial neural networks- a framework of study. *Proc. of the 2008 Int'l Joint Conf. on Neural Networks*, 1, 489-494.

Chan, T. F., Sandberg, B., & Vese, L. A. (2000). Active contours without edges for vector-valued images. *Journal of Visual Communication and Image Representation, 11*(2), 130–141. doi:10.1006/jvci.1999.0442

Chan, T. F., & Vese, L. A. (1990). An active contour model without edges. *Lecture Notes in Computer Science, 1682*, 141–151. doi:10.1007/3-540-48236-9_13

Chan, T. F., & Vese, L. A. (2001). Active contours without edges. *IEEE Transactions on Image Processing, 10*(1), 266–277. doi:10.1109/83.902291 PMID:18249617

Chaturvedi, R. S. (1973). A note on the investigation of ground water resources in western districts of Uttar Pradesh. In *Annual Report* (pp. 86–122). U.P. Irrigation Research Institute.

Chellappa, R., Chatterjee, S., & Bagdazian, R. (1985). Texture Synthesis and Compression Using Gaussian-Markov Random Field Models. *IEEE Transactions on Systems, Man, and Cybernetics*, *15*(2), 298–303. doi:10.1109/TSMC.1985.6313361

Chenouard, N., Smal, I., de Chaumont, F., Maska, M., Sbalzarini, I. F., Gong, Y., ... Meijering, E. (2014). Objective comparison of particle tracking methods. *Nature Methods*, *11*(3), 281–289. doi:10.1038/nmeth.2808 PMID:24441936

Chen, Y., & Wang, J. Z. (2004). Image categorization by learning and reasoning with regions. *JMLR*, *5*, 913–939.

Chib, S., & Greenberg, E. (1995). Understanding the Metropolis Hastings algorithm. *American Statistical Association*, *49*(4), 327–335.

Choi, J. G., Lee, S. W., & Kim, S. D. (1997). Spatio-temporal video segmentation using a joint similarity measure. *IEEE Transactions on Circuits and Systems for Video Technology*, *7*(2), 279–286. doi:10.1109/76.564107

Chung, C. L., Huang, K. J., Chen, S. Y., Lai, M. H., Chen, Y. C., & Kuo, Y. F. (2016). Detecting Bakanae disease in rice seedlings by machine vision. *Computers and Electronics in Agriculture*, *121*, 404–411. doi:10.1016/j.compag.2016.01.008

Ciesla, B. (2012). *Hematology in practice* (2nd ed.). Davis Company.

Coelho, A. M., & Estrela, V. V. (2012c). Data-driven motion estimation with spatial adaptation. *Intl. J. of Image Proc*, *6*(1), 53–67. http://www.cscjournals.org/csc/manuscript/Journals/IJIP/volume6/Issue1/IJIP-513.pdf

Coelho, A.M., & Estrela, V. V. (2012a). EM-based mixture models applied to video event detection. In *Principal Component Analysis - Engineering Applications*. Intech Open. doi:10.5772/38129

Coelho, A. M., Estrela, V. V., do Carmo, F. P., & Fernandes, S. R. (2012b). Error concealment by means of motion refinement and regularized Bregman divergence. In H. Yin, J. A. F. Costa, & G. Barreto (Eds.), Lecture Notes in Computer Science: Vol. 7435. *Proc. Intelligent Data Engineering and Automated Learning - IDEAL 2012. IDEAL 2012*. Berlin: Springer. doi:10.1007/978-3-642-32639-4_78

Cotter, S. M. (2018). *Merck and the Merck veterinary manual*. Retrieved October 13, 2018, from https://www.merckvetmanual.com/circulatory-system/hematopoietic-system-introduction/white-blood-cells

Cundiff, J. S., & Grisso, R. D. (2008). Containerized handling to minimize hauling cost of herbaceous biomass. *Biomass and Bioenergy*, *32*(4), 308–313. doi:10.1016/j.biombioe.2007.10.009

de Jesus, M. A., Estrela, V. V., & Saotome, O. (2016) Super-resolution in a nutshell. In *Proceedings of the Brazilian Technology Symposium 2016 (BTSym 2016)*. Academic Press.

de Jesus, M. A., Estrela, V. V., & Saotome, O. (2016) Super-resolution in a nutshell. *Proceedings of the Brazilian technology symposium 2016* (BTSym 2016).

de Jesus, M. A., Estrela, V. V., Saotome, O., & Stutz, D. (2018). Super-Resolution via Particle Swarm Optimization Variants. In J. Hemanth & V. Balas (Eds.), *Biologically Rationalized Computing Techniques For Image Processing Applications. Lecture Notes in Computational Vision and Biomechanics* (Vol. 25). Cham: Springer. doi:10.1007/978-3-319-61316-1_14

de Jesus, M. A., Estrela, V. V., Saotome, O., & Stutz, D. (2018). Super-resolution via particle swarm optimization variants. In J. Hemanth & V. Balas (Eds.), *Lecture Notes in Computational Vision and Biomechanics* (Vol. 25). Cham: Springer.

DeChant, C., Wiesner-Hanks, T., Chen, S., Stewart, E. L., Yosinski, J., Gore, M. A., & Lipson, H. (2017). Automated identification of northern leaf blight-infected maize plants from field imagery using deep learning. *Phytopathology*, *107*(11), 1426–1432. doi:10.1094/PHYTO-11-16-0417-R PMID:28653579

Deever, A. J., & Hemami, S. S. (2003). Lossless image compression with projection-based and adaptive reversible integer Wavelet transforms. *IEEE Transactions on Image Processing*, *12*(5), 489–499. doi:10.1109/TIP.2003.812374 PMID:18237926

De, I., & Sil, J. (2015). *A Fuzzy Regression Analysis Based No Reference Image Quality Metric. In Advances in Intelligent Informatics* (pp. 87–95). Springer.

Deng, Y., Kenney, C., Moore, M. S., & Manjunath, B. S. (1999). Peer group filtering and perceptual color image quantization. *Proc. of IEEE Intl. Symposium on Circuits and Systems*, *4*, 21-24.

Devi, C. J., Reddy, B. S. P., Kumar, K. V., Reddy, B. M., & Nayak, N. R. (2012). ANN approach for weather prediction using back propagation. *International Journal of Engineering Trends and Technology*, *3*(1), 19–23.

Dey, N., Rajinikanth, V., Amira, S., Ashour, A. S., & Tavares, J. M. R. S. (2018). Social Group Optimization Supported Segmentation and Evaluation of Skin Melanoma Images. *Symmetry*, *10*(2), 51. doi:10.3390ym10020051

Díaz, G., & Manzanera, A. (2009). Automatic analysis of microscopic images in hematological cytology applications. In F. A. González & E. Romero (Eds.), Biomedical image analysis and machine learning technologies: Applications and techniques. Academic Press. doi:10.4018/978-1-60566-956-4.ch008

Dimililer, K. (2013). Backpropagation neural network implementation for medical image compression. *Journal of Applied Mathematics*, *2013*, 1–8. doi:10.1155/2013/453098

Dogra, A., & Bhalla, P. (2014). Image sharpening by gaussian and butterworth high pass filter. *Biomedical & Pharmacology Journal*, *7*(2), 707–713. doi:10.13005/bpj/545

Dripps, W. R., & Bradbury, K. R. (2010). The spatial and temporal variability of groundwater recharge in a forested basin in northern Wisconsin. *Hydrological Processes*, *24*, 383–392.

Drzymala, Z. (1993). *Industrial briquetting-fundamentals and methods. Studies in mechanical engineering* (Vol. 13). Warszawa: PWN-Polish Scientific Publishers.

Durmus, H., Günes, E. O., & Kirci, M. (2017, August). Disease Detection on the Leaves of the Tomato Plants by Using Deep Learning. In *Agro-Geoinformatics, 2017 6th International Conference on* (pp. 1-5). IEEE. 10.1109/Agro-Geoinformatics.2017.8047016

Dzyubachyk, O., van Cappellen, W. A., Essers, J., Niessen, W. J., & Meijering, E. (2010). Advanced level-set-based cell tracking in time-lapse fluorescence microscopy. *IEEE Transactions on Medical Imaging*, *29*(3), 852–867. doi:10.1109/TMI.2009.2038693 PMID:20199920

Eclinpath. (2013). *Atlas of the Cornell University College of Veterinary Medicine*. Retrieved October 23, 2018, from http://www.eclinpath.com/ngg_tag/bovine/nggallery/page/2

Emami, S., & Tabil, L. G. (2008). Friction and compression characteristics of chickpea flour and components. *Powder Technology*. doi:10.1016/j.powtec.2007.07.012

Erichsen, E. S., Viana, L. G., Faria, R. M. D., & Santos, S. M. E. (2015). *Medicina laboratorial para o clínico*. Belo Horizonte, Brazil: Coopmed.

Estrela, V. V., Franz, M. O., Lopes, R. T., & de Araújo, G. P. (2005). Adaptive mixed norm optical flow estimation. Proc. SPIE 5960, Visual Communications and Image Processing 2005, 59603W. doi:10.1117/12.632674

Estrela, V. V., Monteiro, A. C. B., França, R. P., Iano, Y., Khelassi, A., & Razmjooy, N. (2019). Health 4.0 as an application of Industry 4.0 in healthcare services and management. *Medical Technologies Journal*, *2*(4), 262-76. Retrieved from http://ichsmt.org/Journals/ojs/index.php/MTJ/article/view/205

Estrela, V. V., Hemanth, J., Saotome, O., Grata, E. G. H., & Izario, D. R. F. (2019). Emergency Response Cyber-Physical System for Flood Prevention with Sustainable Electronics. In *Proceedings of the 3rd Brazilian Technology Symposium. BTSym 2017*. Springer. 10.1007/978-3-319-93112-8_33

Estrela, V. V., Saotome, O., Loschi, H. J., Hemanth, J., Farfan, W. S., Aroma, J., ... Grata, E. G. H. (2018). Emergency response cyber-physical framework for landslide avoidance with sustainable electronics. *Technologies*, *6*(2), 42. doi:10.3390/technologies6020042

Ewers, H., Smith, A. E., Sbalzarini, I. F., Lilie, H., Koumoutsakos, P., & Helenius, A. (2005). Single-particle tracking of murine polyoma virus-like particles on live cells and artificial membranes. *Proceedings of the National Academy of Sciences of the United States of America*, *102*(42), 15,110–15,115. doi:10.1073/pnas.0504407102 PMID:16219700

Failace, R., & Fernandes, F. (2015). *Hemograma. Manual de interpretação* (6th ed.). Porto Alegre, Brazil: Artemed.

Fan, F., Cong, W., & Wang, G. (2018). Generalized backpropagation algorithm for training second-order neural networks. *International Journal for Numerical Methods in Biomedical Engineering*, *34*(5), e2956. doi:10.1002/cnm.2956 PMID:29277960

Fasina, O. O. (2006). Flow and physical properties of switchgrass, peanut hull and poultry litter. *Transactions of the ASABE*, *49*(3), 721–728. doi:10.13031/2013.20470

Ferentinos, K. P. (2018). Deep learning models for plant disease detection and diagnosis. *Computers and Electronics in Agriculture*, *145*, 311–318. doi:10.1016/j.compag.2018.01.009

Fernandcz, W. P., Xian, Y., & Tian, Y. (2017). *Image-Based Barcode Detection and Recognition to Assist Visually Impaired Persons.* Paper presented at the 2017 IEEE 7th Annual International Conference on CYBER Technology in Automation, Control, and Intelligent Systems (CYBER). 10.1109/CYBER.2017.8446388

Fernandes, S. R., Estrela, V. V., Magalhaes, H. A., & Saotome, O. (2014). On improving sub-pixel accuracy by means of B-spline. *Proc. of the 2014 IEEE Int'l Conf. on Im. Syst. and Techn. (IST 2014)*, Santorini, Greece. 10.1109/IST.2014.6958448

Fischbach, F., & Duynning, M. B. A. (2009). *Manual of laboratory and diagnostic tests.* Lippincott Williams e Wilkins.

Flierl, M., & Girod, B. (2003). Generalised B pictures and the draft H.264/AVC video compression standard. *IEEE Transactions on Circuits and Systems for Video Technology*, *13*(7), 587–597. doi:10.1109/TCSVT.2003.814963

Food and Agriculture Organization of the United Nations. (n.d.). Retrieved from http://www.fao.org/faostat/en/#country/100

Fountas, S., Sorensen, C. G., Tsiropoulos, Z., Cavalaris, C., Liakos, V., & Gemtos, T. (2015). Farm machinery management information system. *Computers and Electronics in Agriculture*, *110*, 131–138. doi:10.1016/j.compag.2014.11.011

Fruit Master Database. (2018). *Different Fruits data.* Retrieved from https://github.com/Horea94/Fruit-Images-Dataset

Fu, S., Ruan, Q., & Wang, W. (2008). Remote sensing image data enhancement based on robust inverse diffusion equation for agriculture applications. *2008 9th International Conference on Signal Processing*. doi:10.1109/icosp.2008.4697353

Ganesan, V., Rosentrater, K. A., & Muthukumarappan, K. (2008). Flowability and handling characteristics of bulk solids and powders - a review with implications for DDGS. *Biosystems Engineering*, *101*(4), 425–435. doi:10.1016/j.biosystemseng.2008.09.008

Garces, E., & Reinhard, E. (2018). *Light-Field Surface Color Segmentation with an Application to Intrinsic Decomposition.* Paper presented at the 2018 IEEE Winter Conference on Applications of Computer Vision (WACV). 10.1109/WACV.2018.00166

García-Lamont, F., Cervantes, J., López-Chau, A., & Ruiz, S. (2018). *Contrast Enhancement of RGB Color Images by Histogram Equalization of Color Vectors' Intensities.* Paper presented at the International Conference on Intelligent Computing. 10.1007/978-3-319-95957-3_47

Getreuer, P. (2012). Chan–Vese Segmentation. *Image Processing On Line*, *2*, 214–224. doi:10.5201/ipol.2012.g-cv

Ghaffari, R., Zhang, F., Iliescu, D., Hines, E., Leeson, M., Napier, R., & Clarkson, J. (2010, July). Early Detection of Diseases in Tomato Crops: An Electronic Nose and Intelligent Systems Approach. In *Neural Networks (IJCNN), The 2010 International Joint Conference on* (pp. 1-6). IEEE.

Ghate, V. N., & Dudul, S. V. (2010). Optimal MLP neural network classifier for fault detection of three phase induction motor. *Expert Systems with Applications*, *37*(4), 3468–3481. doi:10.1016/j.eswa.2009.10.041

Godon, A., Genevieve, F., Marteau-Tessier, A., & Zandecki, M. (2012). Anomalies et erreurs de détermination de l'hémogramme avec les automates d'hématologie cellulaire Partie 3. Hémoglobine, hématies, indices érythrocytaires, réticulocytes. *Annales de Biologie Clinique*, *70*(2), 155–168. doi:10.1684/abc.2012.0685 PMID:22484526

Gollakota, A., & Srinivas, M. B. (2011). Agribot — A multipurpose agricultural robot. *2011 Annual IEEE India Conference*. 10.1109/INDCON.2011.6139624

Golomb, S. W. (1996). Run length encoding. *IEEE Transactions on Information Theory*, *12*(3), 399–401. doi:10.1109/TIT.1966.1053907

Grgic, S., Grgic, M., & Mrak, M. (2004). Reliability of objective picture quality measures. *Journal of Electrical Engineering*, *55*(1-2), 3–10.

Gulati, G. L., Alomari, M., Kocher, W., & Schwarting, R. (2002). Criteria for blood smear review. *Laboratory Medicine*, *33*(5), 374–377. doi:10.1309/TVJW-ENEK-G1U3-WE0A PMID:23301216

Habibi, A., & Wintz, P. A. (1971). Image Coding by linear transformation and block quantization. *IEEE Transactions on Communication Technology*, *COM19*(1), 50–62. doi:10.1109/TCOM.1971.1090601

Haghiabi, A. H., Azamathulla, H. M., & Parsaie, A. (2017). Prediction of head loss on cascade weir using ANN and SVM. *ISH Journal of Hydraulic Engineering*, *23*(1), 102–110. doi:10.1080/09715010.2016.1241724

Hamdin, R. I., Lux, S. E., & Stossel, T. P. (2003). *Blood: Principles and practice of hematology*. Lippincott Williams e Wilkins.

Hamzaoui, R. (1999). Fast iterative methods for fractal image compression. *Journal of Mathematical Imaging and Vision*, *11*(2), 147–159. doi:10.1023/A:1008343530356

Hansen, J. S. (2011). *GNU Octave. Beginner's guide*. Packt Publishing.

Hartension, H., Ruhl, M., & Saupe, D. (2000). Region based fractal image compression. *IEEE Transactions on Image Processing*, *9*(7), 1171–1184. doi:10.1109/83.847831 PMID:18262956

Hartmann, A., Lange, J., Weiler, M., Arbel, Y., & Greenbaum, N. (2012). A new approach to model the spatial and temporal variability of recharge to karst aquifers. *Hydrology and Earth System Sciences*, *16*(7), 2219–2231. doi:10.5194/hess-16-2219-2012

Hartmann, A., Weiler, M., Wagener, T., Lange, J., Kralik, M., Humer, F., ... Huggenberger, P. (2013). Process-based karst modelling to relate hydrodynamic and hydrochemical characteristics to system properties. *Hydrology and Earth System Sciences*, *17*(8), 3305–3321. doi:10.5194/hess-17-3305-2013

Hashimoto, Y., Da Silva, P. H., & Bertassoni, H. (2009). *Hematologia laboratorial*. Rio de Janeiro: Revinter.

Hau, C. C. (2015). *Handbook of pattern recognition and computer vision*. World Scientific.

Haug, S., Michaels, A., Biber, P., & Ostermann, J. (2014). Plant classification system for crop/weed discrimination without segmentation. *IEEE Winter Conference on Applications of Computer Vision (WACV)*, 1142-1149. 10.1109/WACV.2014.6835733

Haug, S., & Ostermann, J. (2014). A Crop /Weed field image dataset for the evaluation of computer vision based precision agriculture tasks. *Lecture Notes in Computer Science*, *8928*, 105–116. doi:10.1007/978-3-319-16220-1_8

Haug, S., & Ostermann, J. (2016). Plant classification for field robots: A machine vision approach. *Computer Vision and Pattern Recognition in Environmental Informatics*, *248*, 248–272. doi:10.4018/978-1-4666-9435-4.ch012

Heckmann, D., Schluter, U., & Weber, A. P. (2017). Machine Learning Techniques for Predicting Crop Photosynthetic Capacity from Leaf Reflectance Spectra. *Molecular Plant. Cell Press*, *10*(6), 878–890. PMID:28461269

Helmuth, J. A., Burckhardt, C. J., Koumoutsakos, P., Greber, U. F., & Sbalzarini, I. F. (2007). A novel supervised trajectory segmentation algorithm identifies distinct types of human adenovirus motion in host cells. *Journal of Structural Biology*, *159*(3), 347–358. doi:10.1016/j.jsb.2007.04.003 PMID:17532228

Hemanth, D. J., & Estrela, V. V. (2017). Deep Learning for Image Processing Applications (vol. 31). IOS Press.

Hemanth, D. J., & Estrela, V. V. (2017). *Deep Learning for Image Processing Applications*. IOS Press.

Herman, N., Trumel, C., Geffré, A., Braun, J.-P., Thibault, M., Schelcher, F., & Bourgès-Abella, N. (2018). Hematology reference intervals for adult cows in France using the Sysmex XT-2000iV analyzer. *Journal of Veterinary Diagnostic Investigation*, *30*(5), 678–687. doi:10.1177/1040638718790310 PMID:30027829

Herrera, P. J., Dorado, J., & Ribeiro, A. (2014). A novel approach for weed type classification based on shape descriptors and a fuzzy decision-making method. *Sensors (Basel)*, *14*(8), 15304–15324. doi:10.3390140815304 PMID:25195854

Herrmann, A. E., & Estrela, V. V. (2016). Content Based Image Retrieval (CBIR) in Remote Clinical Diagnosis and Healthcare. In Encyclopedia of E-Health and Telemedicine. Academic Press.

Herrmann, A. E., & Estrela, V. V. (2016). Content-based image retrieval (CBIR) in remote clinical diagnosis and healthcare. In M. Cruz-Cunha, I. Miranda, R. Martinho, & R. Rijo (Eds.), *Encyclopedia of E-Health and Telemedicine* (pp. 495–520). Hershey, PA: IGI Global. doi:10.4018/978-1-4666-9978-6.ch039

Hess, R. J., Wright, C. T., & Kenney, K. L. (2007). Cellulosic biomass feedstocks and logistics for ethanol production. *Biofuels, Bioproducts & Biorefining*, *1*(3), 181–190. doi:10.1002/bbb.26

Hosseini, H., Tousi, B., & Razmjooy, N. (2014). Application of fuzzy subtractive clustering for optimal transient performance of automatic generation control in restructured power system. *Journal of Intelligent & Fuzzy Systems*, *26*(3), 1155–1166.

Hosseini, K., Nodoushan, E. J., Barati, R., & Shahheydari, H. (2016). Optimal design of labyrinth spillways using meta-heuristic algorithms. *KSCE Journal of Civil Engineering*, *20*(1), 468–477. doi:10.100712205-015-0462-5

Howard, P., & Vitter, J. (1992). New methods for lossless image compression using arithmetic coding. *Journal of Information Processing and Management*, *28*(5), 765–779. doi:10.1016/0306-4573(92)90067-A

Hsu, C.-W., Chang, C.-C., & Lin, C.-J. (2010). A Practical Guide to Support Vector Classification. *Technical Report*, 1–16. doi:10.1177/02632760022050997

Huang, D.-S., & Du, J.-X. (2008). A constructive hybrid structure optimization methodology for radial basis probabilistic neural networks. *IEEE Transactions on Neural Networks*, *19*(12), 2099–2115. doi:10.1109/TNN.2008.2004370 PMID:19054734

Huang, K. Y. (2007). Application of artificial neural network for detecting Phalaenopsis seedling diseases using color and texture features. *Computers and Electronics in Agriculture*, *57*(1), 3–11. doi:10.1016/j.compag.2007.01.015

Huang, W., Zhu, W., Ma, C., Guo, Y., & Chen, C. (2018). Identification of group-housed pigs based on Gabor and Local Binary Pattern features. *Biosystems Engineering*, *166*, 90–100. doi:10.1016/j.biosystemseng.2017.11.007

Hu, J., Chen, Z., Yang, M., Zhang, R., & Cui, Y. (2018). A Multiscale Fusion Convolutional Neural Network for Plant Leaf Recognition. *IEEE Signal Processing Letters*, *25*(6), 853–857. doi:10.1109/LSP.2018.2809688

Hung, N. Q., Babel, M. S., Weesakul, S., & Tripathi, N. K. (2009). An artificial neural network model for rainfall forecasting in Bangkok, Thailand. *Hydrology and Earth System Sciences*, *13*(8), 1413–1425. doi:10.5194/hess-13-1413-2009

Hung, Y. Y., Yang, L. X., & Huang, Y. H. (2013). Non-destructive evaluation (NDE) of composites: digital shearography. In V. M. Karbhari (Ed.), *Non-Destructive Evaluation (NDE) of Polymer Matrix Composites* (pp. 84–115). Woodhead Publishing. doi:10.1533/9780857093554.1.84

Hussain, F., & Jeong, J. (2015). Exploiting deep neural networks for digital image compression. *Proc. 2nd World Symp. on Web Appl. and Networking (WSWAN)*, 1-6. 10.1109/WSWAN.2015.7210294

Irshad, H., Veillard, A., Roux, L., & Racoceanu, D. (2014). Methods for nuclei detection, segmentation, and classification in digital histopathology: A review–current status and future potential. *IEEE Reviews in Biomedical Engineering*, *7*, 97–114. doi:10.1109/RBME.2013.2295804 PMID:24802905

Islam, M. R., Anupam, C., Kazi, M., Islam, M., & Ziaur, R. (2014). Development of rainfall recharge model for natural groundwater recharge estimation in Godagari Upazila of Rajshahi district, Bangladesh. *American Journal of Civil Engineering*, *2*(2), 48–52. doi:10.11648/j.ajce.20140202.16

Islam, S., Ram, K. S., & Roohul, A. K. (2015). Methods of Estimating Ground water Recharge. *International Journal of Engineering Associates*, *5*(2), 6–13.

ISO/IEC 15444 I ITU-T Recommendation T.800, Information technology - JPEG 2000 image coding system: Core coding system, November 2015.

Jagadish, H. P., & Lohit, M. K. (2010). A new lossless method of image compression and decompression using Huffman coding techniques. *Journal of Theoretical & Applied Information Technology*, *5*(2), 18–23.

Jain, A. K. (1981). Image data compression: A review. *Proceedings of the IEEE*, *69*(3), 366–406. doi:10.1109/PROC.1981.11971

Jain, L. C., Kandel, A., & Teodorescu, H.-N. L. (2017). *Fuzzy and neuro-fuzzy systems in medicine*. CRC Press.

Jasim, S. S., & Al-Taei, A. A. M. (2018). A Comparison between SVM and K-NN for classification of Plant Diseases. *Diyala Journal for Pure Science*, *14*(02), 94–105. doi:10.24237/djps.1402.383B

Jeon, B., Park, J., & Jeong, J. (1998). Huffman coding of DCT coefficients using dynamic code word assigned and adaptive codebook selection. *Signal Processing Image Communication*, *22*(3), 253–262. doi:10.1016/S0923-5965(97)00041-6

Jiang, L., Lu, H., My, V. D., Koch, A., & Zell, A. (2015). Superpixel segmentation based gradient maps on RGB-D dataset. *Proc. 2015 IEEE Int'l Conference on Robotics and Biomimetics (ROBIO)*. 10.1109/ROBIO.2015.7418960

Jingfu, L., & Zhijun, L. (2010). Study on agriculture image processing based on discrete wavelet transform. *2010 Second IITA International Conference on Geoscience and Remote Sensing*. 10.1109/IITA-GRS.2010.5603095

Jing, J., Huang, M., Li, P., & Ning, X. (2018). Automatic measurement of yarn hairiness based on the improved MRMRF segmentation algorithm. *Journal of the Textile Institute*, *109*(6), 740–749. doi:10.1080/00405000.2017.1368106

JPEG2000 Part I Final Committee Draft Version 1.0. ISO/IECJTC1/SC29 WG1 N1646R, 2000.

Jumadi, Y. P. (2015). Development and Evaluation of Web GIS Application for Groundwater Management of Karanganyar Regency. *Forum Geografi*, *29*(2), 164–173.

Junqueira, L. C., & Carneiro, J. (2004). *Histologia básica* (10th ed.). São Paulo, Guanabara.

Junqueira, L. C., & Carneiro, J. (2012). *Biologia celular e molecular*. São Paulo: Guanabara Koogan.

Kannappan, P. L. (1972). On Shannon's entropy, directed divergence and inaccuracy. *Probability Theory and Related Fields*, *22*(2), 95–100. doi:10.1007/BF00532728

Kapoor, A., Bhat, S. I., Shidnal, S., & Mehra, A. (2016). Implementation of IoT (Internet of Things) and Image processing in smart agriculture. *2016 International Conference on Computation System and Information Technology for Sustainable Solutions (CSITSS)*. 10.1109/CSITSS.2016.7779434

Kashyap, Y., Sharma, T., & Ali, S. S. (2017). Analysis of Plant Diseases with Detection using Image Processing Methods. *International Journal of Computers and Applications*, *166*(7).

Kateris, D., Moshou, D., Pantazi, X.-E., Gravalos, I., Sawalhi, N., & Loutridis, S. (2014). A machine learning approach for the condition monitoring of rotating machinery. *Journal of Mechanical Science and Technology*, *28*(1), 61–71. doi:10.100712206-013-1102-y

Kaundal, R., Kapoor, A. S., & Raghava, G. P. (2006). Machine learning techniques in disease forecasting: A case study on rice blast prediction. *BMC Bioinformatics*, *7*(1), 485. doi:10.1186/1471-2105-7-485 PMID:17083731

Kawthalksr, S. M. (2013). Essentials of haematology (2nd ed.). New Delhi: Jaypee.

Keerthi, S. S., & Lin, C. J. (2003). Asymptotic behaviors of support vector machines with gaussian kernel. *Neural Computation*, *15*(7), 1667–1689. doi:10.1162/089976603321891855 PMID:12816571

Khmag, A., Ramli, A. R., Al-Haddad, S., & Kamarudin, N. (2018). Natural image noise level estimation based on local statistics for blind noise reduction. *The Visual Computer*, *34*(4), 575–587. doi:10.100700371-017-1362-0

Kim, N. C., & So, H. J. (2018). Directional statistical Gabor features for texture classification. *Pattern Recognition Letters*, *112*, 18–26. doi:10.1016/j.patrec.2018.05.010

Kingston, A., & Autrusseau, F. (2008). Lossless image compression via predictive coding of discrete Radon projections. *Signal Processing Image Communication*, *23*(4), 313–324. doi:10.1016/j.image.2008.03.001

Kowsalya, N., Kalyani, A., Chalcedony, C. J., Sivakumar, R., Janani, M., & Rajinikanth, V. (2018). An approach to extract optic-disc from retinal image using k-means clustering. *Fourth International Conference on Biosignals, Images and Instrumentation (ICBSII)*, 206-212. 10.1109/ICBSII.2018.8524655

Koyama, K., Hoshikawa, H., & Kojima, G. (2013). Eddy Current Nondestructive Testing for Carbon Fiber- Reinforced Composites. *Journal of Pressure Vessel Technology*, *135*(4), 041501. doi:10.1115/1.4023253

Kroeger, T. (2014). Thermographic inspection of composites. *Reinforced Plastics*, *58*(4), 42–43. doi:10.1016/S0034-3617(14)70183-3

Kumar, C. P. (2000). *Groundwater Assessment Methodology. 21*. National Institute of Hydrology.

Kumar, C. P., & Seethapathi, P. V. (2002). Assessment of natural ground water recharge in upper Ganga canal command area. *Journal of Applied Hydrology*, *15*(4), 13–20.

Kumar, P., Barrett, D. M., Delwiche, M. J., & Stroeve, P. (2009). Methods of pretreatment of lignocellulosic biomass for efficient hydrolysis and biofuel production. *Industrial & Engineering Chemistry Research*, *48*(8), 3713–3729. doi:10.1021/ie801542g

Kung, W.-J., Yeh, H.-F., Lin, H.-I., Chen, W.-P., & Lee, C.-H. (2013). An approach to evaluate groundwater recharge from streamflow and groundwater records. *Geosciences Journal*, *17*(3), 353–362. doi:10.100712303-013-0027-x

Kyi, T. M., & Zin, K. C. M. (2018). *Color Segmentation Based on Human Perception Using Fuzzy Logic*. Paper presented at the International Conference on Big Data Analysis and Deep Learning Applications.

Lam, P. S., Sokhansanj, S., Bi, X., Mani, S., Lim, C. J., Womac, A. R., … Nayaran, S. (2008). Physical characterization of wet and dry wheat straw and switchgrass – bulk and specific density. *ASABE Annual International Meeting, 300*(7), 23. 10.13031/2013.24490

Lei, Y., He, Z., & Zi, Y. (2008). A new approach to intelligent fault diagnosis of rotating machinery. *Expert Systems with Applications*, *35*(4), 1593–1600. doi:10.1016/j.eswa.2007.08.072

Le-Qing, Z. (2011). Finger knuckle print recognition based on SURF algorithm. In *Fuzzy Systems and Knowledge Discovery (FSKD), 2011 Eighth International Conference on* (Vol. 3, pp. 1879-1883). IEEE.

Lesage, D., Graf, V. de T., & Westphal, K. (2010). *Global energy governance in a multipolar world*. London: Ashgate.

Li, Z., Aaron, A., Katsavounidis, I., Moorthy, A., & Manohara, M. (2016). Toward a practical perceptual video quality metric. *The Netflix Tech Blog*. Retrieved October 13, 2018, https://medium.com/netflixtechblog/toward-a-practical-perceptual-video-quality-metric-653f208b9652

Li, H., Zhang, Y., & Zheng, H. (2009). Gear fault detection and diagnosis under speed-up condition based on order cepstrum and radial basis function neural network. *Journal of Mechanical Science and Technology*, *23*(10), 2780–2789. doi:10.100712206-009-0730-8

Lin, H.-T., & Lin, C.-J. (2003). *A study on sigmoid kernels for SVMand the training of non-PSD kernels by SMO-type methods*. Academic Press.

Liu, C., & Wechsler, H. (2002). Gabor feature based classification using the enhanced fisher linear discriminant model for face recognition. *IEEE Transactions on Image Processing*, *11*(4), 467–476. doi:10.1109/TIP.2002.999679 PMID:18244647

Liu, L., Zhang, W., Shu, S., & Jin, X. (2013, August). Image Recognition of Wheat Disease Based on RBF Support Vector Machine. In *Proceedings of the International Conference on Advanced Computer Science and Electronics Information* (pp 307-310). Atlantis Press. 10.2991/icacsei.2013.77

Londhe, S. N., & Shah, S. (2017). A novel approach for knowledge extraction from artificial neural networks. *ISH Journal of Hydraulic Engineering*, 1–13. doi:10.1080/09715010.2017.1409667

Lopo, P. (2002). The right grinding solution for you: roll, horizontal or vertical. *Feed Management*, (53), 6–23.

Loveday, R. (1980). *Statistics, A Second Course in Statistics* (2nd ed.). New York: Cambridge University Press.

Lowe, A., Harrison, N., & French, A. P. (2017). Hyperspectral image analysis techniques for the detection and classification of the early onset of plant disease and stress. *Plant Methods*, *13*(1), 80. doi:10.118613007-017-0233-z PMID:29051772

Lu, Q., & Li, D. (1998). Neural network method for diagnosing faults rolling bearing in electrical machines with frequency signatures. *Qinghua Daxue Xuebao/Journal of Tsinghua University*, *38*(4), 94–97.

Lu, H., Kot, A. C., & Shi, Y. Q. (2004). Distance-reciprocal distortion measure for binary document images. *IEEE Signal Processing Letters*, *11*(2), 228–231. doi:10.1109/LSP.2003.821748

Lu, J., Zhou, M., Gao, Y., & Jiang, H. (2017). Using hyperspectral imaging to discriminate yellow leaf curl disease in tomato leaves. *Precision Agriculture*, 1–16.

Luo, Y., & Ward, R. K. (2003). Removing the blocking artifacts of block-based DCT compressed images. *IEEE Transactions on Image Processing*, *12*(7). PMID:18237958

Machacek, M., & Danuser, G. (2006). Morphodynamic profiling of protrusion phenotypes. *Biophysical Journal*, *90*(4), 1439–1452. doi:10.1529/biophysj.105.070383 PMID:16326902

Magli, E., & Olmo, G. (2003). Lossy predictive coding of SAR raw data. *IEEE Transactions on Image Processing*, *41*(5), 977–987.

Mahlein, A. K. (2016). Plant Disease Detection by Imaging Sensors – Parallels and Specific Demands for Precision Agriculture and Plant Phenotyping. *Plant Disease. APS Journals*, *100*(2), 241–251.

Maierhofer, C., Myrach, P., Reischel, M., Steinfurth, H., Röllig, M., & Kunert, M. (2014). Characterizing damage in CFRP structures using flash thermography in reflection and transmission configurations. *Composites. Part B, Engineering*, *57*, 35–46. doi:10.1016/j.compositesb.2013.09.036

Makaremi, M., Razmjooy, N., & Ramezani, M. (2018). A new method for detecting texture defects based on modified local binary pattern. *Signal, Image and Video Processing*, 1–7.

Manic, K.S., Naimi, I.S.A., Hasoon, F.N. & Rajinikanth, V. (2018). Jaya algorithm-assisted evaluation of tooth elements using digital bitewing radiography images. *Computational Techniques for Dental Image Analysis*, 107-128. doi:10.4018/978-1-5225-6243-6.ch005

Manic, K. S., Hasoon, F. N., Shibli, N. A., Satapathy, S. C., & Rajinikanth, V. (2019). An approach to examine brain tumor based on Kapur's entropy and Chan–Vese algorithm. *Advances in Intelligent Systems and Computing*, *797*, 901–909. doi:10.1007/978-981-13-1165-9_81

Mani, S., Tabil, L. G., & Sokhansanj, S. (2004). Grinding performance and physical properties of wheat and barley straws, corn stover and switchgrass. *Biomass and Bioenergy*, *27*(4), 339–352. doi:10.1016/j.biombioe.2004.03.007

Maska, M., Ulman, V., Svoboda, D., Matula, P., Matula, P., Ederra, C., ... Ortiz-de-Solorzano, C. (2014). A benchmark for comparison of cell tracking algorithms. *Bioinformatics (Oxford, England)*, *30*(11), 1609–1617. doi:10.1093/bioinformatics/btu080 PMID:24526711

Meier, T., & Ngan, K. N. (1998). Automatic segmentation of moving objects for video object plane generation. *IEEE Transactions on Circuits and Systems for Video Technology*, *8*(5), 525–538. doi:10.1109/76.718500

Melo, M., & Silveira, C. M. (2015). *Laboratório de hematologia: Teorias, técnicas e atlas*. Rubio.

Meyer, F. (2014). Watersheds on weighted graphs. *Pattern Recognition Letters*, *47*(1), 72–79. doi:10.1016/j.patrec.2014.02.018

Mitchell, P., Kiel, J., Livingston, B., & Dupont-Roc, G. (2007). *Torrefi ed biomass: A foresighting study into the business case for pellets from torrefi ed biomass as a new solid fuel, in All Energy 2007*. University of Aberdeen, ECN, Doosan Babcock, and ITI Energy.

Miucic, R., Mahmud, S., & Popovic, Z. (2009). An enhanced data-reduction algorithm for event-triggered networks. *IEEE Transactions on Vehicular Technology*, *58*(6), 2663–2678. doi:10.1109/TVT.2008.2011361

Moallem, P., Razmjooy, N., & Mousavi, B.S. (2014). Robust potato color image segmentation using adaptive fuzzy inference system. *Iranian Journal of Fuzzy Systems*.

Moallem, P., & Razmjooy, N. (2012). A multi layer perceptron neural network trained by invasive weed optimization for potato color image segmentation. *Trends in Applied Sciences Research*, *7*(6), 445–455. doi:10.3923/tasr.2012.445.455

Moallem, P., & Razmjooy, N. (2012). Optimal threshold computing in automatic image thresholding using adaptive particle swarm optimization. *Journal of Applied Research and Technology*, *10*(5), 703–712.

Moallem, P., Razmjooy, N., & Ashourian, M. (2013). Computer vision-based potato defect detection using neural networks and support vector machine. *International Journal of Robotics and Automation, 28*(2), 137–145. doi:10.2316/Journal.206.2013.2.206-3746

Moallem, P., Razmjooy, N., & Mousavi, B. S. (2014). Robust potato color image segmentation using adaptive fuzzy inference system. *Iranian Journal of Fuzzy Systems, 11*(6), 47–65.

Mogra, M. & Srivastava, V. (2014). A comprehensive review of analysis of counting blood cells using different image processing algorithms. *Int'l J. of Eng. Sc. Invention, 3*(6).

Mohanty, S. P., Hughes, D. P., & Salathé, M. (2016). Using deep learning for image-based plant disease detection. *Frontiers in Plant Science, 7*, 1419. doi:10.3389/fpls.2016.01419 PMID:27713752

Monisha, R., Mrinalini, R., Britto, M. N., Ramakrishnan, R., & Rajinikanth, V. (2019). Social group optimization and Shannon's function-based RGB image multi-level thresholding. *Smart Innovation. Systems and Technologies, 105*, 123–132. doi:10.1007/978-981-13-1927-3_13

Montalvo, M., Guerrero, J. M., Romeo, J., Emmi, L., Guijarro, M., & Pajares, G. (2013). Automatic expert system for weeds/crops identification in images from maize fields. *Expert Systems with Applications, 40*(1), 75–82. doi:10.1016/j.eswa.2012.07.034

Moore, H. (2012). *Matlab for Engineers*. Boston: Pearson.

Moscheni, F., Bhattacharjee, S., & Kunt, M. (1998). Spatio-temporal segmentation based on region merging. *IEEE Transactions on Pattern Analysis and Machine Intelligence, 20*(9), 897–915. doi:10.1109/34.713358

Moshou, D., Kateris, D., Gravalos, I., Loutridis, S., Sawalhi, N., Gialamas, T., … Tsiropoulos, Z. (2010). Determination of fault topology in mechanical subsystems of agricultural machinery based on feature fusion and neural networks. *Conference Proceeding - 4th International Conference, TAE 2010: Trends in Agricultural Engineering 2010.*

Moshou, D., Pantazi, X.-E., Kateris, D., & Gravalos, I. (2014). Water stress detection based on optical multisensor fusion with a least squares support vector machine classifier. *Biosystems Engineering, 117*, 15–22. doi:10.1016/j.biosystemseng.2013.07.008

Motta, G., Storer, J. A., & Carpentieri, B. (2000). Lossless image coding via adaptive linear prediction and classification. *Proceedings of the IEEE, 88*(11), 1790–1796. doi:10.1109/5.892714

Mousavi, B. S., Sargolzaei, P., Razmjooy, N., Hosseinabadi, V., & Soleymani, F. (2011). Digital Image Segmentation Using Rule-Base Classifier. *American Journal of Scientific Researches, 35*, 17–23.

Mousavi, B. S., Soleymani, F., & Razmjooy, N. (2012). Semantic image classification by genetic algorithm using optimised fuzzy system based on Zernike moments. *Signal, Image and Video Processing*. doi:10.100711760-012-0311-7

Mousavi, B. S., Soleymani, F., & Razmjooy, N. (2013). Color image segmentation using neuro-fuzzy system in a novel optimized color space. *Neural Computing & Applications*, *23*(5), 1513–1520. doi:10.100700521-012-1102-3

Murphy, K., Travers, P., & Walport, M. (2010). *Imunibiologia* (7th ed.). Porto Alegre: Artmed, Brazil.

N, M. M. (2017). Solar Powered Digging and Seed Sowing Machine. *International Journal for Research in Applied Science and Engineering Technology, 5*(3), 446-450. doi:10.22214/ijraset.2017.3081

Naemura, N., Fukuda, A., Mizutani, Y., Izumi, Y., Tanaka, Y., & Enami, K. (2000). Morphological segmentation of sports scenes using color information. *IEEE Transactions on Broadcasting*, *46*(3), 181–188. doi:10.1109/11.892154

Naik, A., Satapathy, S. C., Ashour, A. S., & Dey, N. (2016). Social group optimization for global optimization of multimodal functions and data clustering problems. *Neural Computing & Applications*. doi:10.100700521-016-2686-9

Najafzadeh, M., Tafarojnoruz, A., & Lim, S. Y. (2017). Prediction of local scour depth downstream of sluice gates using data-driven models. *ISH Journal of Hydraulic Engineering*, *23*(2), 195–202. doi:10.1080/09715010.2017.1286614

Nasreen, N., Kumar, C., & Nabeel, A. (2015). *Counting of RBC using circular Hough transform with median filtering*. Bonfring.

Nelson, M., & Gailly, J. (1996). *The data compression book* (2nd ed.). New Delhi: BPB Publications.

Nemer, A. S. A., Das Neves, F. J., & Ferreira, J. E. S. (2010). *Manual de solicitação de exames laboratoriais*. Revinter.

Niu, W. J., Feng, Z. K., Cheng, C. T., & Zhou, J. Z. (2018). Forecasting daily runoff by extreme learning machine based on quantum-behaved particle swarm optimization. *Journal of Hydrologic Engineering*, *23*(3), 04018002. doi:10.1061/(ASCE)HE.1943-5584.0001625

Oguma, I., Goto, R., & Sugiura, T. (2012). Ultrasonic inspection of an internal flaw in a ferromagnetic specimen using angle beam EMATs. *Przeglad Elektrotechniczny*, *88*(7B), 78–81.

Oke, Martins, Salako, & Idowu. (2016). Estimation of Groundwater Recharge in the Lower-Ogun River Basin using Two independent Methods. *Nigerian Journal of Hydrological Sciences*, 181 – 200

Oke, I. A. (2007). Reliability and Statistical Assessment of Methods For Pipe Network Analysis. *Environmental Engineering Science*, *24*(10), 1481–1489. doi:10.1089/ees.2007.0001

Oke, I. A., Lukman, S., & Ismail, A. (2017). Development And Performance Evaluation Of A New Numerical Model For Groundwater Recharge Estimation. *Nigeria Journal of Engineering*, *23*(2), 56–65.

Oke, M. O., Martins, O., Idowu, O. A., & Aiyelokun, O. (2015). Comparative Analysis of Groundwater Recharge Estimation Value Obtained Using Empirical Methods In Ogun and Oshun River Basins. *Ife Journal of Science*, *17*(1), 53–64.

Osroosh, Y., Khot, L. R., & Peters, R. T. (2018). Economical thermal-RGB imaging system for monitoring agricultural crops. *Computers and Electronics in Agriculture*, *147*, 34–43. doi:10.1016/j.compag.2018.02.018

Pal, T., Jaiswal, V., & Chauhan, R. S. (2016). DRPPP: A machine learning based tool for prediction of disease resistance proteins in plants. *Computers in Biology and Medicine*, *78*, 42–48. doi:10.1016/j.compbiomed.2016.09.008 PMID:27658260

Pan, X., & Ruan, Q. Q. (2008). Palmprint recognition using Gabor feature-based $(2D)^2$ PCA. *Neurocomputing*, *71*(13-15), 3032–3036. doi:10.1016/j.neucom.2007.12.030

Park, S. I., & Yang, K. (2002). Extended Hamming accumulate codes and modified irregular repeat accumulate codes. *Electronics Letters*, *38*(10), 467–468. doi:10.1049/el:20020316

Parlos, A. G. M. S., & A. A. (1999). Adaptive state estimation using dynamic recurrent neural networks in Neural Networks. In *IJCNN '99. International Joint Conference on. Dept. of Nucl,* (pp. 3361–3364). Academic Press.

Patel, B., & Agrawal, S. (2013). Image compression techniques using artificial neural network. *Int. J. Adv. Res. Comput. Eng. Technol.*, *2*(10), 2725–2729.

Pathak, S., & Sejwar, V. (2017). Optimized noisy image segmentation using genetic algorithm. *Proc. 2017 International Conference on Intelligent Computing and Control Systems (ICICCS)*, 1311-1316. 10.1109/ICCONS.2017.8250681

Patil, D. A., Upadhye, M. Y., Kazi, F. S., & Singh, N. M. (2015). Multi Robot Communication and Target Tracking system with controller design and implementation of SWARM robot using Arduino. *2015 International Conference on Industrial Instrumentation and Control (ICIC)*. 10.1109/IIC.2015.7150777

Paulrud, S. (2004). *Upgraded biofuels d Effects of quality on processing, handling characteristics, combustion, and ash melting*. Swedish University of Agricultural Sciences.

Paul, S., & Bandyopadhyay, B. (2014). A novel approach for image compression based on multi-level image thresholding using Shannon entropy and differential evolution. *Students' Technology Symposium (TechSym)*, 56-61. doi: 10.1109/TechSym.2014.6807914

Paya, B. A., Esat, I. I., & Badi, M. N. M. (1997). Artificial neural network based fault diagnostics of rotating machinery using wavelet transforms as a preprocessor. *Mechanical Systems and Signal Processing*, *11*(5), 751–765. doi:10.1006/mssp.1997.0090

Peng, W., Zhang, Y., Qiu, B., & Xue, H. (2012). A Brief Review of the Application and Problems in Ultrasonic Fatigue Testing. *AASRI Procedia*, *2*, 127–133. doi:10.1016/j.aasri.2012.09.024

Petrellis, N. (2015, September). Plant Disease Diagnosis Based on Image Processing, Appropriate for Mobile Phone Implementation. In *Proceedings of 7th International Conference on Information and Communication Technologies in Agriculture, Food and Environment* (pp. 238-246). Academic Press.

Pham, V. H., & Lee, B. R. (2015). An image segmentation approach for fruit defect detection using k-means clustering and graph-based algorithm. *Vietnam Journal of Computer Science*, *2*(1), 25–33. doi:10.100740595-014-0028-3

Pindoriya, N. M., Singh, S. N., & Singh, S. K. (2008). An Adaptive Wavelet Neural Network-Based Energy Price Forecasting in Electricity Markets. *IEEE Transactions on Power Systems*, *23*(3), 1423–1432. doi:10.1109/TPWRS.2008.922251

PlantVillage. (n.d.). Retrieved from https://plantvillage.org/

Pratt, W. K., Chen, W. H., & Welch, L. R. (1973). Slant transform image coding. *IEEE Transactions on Communications*, *COM22*(8), 1075–1093. doi:10.1109/TCOM.1974.1092335

Provan, D., Baglin, T., Dokal, I., & de Vos, J. (2015). *Oxford handbook of clinical haematology* (4th ed.). Oxford University Press. doi:10.1093/med/9780199683307.001.0001

Qi, Y.-n., Lü, C.-x., Zhang, J.-n., Li, Y.-s., Zeng, Z., Mao, W.-h., . . . Yang, B.-n. (2018). *Image recognition on raw and processed potato detection: a review.* Paper presented at the Fourth Seminar on Novel Optoelectronic Detection Technology and Application. 10.1117/12.2315467

Qinghua, W., Guiling, W., Wei, Z., Haodong, C., & Wei, Z. (2016). Estimation of Groundwater Recharge Using Tracers and Numerical Modeling in the North China Plain. *Water (Basel)*, *8*(8), 353–373. doi:10.3390/w8080353

Rabbani, M., & Joshi, R. (2002). An Overview of the JPEG 2000 Still Image Compression Standard. *Signal Processing Image Communication*, *17*(1), 3–48. doi:10.1016/S0923-5965(01)00024-8

Raja, N.S.M., Arunmozhi, S., Lin, H., Dey, N. & Rajinikanth, V. (2019). A study on segmentation of leukocyte image with Shannon's entropy. *Histopathological Image Analysis in Medical Decision Making*, 1-27. doi:10.4018/978-1-5225-6316-7.ch001

Rajinikanth, V., Raja, N. S. M., Satapathy, S. C., & Fernandes, S. L. (2017). Otsu's multi-thresholding and active contour snake model to segment dermoscopy images. *Journal of Medical Imaging and Health Informatics*, *7*(8), 1837–1840. doi:10.1166/jmihi.2017.2265

Rajinikanth, V., & Satapathy, S. C. (2018). Segmentation of ischemic stroke lesion in brain MRI based on social group optimization and fuzzy-Tsallis entropy. *Arabian Journal for Science and Engineering*, *43*(8), 4365–4378. doi:10.100713369-017-3053-6

Rajinikanth, V., Satapathy, S. C., Dey, N., Fernandes, S. L., & Manic, K. S. (2019). Skin melanoma assessment using Kapur's entropy and level set—A study with bat algorithm. *Smart Innovation. Systems and Technologies*, *104*, 193–202. doi:10.1007/978-981-13-1921-1_19

Rajinikanth, V., Satapathy, S. C., Dey, N., & Vijayarajan, R. (2018). DWT-PCA image fusion technique to improve segmentation accuracy in brain tumor analysis. *Lecture Notes in Electrical Engineering*, *471*, 453–462. doi:10.1007/978-981-10-7329-8_46

Rajinikanth, V., Satapathy, S. C., Fernandes, S. L., & Nachiappan, S. (2017). Entropy based segmentation of tumor from brain MR images–A study with teaching learning based optimization. *Pattern Recognition Letters*, *94*, 87–94. doi:10.1016/j.patrec.2017.05.028

Rajinikanth, V., Thanaraj, K. P., Satapathy, S. C., Fernandes, S. L., & Dey, N. (2019). Shannon's entropy and watershed algorithm based technique to inspect ischemic stroke wound. *Smart Innovation. Systems and Technologies*, *105*, 23–31. doi:10.1007/978-981-13-1927-3_3

Ramesh, V., & Ramar, K. (2011). Classification of agricultural land soils: A data mining approach. *Agricultural Journal*, *6*(3), 82–86. doi:10.3923/aj.2011.82.86

Ramteke, P. R., & Mahmud, S. M. (2005). *An adaptive data-reduction protocol for the future in-vehicle networks. Tech. Rep.* SAE.

Rana, C., & Ray, R. K. (2014). *Assessment of Ground Water Resources: A Review of International Practices A Review of International Practices*. Faridabad: Govt. of India Ministry of Water Resources Central Ground Water Board.

Raza, S. A., Prince, G., Clarkson, J. P., & Rajpoot, N. M. (2015). Automatic detection of diseased tomato plants using thermal and stereo visible light images. *PLoS One*, *10*(4), e0123262. doi:10.1371/journal.pone.0123262 PMID:25861025

Razmjooy, N. and Ramezani m. (2016). Training Wavelet Neural Networks Using Hybrid Particle Swarm Optimization and Gravitational Search Algorithm for System Identification, International Journal of Mechatronics. *Electrical and Computer Technology*, *6*(21), 2987–2997.

Razmjooy, N., Mousavi, B. S., Khalilpour, M., & Hosseini, H. (2014). Automatic selection and fusion of color spaces for image thresholding. *Signal, Image and Video Processing*, *8*(4), 603–614. doi:10.100711760-012-0303-7

Razmjooy, N., Mousavi, B. S., Sargolzaei, P., & Soleymani, F. (2011). Image thresholding based on evolutionary algorithms. *International Journal of Physical Sciences*, *6*(31), 7203–7211.

Razmjooy, N., Mousavi, B. S., & Soleymani, F. (2012). A real-time mathematical computer method for potato inspection using machine vision. *Computers & Mathematics with Applications (Oxford, England)*, *63*(1), 268–279. doi:10.1016/j.camwa.2011.11.019

Razmjooy, N., Mousavi, B. S., & Soleymani, F. (2013). A hybrid neural network Imperialist Competitive Algorithm for skin color segmentation. *Mathematical and Computer Modelling*, *57*(3), 848–856. doi:10.1016/j.mcm.2012.09.013

Razmjooy, N., Mousavi, B. S., Soleymani, F., & Khotbesara, M. H. (2013). A computer-aided diagnosis system for malignant melanomas. *Neural Computing & Applications*, *23*(7-8), 2059–2071. doi:10.100700521-012-1149-1

Razmjooy, N., Ramezani, M., & Ghadimi, N. (2017). Imperialist competitive algorithm-based optimization of neuro-fuzzy system parameters for automatic red-eye removal. *International Journal of Fuzzy Systems, 19*(4), 1144–1156. doi:10.100740815-017-0305-2

Razmjooy, N., Sheykhahmad, F. R., & Ghadimi, N. (2018). A hybrid neural network–world cup optimization algorithm for melanoma detection. *Open Medicine: a Peer-Reviewed, Independent, Open-Access Journal, 13*(1), 9–16. doi:10.1515/med-2018-0002 PMID:29577090

Reyes-Aldosoro, C. C. (2015). *Biomedical image analysis recipes in Matlab: For life and engineers*. London: Wiley Blackweel. doi:10.1002/9781118657546

Roerdink, J. B. T. M., & Meijster, A. (1996). Computation of watersheds based on parallel graph algorithms. *Proc of Math. Morphology and Its Application to Image and Signal Processing.*

Roerdink, J. B. T. M., & Meijster, A. (2000). The watershed transform: Definitions, algorithms and parallelization strategies. *Fundamenta Informaticae, 41*(1-2), 187–228.

Roland, L., Drillich, M., & Iwersen, M. (2014). Hematology as a diagnostic tool in bovine medicine. *Journal of Veterinary Diagnostic Investigation, 26*(5), 592–598. doi:10.1177/1040638714546490 PMID:25121728

Romero-Záliz, R., & Reinoso-Gordo, J.F. (2017). *An Updated Review on Watershed Algorithms*. Academic Press.

Roushangar, K., Alami, M. T., Majedi Asl, M., & Shiri, J. (2017). Modeling discharge coefficient of normal and inverted orientation labyrinth weirs using machine learning techniques. *ISH Journal of Hydraulic Engineering*, 1–10.

Roux, P. (2016). Scilab: I. Fundamentals, from theory to practice. Paris, France: Academic Press.

Rumpf, T., Mahlein, A. K., Steiner, U., Oerke, E. C., Dehne, H. W., & Plümer, L. (2010). Early detection and classification of plant diseases with Support Vector Machines based on hyperspectral reflectance. *Computers and Electronics in Agriculture, 74*(1), 91–99. doi:10.1016/j.compag.2010.06.009

Rupanagudi, S. R., Ranjani, B. S., Nagaraj, P., Bhat, V. G., & Thippeswamy, G. (2015, January). A Novel Cloud Computing based Smart Farming System for Early Detection of Borer Insects in Tomatoes. In *Communication, Information & Computing Technology (ICCICT), 2015 International Conference on* (pp. 1-6). IEEE. 10.1109/ICCICT.2015.7045722

Ruprecht, V., Axmann, M., Wieser, S., & Schutz, G. J. (2011). What can we learn from single molecule trajectories? *Current Protein & Peptide Science, 12*(8), 714–724. doi:10.2174/138920311798841753 PMID:22044145

Russ, J. C. (2016). *The image processing handbook*. CRC Press.

Sabrol, H., & Kumar, S. (2015). Recent studies of image and soft computing techniques for plant disease recognition and classification. *International Journal of Computers and Applications, 126*(1).

Sabrol, H., & Kumar, S. (2016). Fuzzy and Neural Network based Tomato Plant Disease Classification using Natural Outdoor Images. *Indian Journal of Science and Technology*, *9*(44), 1–8. doi:10.17485/ijst/2016/v9i44/92825

Sabrol, H., & Kumar, S. (2016). Intensity based feature extraction for tomato plant disease recognition by classification using decision tree. *International Journal of Computer Science and Information Security*, *14*(9), 622–626.

Sa, I., Ge, Z., Dayoub, F., Upcroft, B., Perez, T., & McCool, C. (2016). Deep fruits: A fruit detection system using deep neural networks. *Sensors (Basel)*, *16*(8), 1222. doi:10.339016081222 PMID:27527168

Salaman, R. N. (2014). *Potato varieties*. Cambridge University Press.

Samanta, B., Al-Balushi, K. R., & Al-Araimi, S. A. (2006). Artificial neural networks and genetic algorithm for bearing fault detection. *Soft Computing*, *10*(3), 264–271. doi:10.100700500-005-0481-0

Sankaran, S., Mishra, A., Ehsani, R., & Davis, C. (2010). A review of advanced techniques for detecting plant diseases. *Computers and Electronics in Agriculture*, *72*(1), 1–13. doi:10.1016/j.compag.2010.02.007

Sankaran, S., Mishra, A., Maja, J. M., & Ehsani, R. (2011). Visible-near infrared spectroscopy for detection of Huanglongbing in citrus orchards. *Computers and Electronics in Agriculture*, *77*(2), 127–134. doi:10.1016/j.compag.2011.03.004

Satapathy, S. C., Raja, N. S. M., Rajinikanth, V., Ashour, A. S., & Dey, N. (2018). Multi-level image thresholding using Otsu and chaotic bat algorithm. *Neural Computing & Applications*, *29*(12), 1285–1307. doi:10.100700521-016-2645-5

Satapathy, S., & Naik, A. (2016). Social group optimization (SGO): A new population evolutionary optimization technique. *Complex & Intelligent Systems*, *2*(3), 173–203. doi:10.100740747-016-0022-8

Savkare, S.S., & Narote, S.P. (2011). Automatic classification of normal and infected blood cells for parasitemia detection. *Int'l J. of Comp. Science and Network Security*, *11*(2).

Sawaitul, S. D., Wagh, K. P., & Chatur, P. N. (2012). Classification and prediction of future weather by using back propagation algorithm-an approach. *International Journal of Emerging Technology and Advanced Engineering*, *2*(1), 110–113.

Sbalzarini, I. F. (2016). Seeing Is believing: Quantifying Is convincing: computational image analysis in biology. *Advances in Anatomy, Embryology, and Cell Biology*, *219*, 1–39. doi:10.1007/978-3-319-28549-8_1 PMID:27207361

Scanlon, B. R., Healy, R. W., & Cook, P. G. (2002). Choosing appropriate techniques for quantifying groundwater recharge. *Hydrogeology Journal*, *10*(1), 18–39. doi:10.100710040-001-0176-2

Scanlon, B. R., Keese, K. E., Flint, A. L., Flint, L. E., Gaye, C. B., Edmunds, W. M., & Simmers, I. (2006). Global synthesis of groundwater recharge in semiarid and arid regions. *Hydrological Processes*, *20*(15), 3335–3370. doi:10.1002/hyp.6335

Sethian, J. A. (1999). *Level Set Methods and Fast Marching Methods*. Cambridge, UK: Cambridge University Press.

Seyed R.; Saghravani, I. Y., Sa'ar I, M., & Seyed, F. S. (2013). Estimating Groundwater Recharge Using Empirical Method: A Case Study in the Tropical Zone. *Sains Malaysiana, 42*(5), 553–560.

Shankar, R., Harsha, S., & Bhandary, R. (2014). *A Practical Guide to Identification and Control of Tomato Diseases*. Tropica Seeds Pvt.Ltd.

Sharma, A., & Datta, U. (2007, July). A Weather Forecasting System using concept of Soft Computing: A new approach. In AIP Conference Proceedings: Vol. 923. *No. 1* (pp. 275–287). AIP. doi:10.1063/1.2767047

Shi, J., & Malik, J. (1998). Motion segmentation and tracking using normalized cuts. *Proc. of IEEE 6th Intl. Conf. on Computer Vision*, 1154-1160.

Shi, Y., & Karl, W. (2005). Real-time tracking using level sets. *Proc. IEEE Conf. CVPR*, 2, 34–41. doi: 10.1109/CVPR.2005.294

Singh, C. B., Jayasa, D. S., Paliwala, J., & White, N. D. G. (2010). Identification of insect-damaged wheat kernels using short-wave near-infrared hyperspectral and digital colour imaging. *Computers and Electronics in Agriculture*, *73*(2), 118–125. doi:10.1016/j.compag.2010.06.001

Singh, V., & Misra, A. K. (2017). Detection of plant leaf diseases using image segmentation and soft computing techniques. *Information Processing in Agriculture*, *4*(1), 41–49. doi:10.1016/j.inpa.2016.10.005

Smereka, M., & Duleba, I. (2008). Circular object detection using a modified Hough transform. *International Journal of Applied Mathematics and Computer Science*, *18*(1), 85–91. doi:10.2478/v10006-008-0008-9

Sneha, Abirami, Ankita, Praveena, & Srimeena. (2015). Agricultural Robot for automatic ploughing and seeding. *2015 IEEE Technological Innovation in ICT for Agriculture and Rural Development (TIAR)*. doi:10.1109/tiar.2015.7358525

Sokhansanj, S., & Fenton, J. (2011). *Cost benefit of biomass supply and preprocessing: BIOCAP (Canada) research integration program synthesis paper*. Retrieved from http://www.biocap.ca/rif/reprt/Sokhansanj_S.pdf

Sokolova, M., & Lapalme, G. (2009). A systematic analysis of performance measures for classification tasks. *Information Processing & Management*, *45*(4), 427–437. doi:10.1016/j.ipm.2009.03.002

Soltanzade, R. (2010). Classification of three types of red blood cells in peripheral blood smear based on morphology. *Proceedings of the 2010 ICSP*. 10.1109/ICOSP.2010.5655754

Somvanshi, V. K., Pandey, O. P., Agrawal, P. K., Kalanker, N. V., Prakash, M. R., & Chand, R. (2006). Modeling and prediction of rainfall using artificial neural network and ARIMA techniques. *J. Ind. Geophys. Union*, *10*(2), 141–151.

Song, J., & Lyu, R. M. (2005). A Hough transform based line recognition method utilizing both parameter space and image space. *Pattern Recognition*, *38*(4), 539–552. doi:10.1016/j.patcog.2004.09.003

Souček, J., Kocánová, V., & Novák, M. (2007). Parametres of energy crop biomass handling. *Research in Agricultural Engineering*, *53*(4), 161–165. doi:10.17221/1952-RAE

Srivastva, Neda, & Kriti. (2017). Robotic Farmer. *International Journal of Engineering Development and Research*, *5*(2).

Stiller, C. (1997). Object-based estimation of dense motion fields. *IEEE Transactions on Image Processing*, *6*(2), 234–250. doi:10.1109/83.551695 PMID:18282920

Stone, M., McKee, K., Formwalt, C., & Benneweis, R. (1999). *ISO 11783: An Electronic Communications Protocol for Agricultural Equipment*. Agricultural Equipment Technology Conference, ASAE, Louisville, KY.

Sugumaran, V., & Ramachandran, K. I. (2011). Effect of number of features on classification of roller bearing faults using SVM and PSVM. *Expert Systems with Applications*, *38*(4), 4088–4096. doi:10.1016/j.eswa.2010.09.072

Suh, H. K., Hofstee, J. W., IJsselmuiden, J., & van Henten, E. J. (2018). Sugar beet and volunteer potato classification using Bag-of-Visual-Words model, Scale-Invariant Feature Transform, or Speeded Up Robust Feature descriptors and crop row information. *Biosystems Engineering*, *166*, 210–226. doi:10.1016/j.biosystemseng.2017.11.015

Sullivan, C. J., Ohm, J.-R., Han, W.-J., & Wiegand, T. (2012). Overview of the high efficiency video coding (HEVC) standard. *IEEE Transactions on Circuits and Systems for Video Technology*, *22*(12), 1649–1668. doi:10.1109/TCSVT.2012.2221191

Sun, W., Chen, J., & Li, J. (2007). Decision tree and PCA-based fault diagnosis of rotating machinery. *Mechanical Systems and Signal Processing*, *21*(3), 1300–1317. doi:10.1016/j.ymssp.2006.06.010

Sweldens, W. (1996). The lifting scheme: A custom-design construction of biorthogonal wavelets. *Applied and Computational Harmonic Analysis*, *3*(2), 186–200. doi:10.1006/acha.1996.0015

Ta, D. N., Chen, W. C., Gelfand, N., & Pulli, K. (2009). Surftrac: Efficient tracking and continuous object recognition using local feature descriptors. In *Computer Vision and Pattern Recognition, 2009. CVPR 2009. IEEE Conference on* (pp. 2937-2944). IEEE.

Tellaeche, A., Pajares, G., Burgos-Artizzu, X. P., & Ribeiro, A. (2011). A computer vision approach for weeds identification through Support Vector Machines. *Applied Soft Computing*, *11*(1), 908–915. doi:10.1016/j.asoc.2010.01.011

TetraTech. (2016). *Estimating Monetized Benefits of Groundwater Recharge from Stormwater Retention Practices.* Prepared for United States Environmental Protection Agency Nonpoint Source Control Branch (4503T).

Tumuluru, J., Wright, C., Kenny, K., & Hess, R. (2010). A Riview on Biomass Densification Technologies for Energy Application. *Statewide Agricultural Land Use Baseline, 2015*, 1–59. doi:10.1017/CBO9781107415324.004

Turgeon, M. L. (2004). *Clinical hematology theory and procedures* (4th ed.). Philadelphia: Lippincott Williams and Wilkins.

Umarkar, S., & Karwankar, A. (2016). Automated seed sowing agribot using arduino. *2016 International Conference on Communication and Signal Processing (ICCSP).* 10.1109/ICCSP.2016.7754380

Umaru, A. B., Lukman, S., Oke, I. A., & Umara, B. G. (2018). A New Numerical Method for Groundwater Recharge Estimation In Yola, Nigeria. *Arid Zone Journal of Engineering, Technology and Environment.*

Vaddella, V., & Rama, K. (2010). Artificial neural networks for compression of digital images: A review. *Int. J. Rev. Comput*, *3*, 75–82.

Valgren, C., & Lilienthal, A. J. (2010). SIFT, SURF & seasons: Appearance-based long-term localization in outdoor environments. *Robotics and Autonomous Systems*, *58*(2), 149–156. doi:10.1016/j.robot.2009.09.010

Vapnik, V. N. (1998). *Statistical Learning Theory.* John Wiley & Sons. doi:10.2307/1271368

Vapnik, V. N. (1999). *The Nature of Statistical Learning Theory* (Vol. 8). Springer; doi:10.1109/TNN.1997.641482

Vavilov, V. P., Budadin, O. N., & Kulkov, A. A. (2015). Infrared thermographic evaluation of large composite grid parts subjected to axial loading. *Polymer Testing*, *41*, 55–62. doi:10.1016/j.polymertesting.2014.10.010

Verrastro, T., Lorenzi, T. F., & Neto, S. W. (2005). *Hematologia e hemoterapia: Fundamentos de morfologia, patologia e clínica.* Rio de Janeiro: Atheneu.

Vibhute, A., & Bodhe, S. K. (2012). Applications of Image Processing in Agriculture: A Survey. *International Journal of Computers and Applications*, *52*(2), 34–40. doi:10.5120/8176-1495

Vincent, L., & Soille, P. (1991). Watersheds in digital spaces: An efficient algorithm based on immersion simulations. *IEEE Transactions on Pattern Analysis and Machine Intelligence*, *13*(6), 583–598. doi:10.1109/34.87344

Vitor, G. B., Körbes, A., Lotufo, R. A., & Ferreira, J. V. (2012). Analysis of a step based watershed algorithm using CUDA. In *Nature-Inspired Computing Design, Development, and Applications.* Hershey, PA: IGI Global; doi:10.4018/978-1-4666-1574-8.ch018

Walsh, D., & Raftery, A. E. (2002). Accurate and efficient curve detection in images: The importance sampling Hough Transform. *Pattern Recognition*, *35*(7), 1421–1431. doi:10.1016/S0031-3203(01)00114-5

Wang, G., Sun, Y., & Wang, J. (2017). Automatic Image-Based Plant Disease Severity Estimation Using Deep Learning. *Hindawi Computational Intelligence and Neuroscience*, 1–8.

Wang, X., Ding, X., & Liu, C. (2005). Gabor filters-based feature extraction for character recognition. *Pattern Recognition*, *38*(3), 369–379. doi:10.1016/j.patcog.2004.08.004

Wang, X., Zhang, M., Zhu, J., & Geng, S. (2008). Spectral prediction of Phytophthora infestans infection on tomatoes using artificial neural network (ANN). *International Journal of Remote Sensing*, *29*(6), 1693–1706. doi:10.1080/01431160701281007

Watkinsa, Y. Z., & Sayeh, M. R. (2016). Image data compression and noisy channel error correction using deep neural network. *Procedia Computer Science*, *95*, 145–152. doi:10.1016/j.procs.2016.09.305

Watson, A. B. (1998). Toward a perceptual video quality metric. *Human Vision, Visual Processing, and Digital Display. VIII*, *3299*, 139–147.

Wei, L., Pordesimo, L. O., Igathinathane, C., & Batchelor, W. D. (2009). Process engineering evaluation of ethanol production from wood through bioprocessing and chemical catalysis. *Biomass and Bioenergy*, *33*(2), 255–266. doi:10.1016/j.biombioe.2008.05.017

Widodo, A., Kim, E. Y., Son, J.-D., Yang, B.-S., Tan, A. C. C., Gu, D.-S., ... Mathew, J. (2009). Fault diagnosis of low speed bearing based on relevance vector machine and support vector machine. *Expert Systems with Applications*, *36*(3), 7252–7261. doi:10.1016/j.eswa.2008.09.033

Wieser, S., Axmann, M., & Schutz, G. J. (2008). Versatile analysis of single-molecule tracking data by comprehensive testing against monte carlo simulations. *Biophysical Journal*, *95*(12), 5988–6001. doi:10.1529/biophysj.108.141655 PMID:18805933

Wieser, S., & Schutz, G. J. (2008). Tracking single molecules in the live cell plasma membrane–do's and don't's. *Methods (San Diego, Calif.)*, *46*(2), 131–140. doi:10.1016/j.ymeth.2008.06.010 PMID:18634880

Wintz, P. A., & Tasto, M. (1971). Image coding by adaptive block quantization. *IEEE Transactions on Communication Technology*, *COM19*, 957–972.

Wohlberg, B., & Jager, G. D. (1999). A review of the fractal image coding literature. *IEEE Transactions on Image Processing*, *8*(12), 1716–1729. doi:10.1109/83.806618 PMID:18267449

Wouwer, A. V., Saucez, P., & Vilas, C. (2014). *Simulation of ODE/PDE models with MATLAB, Octave and Scilab: Scientific and engineering applications*. Springer.

Wu, X., Zhao, R., Bean, S. R., Seib, P. A., McLaren, J. S., & Madl, R. L. … Wang, D. (2007). Factors impacting ethanol production from grain sorghum in the dry-grind process. *Cereal Chemistry*. doi:10.1094/CCHEM-84-2-0130

Wu, Z., Lin, D., & Tang, X. (2016). Deep Markov random field for image modeling. In B. Leibe, J. Matas, N. Sebe, & M. Welling (Eds.), Lecture Notes in Computer Science: Vol. 9912. *Proc. Computer Vision – ECCV 2016. ECCV 2016*. Cham: Springer. doi:10.1007/978-3-319-46484-8_18

Xavier, C. A. N., Moset, V., Wahid, R., & Møller, H. B. (2015). The efficiency of shredded and briquetted wheat straw in anaerobic co-digestion with dairy cattle manure. *Biosystems Engineering*, *139*, 16–24. doi:10.1016/j.biosystemseng.2015.07.008

Xavier, R. M., Dora, J. M., & Barros, E. (2016). *Laboratório na prática clínica: Consulta rápida* (3rd ed.). Porto Alegre: Artmed.

Xie, C., Shao, Y., Li, X., & He, Y. (2015). Detection of early blight and late blight diseases on tomato leaves using hyperspectral imaging. *Scientific Reports*, *5*(16564), 1–11. PMID:26572857

Xu, H., Zhu, S., Ying, Y., & Jiang, H. (2006, October). Application of multispectral reflectance for early detection of tomato disease. In Optics for Natural Resources, Agriculture, and Foods (vol. 6381, p. 63810R). International Society for Optics and Photonics. doi:10.1117/12.685531

Xu, H. R., Ying, Y. B., Fu, X. P., & Zhu, S. P. (2007). Near-infrared Spectroscopy in detecting Leaf Miner Damage on Tomato Leaf. *Biosystems Engineering*, *96*(4), 447–454. doi:10.1016/j.biosystemseng.2007.01.008

Xu, X., Xu, S., Jin, L., & Song, E. (2011). Characteristic analysis of Otsu threshold and its applications. *Pattern Recognition Letters*, *32*(7), 956–961. doi:10.1016/j.patrec.2011.01.021

Yamauchi, Y., Boukari, H., Banerjee, I., Sbalzarini, I.F., Horvath, P., & Helenius, A. (2011). Histone deacetylase 8 is required for centrosome cohesion and influenza A virus entry. *PLoS Pathog.*, *7*(10).

Yang, J., Liu, L., Jiang, T., & Fan, Y. (2003). A modified Gabor filter design method for fingerprint image enhancement. *Pattern Recognition Letters*, *24*(12), 1805–1817. doi:10.1016/S0167-8655(03)00005-9

Yan, W., & Gelan, Y. (2018). A new brain magnetic resonance imaging segmentation algorithm based on subtractive clustering and fuzzy C-means clustering. *Journal of Medical Imaging and Health Informatics*, *8*(3), 602–608. doi:10.1166/jmihi.2018.2309

Yeager, M., Gregory, B., Key, C., & Todd, M. (2018). On using robust Mahalanobis distance estimations for feature discrimination in a damage detection scenario. *Structural Health Monitoring*, 1475921717748878.

Yeh, H. Y., Cheng-Haw, L., Kuo-Chin, H., & Po-Hsun, C. (2009). GIS for the assessment of the groundwater recharge potential zone. *Environ Geol.*, *58*(1), 185–195. doi:10.100700254-008-1504-9

Ye, X., Sakai, K., Okamoto, H., & Garciano, L. O. (2008). A ground-based hyperspectral imaging system for characterizing vegetation spectral features. *Computers and Electronics in Agriculture*, *63*(1), 13–21. doi:10.1016/j.compag.2008.01.011

Yu, M., Womac, A. R., & Pordesimo, L. O. (2003). Review of Biomass Size Reduction Technology. *2003 ASABE Annual International Meeting*, 1–12. 10.13031/2013.15454

Zadeh, L. A. (1994). Fuzzy logic, neural networks, and soft computing. *Communications of the ACM*, *37*(3), 77–84. doi:10.1145/175247.175255

Zaremba, W., Sutskever, I., & Vinyals, O. (2015). Recurrent Neural Network Regularization. *ArXiv Preprint*. Retrieved from http://arxiv.org/abs/1409.2329

Zhang, L., Zhang, L., Zhang, D., & Zhu, H. (2011). Ensemble of local and global information for finger–knuckle-print recognition. *Pattern Recognition*, *44*(9), 1990–1998. doi:10.1016/j.patcog.2010.06.007

Zhang, M., Qin, Z., & Liu, X. (2005). Remote Sensed Spectral Imagery to Detect Late Blight in Field Tomatoes. *Precision Agriculture*, *6*(6), 489–508. doi:10.100711119-005-5640-x

Zhang, M., Qin, Z., Liu, X., & Ustin, S. L. (2003). Detection of stress in tomatoes induced by late blight disease in California, USA, using hyperspectral remote sensing. *International Journal of Applied Earth Observation and Geoinformation*, *4*(4), 295–310. doi:10.1016/S0303-2434(03)00008-4

Zhang, S., Wang, H., & Huang, W. (2017). Two-stage plant species recognition by local mean clustering and Weighted sparse representation classification. *Cluster Computing*, *20*(2), 1517–1525. doi:10.100710586-017-0859-7

Zhang, S., Wu, X., You, Z., & Zhang, L. (2017). Leaf image based cucumber disease recognition using sparse representation classification. *Computers and Electronics in Agriculture*, *134*, 135–141. doi:10.1016/j.compag.2017.01.014

Zhao, F., Liu, H., Fan, J., Chen, C. W., Lan, R., & Li, N. (2018). Intuitionistic fuzzy set approach to multi-objective evolutionary clustering with multiple spatial information for image segmentation. *Neurocomputing*.

Zhao, Y. R., Li, X., Yu, K. Q., Cheng, F., & He, Y. (2016). Hyperspectral Imaging for Determining Pigment Contents in Cucumber Leaves in Response to Angular Leaf Spot Disease. *Scientific Reports*, *6*(27790), 1–9. PMID:27283050

Zhu, X., Zhu, M., & Ren, H. (2018). Method of plant leaf recognition based on improved deep convolutional neural network. *Cognitive Systems Research*, *52*, 223–233. doi:10.1016/j.cogsys.2018.06.008

About the Contributors

Navid Razmjooy has a Ph.D. in Electrical Engineering/Control from Tafresh University (formerly Amirkabir University-Tafresh paradise). He born in 1988; he received his B.Sc. in electrical engineering from the Islamic Azad University of Ardabil Branch, Ardabil, Iran in 2009 and M.Sc. with honor in Mechatronics Engineering from Islamic Azad University, Majlesi Branch, Isfahan, Iran in 2011. He is a senior member of IEEE and young researchers club in the IA University (Majlesi Industrial Unit). His research interests include Control; Optimization; Image Processing, Machine Vision, Soft Computing, Data Mining, Evolutionary Algorithms, and System Control. He published about 74 papers and 3 books in Farsi and English in peer-reviewed journals and conferences and is now a reviewer in the national and international journals.

Vania Estrela has a B.S. degree from Federal University of Rio de Janeiro (UFRJ) in Electrical and Computer Engineering (ECE); M.Sc. from the Instituto Tecnológico de Aeronáutica (ITA), Sao Jose dos Campos; M.Sc. degree in ECE at Northwestern University, Evanston, Illinois (IL), USA; and Ph.D. in ECE from the Illinois Institute of Technology (IIT), Chicago, IL, USA. Taught at DeVry University; DePaul University; Universidade Estadual do Norte Fluminense (UENF), Campos de Goytacazes, Rio de Janeiro (RJ), Brazil; and for the Universidade Estadual da Zona Oeste (UEZO), Rio de Janeiro, RJ, Brazil. Visiting professor at the Polytechnic Institute of Rio de Janeiro (IPRJ)/State University of Rio de Janeiro (UERJ) in Nova Friburgo, RJ. Currently working at Universidade Federal Fluminense (UFF), Niterói, RJ for the Department of Telecommunications. Research interests include signal/image/video processing, inverse problems, computational and mathematical modeling, stochastic models, multimedia, communications, motion estimation, machine learning and geoprocessing. Reviewer for the following journals/magazines: IMAVIS (Elsevier); Pattern Recognition (Elsevier); Computers & Electrical Engineering (Elsevier); IET Image Processing; IET Computer Vision and International Journal of Image Processing (IJIP). Engaged in topics such as technology transfer, engineering/math/computer education, environmental issues and digital inclusion. Member of IEEE, and ACM. Editor of IJIP.

* * *

Meena Arunachalam completed B.E. in Computer science Engineering from VCE, Anna University with First Class Distinction in 2005. Following that, she completed M.E. in Network Engineering from AKCE, Anna University in 2008 with gold medal. Currently, she is working as assistant professor in Electronics and Communication Engineering department, KLN college of Engineering, India, where she has been working since 2011. She has more than 10 years of teaching experience in academic. Her area of interest is Image processing and wireless communication. He is a life member of ISTE, IE. He has successfully guided 16 batch B.Tech students.

Muthukumar Arunachalam received his B.E degree from Madurai Kamaraj University in 2004, and M.E. degree from Anna University, Chennai, Tamil Nadu in 2006. He received his Ph.D degree from Kalasalingam University in 2014.He is working as an Associate Professor in Electronics and Communication Engineering department of Kalasalingam University, Tamilnadu, India. His research interest includes image processing with cryptography.

S. Arunmozhi is currently working as as Associate Professor in the department of ECE. He is having a total experience of 15 years in teaching Electronics and communication engineering. His areas of interest include wireless Communication and signal processing techniques.

Anuradha Chug has long teaching experience of almost 20 years to her credit as faculty and in administration at various educational institutions in India. She has worked as guest faculty in Netaji Subhash Institute of Information and Technology, Dwarka, New Delhi and Regular Faculty at Government Engineering College, Bikaner. Before picking the current assignment as Assistant Professor at USICT, GGSIP University, she has also worked as Academic Head, Aptech, Meerut and Program Coordinator at Regional Centre, Indira Gandhi National Open University (IGNOU), Meerut. In academics, she has earned her doctorate degrees in software engineering from the Delhi Technological University, Delhi, India. Before pursuing PhD, she has achieved top rank in her M.Tech (IT) degree and conferred the University Gold Medal in 2006 from Guru Gobind Singh Indraprastha University. Previously she has acquired her Master's degree in Computer Science from Banasthali

Vidyapith, Rajasthan in the year 1993. Her H-index as reported by Google Scholar is 5. She has published more than 35 research papers in international and national journals and conferences. She has also served as reviewer of several national and international journals and conferences in the area of software engineering (ACM transaction, IJKESE, Informatica, Inderscience, JOT, SEED, WCI).

Reinaldo França has a B.Sc. in Computer Engineering by the Regional University Center of Espírito Santo de Pinhal in 2014. Master graduated by the Department of Semiconductors, Instruments and Photonics of the Faculty of Electrical and Computer Engineering of LCV-UNICAMP, (2017). He is currently a PhD student by the Department of Communications of the Faculty of Electrical Engineering and Computing of LCV - UNICAMP. He works in technological and scientific research, as well as programming and development in C / C ++, Java and .NET languages. His main topics of interest are simulation, operating systems, software engineering, wireless networks, internet of things, systems transmission and telecommunications.

Theodoros Gialamas has took his Ph.D. at Faculty of Agriculture, Aristotle University of Thessaloniki. He holds a postgraduate degree (M.Sc. & Ph.D.) in "Agricultural Engineering" from the Agricultural University of Thessaloniki. He is Associate Professor at the Department of Biosystems Engineering of the Faculty of Agricultural Technology, Technological Educational Institute of Thessaly. His research activities include experimental tests to determine the allowable angle stability for movement of farm machinery on sloping and uneven ground, and measurements of the impact on the health of the operator from vibration and noise.

Ioannis Gravalos is a Professor at Department of Biosystems Engineering, Technological Educational Institute of Thessaly. He obtained Diploma (Dipl.-Ing.) in "Agricultural Engineering" from Agriculture University of Prague, M.Sc. in "Biotechnology - Quality Assessment in Nutrition and the Environment" from Department of Biochemistry & Biotechnology, University of Thessaly, and Ph.D. in "Automation & Control Technology" from Department of Electrical Engineering and Automation, Faculty of Engineering, Agriculture University of Prague. His research activities include: Agricultural Mechatronics & Automation, Farm Machinery Design, Intelligent Fault Diagnosis Systems, and Biofuel Combustion Engines. He has more than 150 publications in scientific journals and conferences and two patents.

Yuzo Iano has B.Sc. (1972), M.Sc. (1974) and Ph.D. degrees (1986) in Electrical Eng. at UNICAMP, Brazil. Since then he has been working in the technological production field, with 1 patent granted, 8 filed patent applications and 36 projects completed with research and development agencies. He has supervised 29 doctoral theses, 49 master's dissertations, 74 undergraduate and 48 scientific initiation works. He has participated in 100 master's examination boards, 50 doctoral degrees, author of 2 books and more than 250 published articles. He is currently Professor at UNICAMP, Editor-in-Chief of the SET International Journal of Broadcast Engineering and General Chair of the Brazilian Symposium on Technology (BTSym). He has experience in Electrical Engineering, with knowledge in Telecommunications, Electronics and Information Technology, mainly in the field of audio-visual communications and multimedia.

Dimitrios Kateris received his Ph.D. in Agricultural Engineering from Aristotle University of Thessaloniki in 2015, for his dissertation on "Intelligent system for fault diagnosis in agricultural tractor mechanical subsystems". He also holds a M.Sc. in Agricultural Engineering and Water Resources from the same institute in 2006 and two B.Sc. in Agricultural Engineering and Water Resources from Aristotle University of Thessaloniki (2006) and Technological Educational Institute of Thessaly (1999), respectively. His main research interests lie in the areas of Computational intelligence in agricultural applications, Sensor based non - destructive testing and evaluation, Fault diagnosis techniques in mechanical systems, Automatic control of vibration in agricultural machinery, Vibration Analysis, Acoustic Emission Analysis, Machine Condition Monitoring, Automation and New Technologies in Agricultural Machinery, Neural Networks, Self-Organizing Maps and Support Vector Machines (SVM) in agricultural applications, Safety and Reliability in Agricultural Machinery.

Seetharaman Krishnamoorthy received the M.Sc. degree in Statistics, and the Ph.D. in Computer Science and Engineering from Annamalai University, Tamil Nadu, India, in 1990 and 2006 respectively, and the M.S. degree in Software Systems from BITS, Pilani, India, in 1995; and M.Tech. degree in Computer and Information Technology from ManonmaniamSundaranar University, Tamil Nadu, India, in 2013. Currently, he is working as Associate Professor in the Department of Computer and Information Science, Annamalai University. His research interests include Statistical Pattern Recognition, Image and Video Retrieval, Digital Image Analysis, Data Mining and Knowledge Discovery, and Big Data Analytics. He is actively involved in professional activities, viz. reviewer in the journals: IET Image Processing, Pattern Recognition (Elsevier), Swarm and Evolutionary Computation (Elsevier), Knowledge–Based SYSTEMS (Elsevier), Computer and Electronics in Agriculture (Elsevier). He is currently serving as an Associate Editor of the Aloy Journal of Soft Computing and Applications, which is an International Journal, and

Editorial Board Member of the International Journal of Information Technology and Management Sciences. He is also a member of the professional societies – IEEE, ACM, IAENG, and Indian Science Congress Association. He has received Two R & D Major Research Projects from UGC and DST, India. He has authored the book entitled Image Processing: Stochastic Model-Based Approach; published more than 75 articles in journals of the Elsevier, Springer, Taylor and Francis, IEEE, etc. He has successfully supervised 9 Ph.D. scholars and 8 M.Phil. Scholars. Dr. Seetharaman has received an Outstanding Scientist award for the contribution and achievement in the field of Image Processing, awarded by the Centre for Advanced Research and Design, Venus International Foundation, India, in November 2017.

Hermes José Loschi graduated in Control and Automation Engineering at Paulista University (UNIP). He holds an M.Sc in Electrical Engineering by the State University of Campinas (UNICAMP). Currently a Ph.D. Candidate by Department of Communications (DECOM), Faculty of Electrical and Computer Engineering (FEEC) at UNICAMP. Professional with experience in applications engineering, processes and definition of technical scope of Diesel engines and Diesel generators, in particular MTU and Mercedes Diesel engines. Technical skills conquered with high level of commitment on all projects assigned for, self motivated, with straight focus on activities under responsibility, effective team work skills and leadership potential on Photovoltaic Solar Generation R&D. Currently, Technical Program and Finance Chair of Brazilian Technology Symposium (BTSym) and researcher at the Laboratory of Visual Communications (LCV). The main topics of interest are Wireless Sensor Network, Internet of Things, Smart Grid, Broadcasting, Biomass, Photovoltaic Systems Applications, Solar Energy, Photovoltaic Solar Generation Prediction Systems, Solar Tracking, Open Access, and STEM. Reviewer for the following publishers: SCIENCEDOMAIN International, Elsevier, AIP Publishing LLC, International Knowledge Press, Scientific Research Publishing, David Publishing Company, Canadian Center of Science and Education, PIBIC-PRP- UNICAMP. Student member of Institute of Electrical and Electronics Engineers (IEEE) and Member of International Society for Development and Sustainability (ISDS).

Ana Carolina Monteiro has a B.Sc. in Biomedicine from Amparense University Center - UNIFIA in 2015. Currently, she is an M.Sc. candidate by Department of Communications, Faculty of Electrical and Computer Engineering at LCV-UNICAMP. She has expertise in clinical analysis automation (emphasis on hematology), digital image processing, biomedical engineering, and MATLAB © software. She currently works in research lines related to the development of algorithms for medical laboratory applications.

Dasari Naga Vinod has Studied Masters Degree (Masters in Technology) with specialization in Embedded Systems from Department of Compute Science and Engineering, Amrita University, Bangalore. He has special interest in developing Robots for various agricultural needs. He received grants for his work. He perused Degree in Bachelor of Technology (B.Tech.) in the Field Of Study Electronics and Communication Engineering from Jawaharlal Nehru Technological University, Kakinada . He comes as a confident student with excellent presentation and quick learning skills. He is very affable, adaptable and cooperative with a good demeanor. He has many publications to his credit.

Isaiah Oke holds a PhD degree in Civil Engineering. His main research Interest is in water Engineering. He has published a number of articles in reputable journals and Conference Proceedings. He is an academic in the Department of Civil Engineering, Obafemi Awolowo University, Ile-Ife, Nigeria.

K. Palani Thanaraj is currently working as Assistant Professor in the department of Electronics & Instrumentation Engineering. He completed his PhD in 2018 in the Faculty of Information and Communication Engineering from Anna University, Chennai. His research areas includes Image Processing, Advanced signal processing, Image segmentation, Machine learning, and Deep learning.

B. Parvatha Varthini is currently the Dean (Research) in St. Joseph's College of Engineering, Chennai, India. She is having a vast teaching and research experience and her area of research includes computer networks, image processing and computational algorithm.

V. Rajinikanth is working as a Professor in Electronics and Instrumentation Engineering Department. He completed his PhD in 2013 from MIT Campus, Anna University, Tamilnadu, India. His area of research includes Heuristic algorithm applications, Medical Image processing and Signal processing.

Shubham Sharma is currently a student of Bachelor of Technology (Undergraduate degree course) in the field of Information Technology at Guru Gobind Singh Singh Indraprastha University. His work focuses specifically on application of deep learning for the purpose of Image recognition and object detection. He also makes machine learning based web applications and android apps.

Amit Prakash Singh is working as Professor in University School of Information Communication & Technology, GGSIPU since July 2001. He obtained Ph.D. in Information Technology from Guru Gobind Singh Indraprastha University, Delhi in

2011. He has done his doctoral research in the area of Artificial Neural Networks. He has earlier worked as Lecturer in University of Hyderabad and Banasthali Vidyapith. His area of interest are Artificial Neural Network, Embedded System Design and Data Science. He is a member of IEEE, IETE, ACM & CSI. He has published more than 40 Research papers in International / National Journals and attended various National / International Workshops and Seminar. Dr. Singh visited UK, Portugal, Singapore, Thailand and Italy to present his research work in international conferences. Dr. Singh is also working on the project funded by Department of Science & Technology, Government of India.

Tripty Singh is an Associate Prof in the Department of Computer Science at Amrita School of Engineering, Bengaluru. She holds her Engineering graduation in Electronics and Instrumentation (1999) from the SGSITS Indore DAVV University, Masters in Information Technology (2002-2004) from the IIITM- Gwalior and Doctorate in Medical Image Processing (2008-2013)from MITS, Gwalior. Her area of interest include biomedical Image Processing, Artificial Intelligence, Computer and Information Sciences, Parallel and Distributed Systems, Ad-Hoc Networks, Network Security and Digital System. She has secured several International awards that include Best Idea Award in 3Bs, Young international investigator award and Best paper presentation and best contains in the track (ICEEE). Moreover she carries the Editorship and Reviewer of several Journals such as International Journal of Computer Engineering, International Journal of Computer and Information etc. Also she has membership in Professional bodies like MIAENG, LMISOI etc . Moreover she has Book Chapter Publication in: Medical Imaging: Concepts, Methodologies, Tools, and Applications, chapter: Computer aided diagnosis system for breast cancer detection published by IGI global. With her 15 years career experience she has secured more than 15 journals and 40 conference publications. She was Program Chair " The Fifth International Conference on the Applications of Digital Information and Web Technologies (ICADIWT 2013)" .Also she was Program Co-Chair for The Third international conference on Innovative Computing Technology was held at London U.K INTECH 2013.Currently she is guiding four research scholars in the area of Medical imaging in collaboration with various reputed hospitals and research center.

N. Sri Madahava Raja is currently working as the associate professor in the department of EIE, St. Joseph's college of Engineering. He completed his PhD in the field of Medical Image Analysis from MIT Campus, Anna University Chennai. His area of interest includes the heuristic algorithms, image and signal processing.

Shradha Verma is currently pursuing her doctorate from USICT, GGSIP University, New Delhi, India. She has a teaching experience of more than 8 years. Her areas of interest include Machine Learning, Artificial Neural Networks, Analysis and Design of Algorithms. She is extremely dedicated and honest towards her work.

Vivek Verma is an experienced Assistant Professor with a demonstrated history of working in the Technical education industry. Skilled in Research, E-Learning, Lecturing, Teaching, and Higher Education. Strong education professional with a Doctor of Philosophy (Ph.D.) focused in Natural Language Processing from Manipal University Jaipur. Working as Assistant Professor at the School of Computing & Information Technology Manipal University Jaipur.

Index

R

S

T

U

V

W

www.ingramcontent.com/pod-product-compliance
Lightning Source LLC
LaVergne TN
LVHW081532300926
849161LV00006B/1392